Lecture Notes in Biomathematics

ctd. on inside back cover

Lecture Notes in Biomathematics

Managing Editor: S. Levin

72

Modeling and Management of Resources under Uncertainty

Proceedings of the Second U.S.-Australia
Workshop on Renewable Resource Management
held at the East-West Center, Honolulu,
Hawaii, December 9–12, 1985

Edited by T. L. Vincent, Y. Cohen, W. J. Grantham,
G. P. Kirkwood, and J. M. Skowronski

Springer-Verlag
Berlin Heidelberg New York Tokyo

Editors

Thomas L. Vincent
Aerospace and Mechanical Engineering
University of Arizona
Tucson, AZ 85721, USA

Yosef Cohen
Department of Fisheries and Wildlife
University of Minnesota
St. Paul, MN 55108, USA

Walter J. Grantham
Department of Mechanical and Materials Engineering
Washington State University
Pullman, WA 99164-2920, USA

Geoffrey P. Kirkwood
Division of Fisheries Research
CSIRO Marine Laboratories
GPO Box 1538
Hobart, Tasmania, Australia

Jan M. Skowronski
Department of Mathematics
University of Queensland
St. Lucia Queensland 4067, Australia

Mathematics Subject Classification (1980): 34 C 11, 34 C 35, 34 D 20, 49 A 10,
49 E 99, 60 H 99, 90 D 99, 93 D 05, 93 C 15

ISBN-13: 978-3-540-17999-3   e-ISBN-13: 978-3-642-93365-3
DOI: 10.1007/978-3-642-93365-3

PREFACE

This volume contains the proceedings of the second U.S.-Australia workshop on Renewable Resource Management held at the East-West Center, Honolulu, Hawaii, December 9-12, 1985. The workshop was jointly sponsored by the National Science Foundation (USA) and the Department of Science and Technology (Australia) under the U.S.-Australia Cooperative Science Program.

The objective of the workshop was to focus on problems associated with the management of renewable resource systems. A particular emphasis was given to methods for handling uncertain elements which are present in any real system. Toward this end, the participants were chosen so that the collective expertise included mathematical modeling, dynamical control/game theory, ecology, and practical management of real systems.

Each participant was invited to give an informal presentation in his field of expertise as related to the overall theme. The formal papers (contained in this volume) were written after the workshop so that the authors could utilize the workshop experience in relating their own work to others. To further encourage this exchange, each paper contained in this volume was reviewed by two other participants who then wrote formal comments. These comments (with author's reply in some cases) are attached to the end of each paper. I feel that these comments (and replies) form a very valuable part of this volume in that they provide to the reader a share of the workshop experience. They also provide a thread which helps tie many of the papers together. Indeed the reader may wish to peruse the comments first.

For review purposes, the papers were originally grouped according to the classifications Modelling/Biology, Controls/Techniques, and Management/Real Problems. These headings give a flavor of the material presented here and the same headings are used to group the papers in this volume. The fact that many of the papers cross these boundaries reflects the focus and interdisciplinary nature of the workshop.

Since modelling is the key to analysis, the majority of papers in this volume deal either directly or indirectly with system modelling. Papers dealing more directly with the modelling question are contained in Part I. Topics include scale and predictability, stochastic dynamic programming, stage structure, uncertain stock sizes, random environmental variables, identification and verification, evolutionary effects, and habitat selection.

A number of techniques particularly applicable to the determination of management policy from models subject to uncertainty are contained in Part II. Both discrete and continuous system models are used. The first two papers use a statistical approach to examine adaptive control and parameter estimation. The next four papers use deterministic methods to deal with system uncertainty. The first two of these use reachable set methods and the last two use Liapunov methods.

The remaining two papers in this section use methods from optimal control and game theory to arrive at management policy.

Part III is devoted to those papers dealing more directly with the management of real problems. These include the management of the Ogalalla Aquifer, the Great Lakes, California sardines, Oregon groundfishery, tuna in the Western Pacific, the California river system, and Western King prawns in Spencer Gulf.

I would like to acknowledge the service that coeditor Jan Skowronski provided in helping to organize the workshop. I am also indebted to coeditors Yosef Cohen, Walt Grantham, and Geoff Kirkwood who did most of the work in organizing the reviews and comments.

T.L. Vincent
Tucson, Arizona
March 2, 1987

PARTICIPANTS

Dr. Ann Lowes Blackwell. Mechanical Engineering, University of Texas, Arlington, Texas 76019 USA.

Dr. Louis W. Botsford. Department of Wildlife and Fisheries Biology, University of California, Davis, California 95616 USA.

Dr. Colin W. Clark. Institute of Applied Mathematics, University of British Columbia, Vancouver, B.C., Canada V6T 1W5.

Dr. Yosef Cohen. Department of Fisheries and Wildlife, University of Minnesota, St. Paul, Minnesota 55108 USA.

Dr. Michael E. Fisher. Department of Mathematics, University of Western Australia, Nedlands, Western Australia 6009.

Dr. Wayne M. Getz. Division of Biological Control, University of California, Berkeley, California 94720 USA.

Dr. Walter J. Grantham. Department of Mechanical Engineering, Washington State University, Pullman, Washington 99164-2920 USA.

Dr. Susan Hanna. Department of Agricultural and Resource Economics, Oregon State University, Corvallis, Oregon 97331-3601 USA.

Dr. Ray Hilborn. Tuna and Billfish Assessment Programme, South Pacific Commission, Post Box D-5, Noumea Cedex, New Caledonia.

Dr. Geoff P. Kirkwood. Division of Fisheries Research, CSIRO Marine Laboratories, GPO Box 1538, Hobart, Tasmania, Australia.

Dr. Cho Seng Lee. Department of Mathematics, University of Malaya, Kuala Lumpur 22-11, Malaysia.

Dr. George Leitmann. Department of Mechanical Engineering, University of California, Berkeley, California 94720 USA.

Dr. Paul F. Lesse. CSIRO, Building Research Division, Graham Road, Highett, Victoria 3190, Australia.

Dr. Simon A. Levin. Ecology and Systematics and Ecosystems Research Center, Cornell University, Ithaca, New York 14853 USA.

Dr. Donald Ludwig. Institute of Animal Resource Ecology, The University of British Columbia, 2204 Main Mall, Vancouver, British Columbia, Canada V6T 1W5.

Dr. Marc Mangel. Department of Mathematics, University of California, Davis, California 95616 USA.

Dr. Robert McKelvey. Department of Mathematical Sciences, University of Montana, Missoula, Montana 59812 USA.

Mr. Brodie Nicol. Department of Mathematics, University of Queensland, St. Lucia, Brisbane, Queensland 4067 Australia.

Dr. Michael L. Rosenzweig. Department of Ecology and Evolutionary Biology, The University of Arizona, Tucson, Arizona 85721 USA.

Dr. Jan M. Skowronski. Department of Mathematics, University of Queensland, St. Lucia, Brisbane, Queensland 4067, Australia.

Dr. Philip R. Sluczanowski.   Department of Fisheries, Box 1625, G.P.O., Adelaide, South Australia, 5001.

Dr. George R. Spangler.   Department of Fisheries and Wildlife, 200 Hodson Hall, University of Minnesota, St. Paul, Minnesota 55455 USA.

Dr. Russel J. Stonier.   Department of Mathematics and Computing, Capricornia Institute of Advanced Education, Rockhampton, Queensland, Australia.

Dr. Thomas L. Vincent, Department of Aerospace and Mechanical Engineering, The University of Arizona, Tucson, Arizona 85721 USA.

Dr. Carl J. Walters. Institute of Animal Resource Ecology, The University of British Columbia, Vancouver, B.C., Canada V6T 1W5.

CONTENTS

CONTENTS (continued)

PART III   MANAGEMENT/REAL PROBLEMS

PART I

MODELLING/BIOLOGY

SCALE AND PREDICTABILITY
IN ECOLOGICAL MODELING

Simon A. Levin
Ecology and Systematics and
Ecosystems Research Center
Cornell University
Ithaca, New York 14853

Fundamental to sound environmental management is an understanding of the relevant scales of interest, and the interrelationships among scales. Predictability is intertwined with variability, and with the choice of temporal and spatial scales. Overly detailed models are useless as predictive devices, and techniques for aggregation and simplification are essential.

INTRODUCTION

In the late 1960's and early 1970's, as a result of environmental activism and concern over the effects of chemicals in the environment, a suite of environmental laws were passed that, in one form or another, called for the protection of the ecosystems. In general, those laws tended to be vague when it came to establishing criteria for measuring ecological effects, since they were designed to address broad objectives without being hampered in applicability by specific references to particular systems. Much legislation, and certainly the conceptual and mathematical models that they spawned, emphasized ideas such as equilibrium, constancy, homogeneity, stability, and predictability. This in turn led to the development of simple-minded and misguided predictive approaches, built upon large-scale and highly detailed models.

PATTERN AND SCALE AS GUIDING PRINCIPLES IN ECOLOGICAL MODELING

The description of the behavior of any system can be carried out blindly, or it can be based on some degree of understanding of the basic mechanisms underlying system dynamics. The search to understand any complex system is a search for pattern; that is, for the reduction of complexity to a few simple rules, principles that allow abstraction of the essence from the noise. This is also the key to management: There exists pattern at all levels and at all scales, and recognition of this multiplicity of scales is fundamental to describing and understanding ecosystems. It is essential to strip away what is irrelevant detail, and to determine those processes and components that are central to the integrity of the system: the keystone species of Paine (1966), or the factors controlling recruitment and larval survival (see, for example, the paper by Getz, this volume). The emphasis on equilibrium and homogeneity obscures the search for pattern, and represents the baggage of historical tradition.

In the early 20th century, as attention in the ecological literature turned to community organization, Gleason's individualistic approach, emphasizing contingency and stochasticity, lost out to Clements' notion of the climax, the stable state to be attained inexorably given any particular set of local environmental conditions. The community came to be viewed as having an individuality of its own; it was a superorganism that behaved as a unit both ecologically and evolutionarily. Only in the middle of this century, as the writings of A.W. Watt (1947) and R.H. Whittaker (1975) gained prominence, did the importance of variation in space and time become recognized as being central to how species coexist, to their evolution, and to concepts of ecosystem structure and function. More recent work (e.g., Levin and Paine, 1974; Paine and Levin, 1981; Pickett and White, 1985), extending Watts arguments for the importance of gaps and mosaic phenomena, has demonstrated the inseparability of the concepts of equilibrium and scale. As one moves to finer and finer scales of observation, systems become more and more variable over time and space, and the degree of variability changes as a function of the spatial and temporal scales of observation. Such a realization has long been part of the thinking of oceanographers, who observe patchiness and variability on virtually every scale of investigation. The situation is summarized elegantly in the Stommel diagram (Fig. 1, taken from Haury et al., 1978; original sources Stommel, 1963, 1965), a cornerstone of the oceanographic literature. A major conclusion is that there is no single correct scale of observation, and that the insights one achieves from any investigation are contingent on the choice of scales. Pattern is neither a property of the system alone nor of the observer, but of an interaction between them.

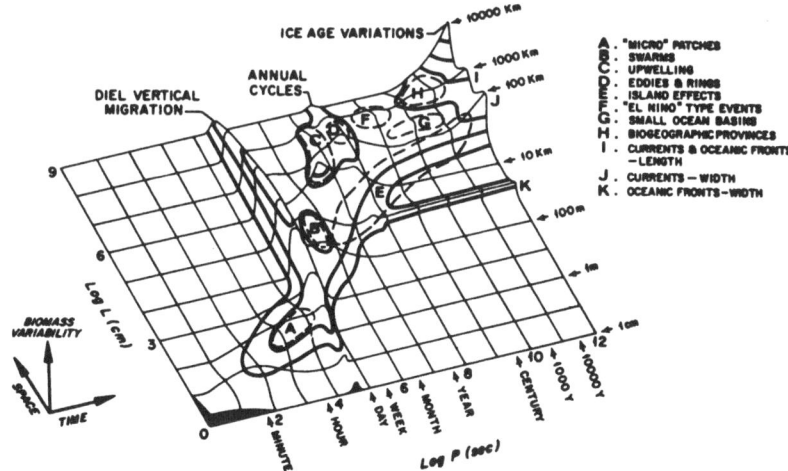

Figure 1.   The Stommel Diagram, a conceptual model of the time-space scales of zooplankton biomass variability and the factors contributing to these scales. I, J and K are bands centered about 1000's 100's and several kilometers in space scales, with time variations between weeks and geological time scales. Figure and caption taken from Haury et al. (1978).

The importance of scales also becomes apparent from an examination of population models, both in terms of their general dynamic properties and in terms of their applicability to real populations. Much recent mathematical work (see, for example, May, 1974) has demonstrated that even the simplest models of populations can exhibit oscillatory and even chaotic behavior; and that, as a consequence, it is impossible to predict accurately the precise dynamics of populations governed by such equations.

To some extent, such investigations render moot the classical debate over whether populations are controlled by density-dependent or density-independent factors. Close to the theoretical equilibrium, the dynamics of such populations may be indistinguishable from those of appropriately chosen stochastic density-independent models; near the equilibrium, density dependence is very weak, and will be obscured by any overriding density-independent variation. On the other hand, far from equilibrium, density-dependent factors assume more importance because the nonlinearities are stronger. Thus, density dependence is the primary mechanism constraining major excursions in population density and keeping populations within bounds; but within those bounds, density independent phenomena predominate. Concepts of stability that rely on asymptotic return to an equilibrium state are seen to be irrelevant, and more general concepts such as boundedness and resiliency replace them (see Botkin and Sobel, 1975; and other papers in Levin, 1975; Holling, 1973).

The major conclusions of such studies are that there are inherent limits to predictability, that predictability is a property in part of the scale of investigation, and that there is no single correct scale of inquiry. Moreover, overly detailed and reductionistic models of populations and systems obscure any pattern by introducing irrelevant detail, often on the specious premise that somehow more detail and more reduction assures greater truth. However, this is predicted in part on the fallacious notion that there is some exact system description possible, and that any model is an approximation to that exact description. In reality, there can be no such ideal, because there can be no "correct" level of aggregation. The taxonomic species, for example, is an imperfect tool of classification, and ignores the differences among the individuals of the species with regard to demographic and phenotypic properties; thus it is just one possible grouping within a particular nested hierarchy. More importantly, the particular hierarchical decomposition of a system that arises from a taxonomic classification system bears much less relevance for some ecological descriptions than would one that was functionally based, and there are other possible schemes. Thus, as with Mandelbrot's fractal curves (Mandelbrot, 1983), there is no natural level of choice within any nesting, and the problem is complicated further by the existence of multiple overlapping nested and non-nested hierarchical schemes.

SIMPLIFICATION

For any particular application, one must be concerned with levels of description that best relate to the phenomena of interest and the scales of observation. One needs to simplify system descriptions such that the essential features are preserved and the obfuscating details are excised. This argues for simplified mathematical models that deliberately eliminate detail by abstracting the essential features. The importance of such simplification becomes obvious in another way from consideration of the problem of parameter estimation. Andreasen (unpublished) has pointed out that in some highly detailed models, such as the Nordsee fisheries model, the number of parameters to be estimated may be of the same order or larger than is the number of observations available. Such models may still be useful as sketches of the systems, as tableaus on which all of the interactions of interest may be displayed; but they do not serve as bases for prediction. Instead, one must find techniques for lumping and simplification, for example based on size classes or habitat, that render models manageable.

More generally, even when the number of parameters to be estimated is given, some models (especially those that can exhibit chaotic behavior and high parametric sensitivity) pose much greater estimation problems than do others (Ludwig and Walters, 1985). A consequence is that the choice of a useful model must be governed by knowledge of the dynamic behavior of candidate models, by their parametric sensitivity, and by the tradeoffs between uncertainty in knowledge of model structure and uncertainty in acquiring data (see the papers by Ludwig and Walters, this volume).

Given the necessity of finding ways to simplify systems, attention must turn to the issue of techniques of simplification. The most successful are likely to be direct methods, for example those that capitalize on the explicit separation of temporal and spatial scales caused by patch dynamics (e.g., Levin and Paine, 1974; Paine and Levin, 1981), or that separate scales by building on knowledge of the dynamics of key physical and biological processes (e.g., Ludwig, Jones, and Holling, 1978). In community and ecosystem models, the most usual candidates for simplification are based on aggregations on the basis of trophic status or functional equivalence, or system separation based on tight interactions (modules, sensu Paine, 1980). Hierarchical decomposition according to tightness of interaction is a familiar technique in general systems theory, especially as applied in economics. Simon and Ando (1961) write that "hierarchy is a common characteristic of real dynamical systems," and even were this not true, "research is conducted as if this were the case." Time scale separation is one of the most powerful tools available, but often emerges indirectly from separation according to tightness of interaction, according to such concepts as near decomposability (Simon and Ando, 1961).

In many situations, it is not at all clear what represent natural simplifications, and one must turn to indirect methods. Matrix methods provide ways to analyze interaction networks and to find new vector space bases that render the systems almost separated; but such methods are inherently linear, and do not work well for systems with strong nonlinearities. A number of direct methods are available for gaining insight into underlying dimensionality; but these remain unproven in ecological applications. They include techniques borrowed from dynamical systems theory, particularly the method of Poincaré maps as applied by Takens (1981) and used with success by Schaffer and his coworkers in analyzing data sets in ecology and epidemiology (see, for example, Schaffer and Kot, 1985; Schaffer, Ellner, and Kot, 1987). Such methods in principle could also be used to simplify highly detailed models, by analysis of the "phony" data sets generated by simulations of the models.

There exists a formal theory of aggregation and simplification (for example, Iwasa, Andreasen, and Levin, 1987); but it remains open to question whether such formal methods will ever prove very useful in specific applications. The general approach is to begin with some particular system description, to study the behavior of the dynamics associated with that description, and to compare the behavior with that of reduced (e.g., aggregated) descriptions, with reference to some set of indicator variables. This basically is the method employed by Ludwig and Walters (1985), although their approach was the more productive one of making explicit comparisons rather than trying to rely on theorems arising from general theory.

MODELS, MONITORING, AND ADAPTIVE ENVIRONMENTAL MANAGEMENT

What society values, and what environmental laws are designed to protect, are not the intricate details of systems, but rather the essential features that make those systems recognizable to biologists and that are somewhat predictable over time. We should not expect to make systems more predictable in our models than natural variability dictates; in fact, variability, such as that associated with seasonality, gap phase dynamics, or fire, is one of the most predictable features of many systems, and is essential to the maintenance of resiliency and to the persistence of most species.

Furthermore, predictability is inextricably intertwined with variability, and with the temporal and spatial scales of interest. Thus, a central challenge in ecological theory must be an elaboration of the understanding of how scales relate, how systems behave on multiple scales, and how the measurement and dynamics of particular phenomena vary across scales. The inherent limits to predictability on long time scales emphasize the importance of monitoring, and of coupling any management action with some mechanism for modification based on analysis of the data obtained from monitoring. Such adaptive management (Holling, 1978) recognizes explicitly the limits to predictability, and places emphasis on short-term prediction in which nonlinear phenomena have diminished importance.

In this regard, renewable resource management presents a number of problems not fundamentally different in kind from those presented by other environmental stresses. The prediction of the dynamics of natural populations has never been adequately resolved, even for renewable resources such as fisheries that have been the objects of scientific scrutiny for decades. The problems relate to the inability to identify and predict changes in the factors controlling dynamics, to the spatial and temporal variability of parameters and even mechanisms of control, and to the inherent propensity of nonlinear dynamical models to exhibit turbulent dynamics that make parameter estimation a daunting challenge. When communities and ecosystems are considered, with the consequent multiplication of pathways of interaction, the problems are similarly multiplied. Some relief can be achieved by judicious simplification that properly recognizes the shift in detail appropriate to the shifts in levels of organization, but a core of irremovable uncertainty will always remain. Scientists and modelers advising managers must make clear the levels of uncertainty, and not give in to the temptation to seem to present the managers with the certainty that they seek. Model outputs must be presented not just in terms of means, but with associated variances in relation to stochastic effects and uncertainties in parameter estimation. Finally, an inescapable conclusion from the existence of such uncertainty is that there will be surprises associated with virtually any management action (Holling, 1978), and that any management strategy must have some potential for (adaptive) modification when experience and monitoring so dictate.

ACKNOWLEDGMENT

This publication is ERC-120 of the Ecosystems Research Center (ERC), Cornell University, and was supported by the U.S. Environmental Protection Agency Cooperative Agreement Number CR811060. Additional funding was provided by Cornell University and by NSF Grant DMS-8406472 to the author. The views expressed herein are those of the author, and do not necessarily represent those of the granting agencies or institutions.

REFERENCES

Botkin, D.B. and M.J. Sobel. 1975. The complexity of ecosystem stability. pp. 144-150 in S.A. Levin (Ed.), Ecosystem Analysis and Prediction. SIAM, Philadelphia.

Haury, L.R., J.A. McGowan, and P.H. Wiebe. 1978. Patterns and processes in the time space scales of plankton distribution. pp. 277-328 in J.H. Steele (Ed.), Spatial Pattern in Plankton Communities. Plenum, New York.

Holling, C.S. 1973. Resilience and stability of ecological systems. Ann. Rev. Ecol. Syst., Vol. 4, pp. 1-23.

Holling, C.S. (Ed.). 1978. Adaptive Environmental Assessment and Management. Wiley, New York, 377 pp.

Iwasa, Y., V. Andreasen, and S.A. Levin. 1987. Aggregation in model ecosystems. I. Perfect aggregation. Ecol. Modeling, in press.

Levin, S.A. 1975. Ecosystem Analysis and Prediction. SIAM, Philadelphia, 337 and xiv pp.

Levin, S.A. and R.T. Paine. 1974. Disturbance, patch formation, and community structure. Proc. Nat. Acad. Sci., U.S.A., Vol. 71, pp. 2744-2747.

Ludwig, D., D.D. Jones, and C.S. Holling. 1978. Qualitative analysis of insect outbreak systems: The spruce budworm and the forest. J. Anim. Ecol., Vol. 47, pp. 315-332.

Ludwig, D. and C.J. Walters. 1985. Are age-structured models appropriate for catch-effort data? Can. J. Fish. Aquat. Sci., Vol. 42, pp. 1066-1072.

Mandelbrot, B. 1983. The Fractal Geometry of Nature. W.H. Freeman & Co., 468 pp.

May, R.M. 1974. Biological populations with non-overlapping generations: Stable points, stable cycles, and chaos. J. Theor. Biol., Vol. 49, pp. 511-524.

Paine, R.T. 1966. Food web complexity and species density. Amer. Natur., Vol. 100, pp. 65-75.

Paine, R.T. 1980. Food webs: Linkage, interaction strength and community infrastructure. J. Animal Ecology, Vol. 49, pp. 667-685.

Paine, R.T. and S.A. Levin. 1981. Intertidal landscapes: Disturbance and the dynamics of pattern. Ecol. Monog., Vol. 51, pp. 145-178.

Pickett, S.T.A. and P.S. White. 1985. The Ecology of Natural Disturbance and Patch Dynamics. Academic Press, Orlando, Florida.

Schaffer, W.M. and M. Kot. 1985. Nearly one dimensional dynamics in an epidemic. J. Theor. Biol., Vol. 112, pp. 403-427.

Schaffer, W.M., S. Ellner and M. Kot. 1987. Effects of noise on some dynamical models in ecology. J. Math. Biol., in press.

Simon, H.A. and A. Ando. 1961. Aggregation of variables in dynamic systems. Econometrica, Vol. 29, pp. 111-138.

Stommel, H. 1963. Varieties of oceanographic experience. Science, Vol. 139, pp. 572-576.

Stommel, H. 1965. Some thoughts about planning the Kuroshio survey. In Proc. Symp. on the Kuroshio, Tokyo, Oct. 29, 1963. Oceanogr. Soc. Japan and UNESCO.

Takens, F. 1980. Detecting strange attractors in turbulence. pp. 366-381 in D.A. Rand and L.S. Young (Eds.), Dynamical Systems and Turbulence. Warwick, 1980. Lecture Notes in Mathematics, 898. Springer-Verlag, 1981.

Watt, A.S. 1947. Pattern and process in the plant community. J. Ecol. Vol. 35, pp. 1-22.

Whittaker, R.H. 1975. Communities and Ecosystems. 2nd edition. MacMillan, New York.

PARTICIPANT'S COMMENTS

Dr. Levin's paper deals with particularly interesting and important issues. In the management of any renewable natural resource there is no substitution for a penetrating and detailed understanding of the managed system, which often emerges after years of experience. Yet, there comes a time where intuition fails us, and this is when one has to rely, for better or worse, on models as aids in management decisions. There are a few points which I find particularly interesting and which I shall briefly amplify (at the risk of being somewhat archaic).

It is my feeling that the purpose of modeling is often overlooked, and managers tend to believe more or less in a particular model not based on its usefulness for a specific problem, but rather based on the salesmanship of its proprietors. This is partly because ecological modeling has become almost a discipline by itself, and managers have to rely on others' expertise. For example, the difference between the so-called strategic and tactical models--with their inherent scale differences--is often overlooked, and managers, in general, do not believe much in strategic models to begin with. Yet, there is a need for both types of models, and for a healthy dose of interaction between them. One way to overcome this difficulty is to emphasize, time and again, the purpose and limitations of a particular model and foster an attitude of constructive criticism among modelers. There is also a need to foster an attitude of mutual respect between modelers and managers.

There is not enough attention paid to modeling for the purpose of pattern recognition. Most modelers, I suspect, approach a modeling problem based on some preconceived notion of how the system operates, forgetting that observations must precede hypotheses (and that hypotheses then generate new observations). With the aid of pattern recognition one might get a better handle on the problem of scale. Closer working relationships between modelers and managers during the early stages of model development might be partly helpful. In this respect, recent work on fractal analysis, where the underlying dimensionality of systems can be investigated, might be particularly useful. Such analysis will not, of course, provide answers to the more important problem of identifying these particular dimensions.

Some of the most compelling desires of any management are predictability and, more importantly, reduction of variability (be it regular or not). The latter usually leads to the former, whereas the reverse is not necessarily true. Predictability is not a particularly difficult problem, and with adequate resources and time may be achieved (or at least we may answer the question of how achievable it is). Reduction of variability requires altering system characteristics, and is achievable to a much lesser extent. Management then should concentrate on predictability, with the recognition that reduction in variability is not only difficult to achieve (at best), but is essential for the proper functioning of many ecological systems. Forest-fire management is a good example where such ideas are recognized and practiced in management.

Finally, I would like to comment on the problem of stiff, as opposed to flexible models. Stiff models are those in which some components change very rapidly compared to others. Flexible models are those in which all components change with the same order of magnitude. Stiff models often lead to the so-called "catastrophic" behavior. If one accepts the attitude that models only approximate reality, and that some of their major purposes are to provide predictions and aid in management decisions, then a stiff model is a good indication that either its scale should be changed or the whole model should be revised. One of the goals of modelers should be to produce flexible models.

Yosef Cohen

Professor Levin emphasizes some critical aspects of the modeling process that are common to the development of mathematical models of ecological systems and their associated biological resources and that have been recognized as crucial in the process of modeling physical/chemical systems as well.

1. The essential elements of a mathematical model depend equally upon the questions to be addressed in the modeling effort and upon the nature of the system being modeled.

2. Levels of temporal and spatial scale to be encompassed in the model format are dictated by (1) and are reflected in appropriate aggregation and simplification schema.

3. Uncertainty in the parameterization of the model, whether due to inherent biological variability or to lumped and/or unmodeled dynamics, produces unavoidable variability and unpredictability in the forecasts of these models.

4. Any proposed management scheme that is based upon the forecasts of these models should be designed to incorporate fully the effects of these aggregated processes, functional uncertainties, and parametric variabilities.

Ann Lowes Blackwell

BEHAVIORAL MODELLING AND RESOURCE MANAGEMENT

Colin W. Clark
Institute of Applied Mathematics
University of British Columbia
Vancouver, B.C. V6T 1W5 Canada

The application of stochastic dynamic programming, or Markov Decision Processes (MDP) to the modelling of behavior is discussed. Two examples are described: the hunting behavior of Serengeti lions, and the oviposition behavior of parasitic wasps. The MDP approach appears to explain the published data for these examples much more satisfactorily than does classical optimal foraging theory. The scope for applying MDP models to resource management problems is discussed briefly.

INTRODUCTION

A fertile area for the application of control theory is the modelling of behavior--both human and animal. The payoffs to research in this area promise to be unusually high, for two reasons. First, an improved understanding of human behavior would be invaluable, for example in the design of more effective resource regulation systems. An improved understanding of animal behavior could have applications in such areas as pest control and fishery management (Bardach et al., 1980). Second, believe it or not, the field is almost untouched.

The latter claim needs some qualification. Over the past 20 years ecologists have developed many optimization models of animal behavior (Krebs and Davies, 1984, provide a good review). The vast majority of these models, however, have been either deterministic or averaged stochastic. The examples discussed later in this paper suggest that dynamic stochastic behavioral models could lead to major new insights into animal behavior. The modelling of human behavior is also well advanced in certain directions (for example, game theory models), but again dynamic stochastic optimization models are scarce. Ray Hilborn has been in the forefront in recognizing the need for such models in order to understand fishing fleet dynamics (Hilborn, 1985).

Of course, stochastic optimization models can be quite unwieldy. One such class of models, however, has been extensively studied, and applied successfully in the field of Operations Research--namely the Markov Decision Process (MDP), or more generally, stochastic dynamic programming models (Heyman and Sobel, 1984). Marc Mangel and I have recently been studying MDP models of animal behavior (Mangel and Clark, 1986), and we are excited about their potential.

MDP MODELS OF ANIMAL BEHAVIOR

Consider an animal that can forage in one of N patches. The patches vary in their productivity, and also in the risk of predation. Patch-choice decisions are made for specific periods--e.g. one day. Many--perhaps most--animals regularly face such decisions; how can their behavior be modelled?

To begin with, a dynamic stochastic model seems indispensible. Let X_t denote the forager's gut contents in terms of energy (X_t is the state variable) at the beginning of period t. If the forager is not killed by a predator, we have

$$X_{t+1} = X_t - \alpha_t + Z_t \tag{1}$$

where α_t denotes the decrease in X_t due to metabolic processes, and Z_t is a random variable denoting food intake. It is reasonable to assume a priori bounds on X_t:

$$x_c < X_t < C \tag{2}$$

where C denotes capacity and x_c is some critical level such that death occurs (by starvation) if X_t falls below x_c. Usually one can assume that $x_c=0$, but in the case of birds foraging in winter, x_c might represent the minimal energy requirement to survive the night (McNamara and Houston, 1982).

Suppose that the i-th patch entails a metabolic cost α_i, and yields forage according to a distribution f_i:

$$\Pr (Z_t=e_{ij}) = f_{ij} \qquad (j=1,2,\ldots,J_i) \quad \text{on patch } i \tag{3}$$

(where $e_{i1}=0$). Finally suppose patch i carries a probability β_i of being killed by a predator. We will treat the objective of maximizing the forager's probability of survival over a given time period (but other objectives, for example involving breeding success, can also be considered). Let $p(x,t)$ denote the maximum probability of survival for t periods, given that $X_0=x$. We then obtain

$$p(x,0) = \begin{cases} 1 & \text{if } x > x_c \\ \\ 0 & \text{if } x < x_c \end{cases} \tag{4}$$

(the forager is alive with t=0 if and only if $x > x_c$). Next, with t+1 periods left, suppose the forager selects the i-th patch in the first period. If it escapes predation and discovers food e_{ij} then its maximum probability of surviving for the remaining t periods is $p(x'_{ij},t)$ where

$$x'_{ij} = \min (x - \alpha_i + e_{ij},C)$$

Hence, the maximum probability of surviving all t+1 periods is given by

$$p(x,t+1) = \max_i (1 - \beta_i) \, \Sigma_j \, f_{ij} p(x'_{ij},t) \tag{5}$$

where

$$x'_{ij} = \min (x - \alpha_i + e_{ij},C) \tag{6}$$

This is a typical MDP model. It is simple enough to be quickly solved by iteration (given numerical parameter values)--an example is worked out in Mangel and Clark (1986). The results are not surprising--given a choice between dangerous, highly productive patches and safer, but less productive patches, a forager should prefer safe patches whenever x is near C, but should choose progressively more dangerous patches as x gets close to x_c. See Milinski (1986) for experimental tests of this and other predictions of the foraging behavior of fish.

Such a result seems trite, but it is worth noting that classical foraging theory *completely ignores the state of the forager* (and usually ignores predation also). Experimentalists usually standardize experimental conditions by working with starved animals, but this clearly misses many interesting features of behavior. (Similarly, most microeconomic models commonly ignore the financial state of the firm or entrepreneur, even though most people probably behave quite differently when they are well off than when they are poor.)

GROUP FORAGING

Many species of animals forage in groups, at least part of the time. Such behavior can be economically important, for example in the case of fish schools (Murphy, 1980). The adaptive significance of group foraging has been extensively analyzed (Pulliam and Caraco 1984). Early models (Thompson et al., 1974) indicated no increase in average feeding rate, but a decrease in variance, as group size increases. Clark and Mangel (1986) describe various conditions under which average feeding rates can be increased--sometimes dramatically--by group foraging. The survival value of reduced variance has been studied by Caraco (1981a), who showed that foragers should be risk averse whenever expected feeding rate exceeds requirements, and vice versa. Several species of birds have been observed to employ such risk-sensitive behavior (Caraco, 1981b, 1982).

The foraging behavior of social carnivores (lions, hyenas, wolves, etc.) has been studied by Schaller (1972), Kruuk (1975), and others. Caraco and Wolf (1975) used Schaller's data on Serengeti lions to calculate the average daily individual intake of food, as a function of hunting group size, for lions hunting zebra, wildebeest, and gazelle under various ecological conditions. In all cases it turned out that groups of size n = 2 would result in the maximum average feeding rate (but see Packer, 1986). Except for the case of gazelle, however, observed feeding group sizes typically ranged from four to eight lions: see Table 1.

Various hypotheses have been proposed to account for this discrepancy: breeding success and competition with hyenas (Caraco and Wolf, 1975); kin selection (Rodman, 1981); game theory (Sibly, 1983; Clark and Mangel, 1984). Excellent numerical predictions are obtained from the ad hoc hypothesis of satisficing behavior -- lions simply feed in the largest groups that will satisfy their daily

Table 1. Lion hunting group sizes, as predicted by game theory (ESS), and by the satisficing hypothesis, as compared with mean observations (Schaller, 1972, Table 40). From Clark (1986).

Prey	Region	Game Theory	Satisficing	Mean Observed
gazelle	Border	2	2	2.0
	E. plains	2	2	1.5
wildebeest	Border	4	7	6.7
	W. woodlands	4	4	4.8
	E. plains	4	4	3.6
zebra	Border	3	7	7.3
	W. woodlands	3	4	4.0
	E. plains	3	4	3.7

meat requirement of 6 kg per lion (see Table 1).

Is satisficing a more "rational" objective than maximizing expected food intake? In order to answer this, I considered the following MDP model (see Clark, 1986, for full details):

$$X_{t+1} = X_t - \alpha + Z_t \tag{7}$$

$$0 < X_t < C \tag{8}$$

$$Pr(Z_t = e/n) = \lambda_n \qquad Pr(Z_t = 0) = 1 - \lambda_n \tag{9}$$

where n denotes group size (the decision variable) and e is the size of food items (this can easily be made random).

With $p(x,t)$ again denoting the probability of survival over t periods, given $X_0 = x$, we have as before

$$p(x,0) = \begin{cases} 1 & \text{if } x > 0 \\ 0 & \text{if } x < 0 \end{cases} \tag{10}$$

$$p(x,t+1) = \max_n \left[\lambda_n \, p(x'_n,t) + (1-\lambda_n)p(x-\alpha,t) \right] \tag{11}$$

$$x'_n = \min (x - \alpha + e/n, C) \tag{12}$$

The following parameter values are taken from Schaller (1972):

$$\alpha = 6 \text{ kg} \qquad C = 30 \text{ kg}$$

$$e = \begin{cases} 164 \text{ kg} & \text{(zebra)} \\ \\ 12 \text{ kg} & \text{(gazelle)} \end{cases} \tag{13}$$

Kill probabilities λ_n were derived from Caraco and Wolf's estimates; e.g., for gazelle, $\lambda_1 = 0.15$ and $\lambda_n = 0.37$ (wet season), $\lambda_n = 0.31$ (dry season) for $n \geqslant 2$.

The numerical results produced by the model (with several minor refinements described in Clark, 1986) are shown in Table 2. Several observations are worth noting. First, the optimal hunting group size is in most cases quite sensitive to the lions' hunger state. Hungry lions need more food, and should hunt in smaller groups than well-fed lions. In fact, well-fed lions should approximately satisfice (e.g., for zebra prey, the carcasses of which can last up to three days, the satisficing group size is $n = 9$, but I restricted n to be at most six). In this regard, note that because of the capacity constraint $x < 30$, two lions can only eat at most 96 kg over a three-day period. If $\lambda_3 > \lambda_2$, clearly a group of three lions does better than a group of two. None of the previous analyses of lion group size considered the effect of a capacity constraint.

Table 2. Thirty-day survival probabilities $p(x,30)$ and optimal group sizes n^* for lions hunting Thomson's gazelle and zebra.

(a) Wet season

initial reserves x(kg)	gazelle P(x,30)	n*	zebra p(x,30)	n*
5	.256	2	.860	3
10	.429	2	.902	4
20	.640	2	.971	4
30	.736	2	.978	6

(b) Dry season

initial reserves x(kg)	gazelle P(x,30)	n*	zebra p(x,30)	n*
5	.063	1	.542	3
10	.128	1	.571	4
20	.256	2	.700	4
30	.352	2	.714	6

For lions hunting gazelle in the dry season, the optimal group size becomes $n^* = 1$ for $x < 10$ kg. Under these conditions, many lions will in fact be hungry, and our model therefore predicts that some lions will hunt as individuals. Note that this also agrees with the observations (Table 1), another feature not discussed in any previous analysis.

The MDP model thus provides a completely different picture of lion foraging behavior than do simple average-rate models. Biologically important constraints and parameters, which are difficult to incorporate in averaging models, are easily, often automatically, included in the MDP model; their effects are easily assessed.

There is more to lion hunting behavior than I have space to discuss here. For example, lions sometimes share their kills with pride mates which converge from several kilometers distance (Packer, 1986). It can be shown that this communal sharing behavior implies smaller optimal hunting group sizes and greater survival probabilities (because of increased average food rates together with reduced variances) than those given in Table 2. Communal sharing behavior does not appear to be commonly found in nature, the main exceptions being the social insects (Oster and Wilson, 1978), and man. To what extent are the social institutions of early man a response to the advantages of variance reduction? This question is of current interest to anthropologists; perhaps Markovian models may help in understanding this interesting possibility.

I have not discussed the dependence of the optimal group size n^* on the time horizon t, or on the form of the terminal payoff function of Eq. (10). For the lions model it turns out that the optimal group sizes n^* shown in Table 2 (for $t = 30$ days) in fact constitute an optimal equilibrium policy, valid for all $t >$ approx 10 days. For shorter time horizons, n^* depends upon T; e.g., for $t = 1$ and $x > 6$ kg clearly $n^* = 0$--the lion will survive one day without hunting at all. As one would expect, such artificial horizon effects tend to disappear as t increases.

Parasitic wasps. Charnov and Skinner (1984) used a classical model of optimal clutch size (originally developed for birds) to calculate the optimal number of eggs that a parasitic wasp should lay on each host. The model failed to explain the data: wasps laid eggs in clutches of all sizes *less than or equal to* the predicted optimum! The following MDP model of Mangel (1986) completely explains the data.

Assume that an insect initially has a fixed supply R of eggs that it can lay over its lifetime (other assumptions can also be modelled). Let X_t denote the number of eggs that it has left to lay at period t; $X_0 = R$. Hosts occur in various types (e.g. sizes) indexed by i. Let

$$p_i = Pr \text{ (wasp encounters host of type i)}$$

where

$$\Sigma \ p_i \ < \ 1$$

(no host may be encountered in a given period). Also let $\phi_i(C)$ denote the increment in parental fitness if a clutch of size C is laid on a host of type i. (Charnov and Skinner estimate ϕ_i in terms of the expected number and size of surviving offspring.) If $p(x,t)$ denotes maximum expected lifetime fitness with t periods left, and $X_0 = x$, we then have

$$p(x,0) = 0 \tag{14}$$

$$p(x,t+1) = \Sigma_i p_i \max_{C \leqslant x} \ [\phi_i \ (C) + p(x-C,t)] + (1 - \Sigma_i p_i) \ p(x,t) \tag{15}$$

(Notice how simple and flexible the MDP approach is!)

Upon running this model, Mangel found that the optimal clutch size, for given t, depended strongly on the "reserves" x. The smaller x (and the longer the horizon t), the smaller the optimal clutch size. Also the optimal clutch size never exceeds C_i^*, the maximand of $\phi(C)$ (the predicted clutch size of Charnov and Skinner). Once it has been obtained, this result is almost obvious! With an infinite time horizon, the wasp should obviously lay *one* egg per host (ϕ_i is a concave function). With lots of eggs and little time in which to lay them, the wasp has to lay more per host, but it never pays to lay more than C_i^*. Since finding hosts in nature is a random process, any population of wasps will have members with a variety of (x,t) pairs at any given time. Hence the spread in the data.

We're convinced that the MDP paradigm vastly outperforms classical foraging theory. It often leads to new testable predictions. For example, it seems that it would be easy to test the prediction that wasps lay larger clutches of eggs after a period of unsuccessful prey hunting. Another testable prediction is that the average clutch size will decrease as the abundance of prey increases. Charnov and Skinner discuss a simple alternative model which also yields the latter prediction, but it's nice to have *one* model framework that does everything.

The study of the behavior of parasitic insects clearly could have applications to biological pest control. Marc tells me that he plans to return to Hawaii to work on this problem.

SUMMARY

It's always risky to claim to have found the philosopher's stone. But I strongly believe that Markov Decision Process models have great potential for the study of animal behavior. The whole question of mean-variance tradeoffs, currently a hot topic in foraging theory (and not neglected in the resource literature, either) can be put into a unified, meaningful framework by using the MDP approach. Multiple behavioral decisions can also be modelled in a consistent way. New experiments are suggested, while old experimental results are explained better or more simply.

MDP models of behavior provide many theoretical problems to attract the mathematician. To what extent can continuous-time models, which are probably more realistic, be used? How can models be linked sequentially? What is the best objective function from an evolutionary standpoint, and how sensitive are predictions to the choice of objective function? What applications are there to economic theory? Perhaps answers to these and other questions will begin to emerge in the near future.

REFERENCES

Bardach, J.E., Magnuson, J.J., May, R.C., and Reinhart, J.M. (eds.) 1980. Fish Behavior and its Use in the Capture and Culture of Fishes. International Center for Living Aquatic Resource Management, Manilla.

Caraco, T. 1981a. Risk sensitivity and foraging groups. Ecology, Vol. 62, pp. 527-531.

Caraco, T. 1981b. Energy budgets, risk, and foraging preferences in dark-eyed juncos (Junco hymelais). Behav. Ecol. Sociobiol., Vol. 8, pp. 213-217.

Caraco, T. 1982. Aspects of risk-aversion in foraging white-crowned sparrows. Anim. Behav., Vol. 30, pp. 719-727.

Caraco, T. and Wolf, L.L. 1975. Ecological determinants of group sizes of foraging lions. Amer. Natur., Vol. 109, pp. 343-352.

Charnov, E.L. and Skinner, S.W. 1984. Evolution of host selection and clutch size in parasitoid wasps. Florida Entomol., Vol. 67, pp. 5-21.

Clark, C.W. 1987. The lazy, adaptable lions: a Markovian model of group foraging. Anim. Behav., Vol. 35, pp. 361-368.

Clark, C.W. and Mangel, M. 1984. Foraging and flocking strategies: information in an uncertain environment. Amer. Nat., Vol. 123, pp. 626-641

Clark, C.W. and Mangel, M. 1986. The evolutionary advantages of group foraging. Theor. Pop. Biol., Vol. 30, pp. 45-75.

Heyman, D.P. and Sobel, M.J. 1984. Stochastic Models in Operations Research. Vol. 2. McGraw Hill, New York, 555 pp.

Hilborn, R. 1985. Fleet dynamics and individual variation. Can. J. Fish. Aquatic Sci., Vol. 42.

Krebs, J.R. and Davies, N.B. 1984. Behavioral Ecology. 2nd Edition. Blackwell Scientific Publications, Sinauer, Sunderland, MA, 494 pp.

Kruuk, H. 1975. Functional aspects of social hunting by carnivores. In Baerends, G.P., Beer, D., and Manning, A. (eds.), Function and Evolution in Behavior. Clarendon, New York, pp. 119-141.

Mangel, M. 1986. Understanding oviposition site selection and clutch size in parasitic insects using Unified Foraging Theory. J. Math. Biol., (in press).

Mangel, M. and Clark, C.W. 1986. Towards a unified foraging theory. Ecology, Vol. 67, pp. 1127-1138.

McNamara, J.M. and Houston, A.I. 1982. Short term behavior and lifetime fitness. In McFarland, D.J. (ed.), Functional Ontogeny. Pitmans, London, pp. 60-87.

Milinski, M. 1986. Constraints placed by predators on feeding behaviour. in Pitcher, T.J. (ed.), The Behavior of Teleost Fishes, Croom Helm, London, pp. 236- 252.

Murphy, G.I. 1980. Schooling and the ecology and management of marine fish. In Bardach, J.E., Magnuson, J.J., May, R.C., and Reinhart, J.M. (eds.) Fish Behavior and its Use in the Capture and Culture of Fishes, International Center for Living Aquatic Resource Management, Manilla.

Oster, G.F. and Wilson, E.O. 1978. Caste and Ecology in the Social Insects. Princeton Univ. Press, Princeton, NJ.

Packer, C. 1986. The ecology of sociality in fields. In Rubenstein, D.I. and Wrangham, R.W., Ecological Aspects of Social Evolution, Princeton University Press (in press).

Pulliam, H.R. and Caraco, T. 1984. Living in groups: is there an optimal group size? Ch. 5 in Krebs, J.R. and Davies, N.B. (eds.), Behavioral Ecology: An Evolutionary Approach. Sinauer, Sunderland MA, pp. 122-147.

Rodman, P.S. 1981. Inclusive fitness and group size, with reconsideration of group sizes of lions and wolves. Amer. Nat., Vol. 118, pp. 275-283.

Schaller, G.B. 1972. The Serengeti Lion. Univ. Chicago Press, Chicago, IL.

Sibly, R.M. 1983. Optimal group size is unstable. Anim. Behav., Vol. 31, pp. 947-948.

Thompson, W.B., Vertinsky, I. and Krebs, J.R. 1974. The survival value of flocking in birds: a simulation model. J. Animal Ecol., Vol. 43, pp. 785-820.

PARTICIPANT'S COMMENTS

Optimality studies of animal behavior have proceeded at two levels in ecology. The first has been an attempt to define the rules governing the optimal behavior of individuals. The second has been to allow those rules to depend upon intraspecific and interspecific interactions. Clark's models have elements of both.

I believe that one of the hallmarks of great science is that it seems obvious once done. We should not, therefore, be misled by the transparency and reasonableness of Clark's conclusions. They are indications of true generality and power, and not of triviality. Bridging the gap between the individual and the group in optimality theory is a fundamental step.

A bonus is the model's ability to deal with individual variation. The importance of doing this is estimable. Pulliam (pers. comm.), for example, has recently noted that because of such variation, a population may regularly be forced to use habitat types in which it cannot survive. Its realized niche may be larger than its fundamental niche. The successful manager is going to need to know that. Otherwise, one might put a refuge in the wrong place or seriously misinterpret the health and exploitability of a population.

The full effects of the model will be felt only after it is fully adapted to situations of density dependence. This, for example, is how Pulliam gets his conclusion. There is no reason this cannot be accomplished. Markov Decision Process models should certainly be useful in redefining out understanding of isoleg models and using them to discern the rules by which communities are organized (Rosenzweig, 1985, in Cody, M.L. (ed.), Habitat selection in birds).

There is an implicit caveat in Clark's approach which should be taken to heart by all ecologists. Many have rejected optimality models entirely because particular models have failed to predict every datum. This is foolishness. MDP models are just as much optimality models as the more primitive models they replace. Optimality models are a strategy of investigation and as such are no more or less valid than statistics or field observation.

A case in point is satisficing. Many ecologists have absorbed this as an alternative to optimization, but it is not. Clark has accurately mirrored the way it is (mis)used by ecologists: lions eat until they need no more; then they sleep or play or engage in omphaloskepsis. But satisficing has its roots in economics where it means something somewhat different. There, it is simply another form of optimization: find tactics that work, i.e., that make you a profit, and then maintain them. The reason this is an optimization strategy (even though you do not know whether they are the best tactics) is that there is often a high cost to collecting information about alternatives. You could go bankrupt on your way to a better set of tactics. Once you have factored in the cost of collecting that information, you conclude rightly that it is suboptimal to obtain it. Perhaps we ecologists could reform so that we stop forcing good papers like Clark's to quote our abominations in order to communicate with us.

Michael L. Rosenzweig

My remarks will focus on two logical extensions of the modelling of animal behavior discussed in this paper to the modelling of human behavior. The models described here raise some fundamental and exciting questions about the concurrence of behavioral dynamics of animals and humans as resource users, and what those dynamics imply for successful resource management.

Models of animal foraging seem particularly suited to the representation of human behavior in the fishers. Fishermen choose patches -- fishing grounds -- for a fixed amount of time which depends in large part on the fishery. Fishing entails area-specific costs and realizes area-specific returns. Fishermen work in a stochastic environment which presents a significant level of physical risk, not of predation but of gear damage, vessel loss, or loss of life. Wide differences in fishing skill lead to different economic states among fishermen. Various objectives may be specified for the fishing process, including maximizing profits, meeting some minimal requirements for long run economic survival ("satisficing"), or minimizing time at sea.

The major deficiencies of classical foraging theory addressed by the Markov Decision Process (MDP) model, namely omission of both forager state and the risk of predation, can also be claimed for much of microeconomic theory, as Clark points out. A notable exception to this nontreatment of wealth effects on economic decision making is found in the literature on agricultural producers and their attitudes toward risk.

In these studies, the financial state of the decision maker is found, along with socioeconomic factors, to be important in determining risk attitudes. The explicit consideration of the influence of the state of the decision maker on behavior has not carried over from agricultural economics to natural resource economics. Economic models of the fishery typically make assumptions of a homogeneous behavioral motivation for all users. The current economic condition of fishermen is not assumed to affect their decision making; rather the structure of property rights is considered the critical element in shaping decisions regarding resource exploitation.

Clark's discussion of the MDP approach suggests to me that there are two immediate possibilities for extension of the theory into the realm of human resource use. MDP models yield some intriguing and testable hypotheses about economic behavior in the fishery, namely the level of risk acceptance by fishermen and the incentives for the formation of fishermen code groups to share information on the location of the stocks of fish.

According to the MDP model, given a choice between risky productive patches and safer less productive patches, foragers prefer safer patches when they are near their capacity for food intake and risky patches when they approach starvation. This result can be directly tested in the fishery by hypothesizing that a poor financial state of the fleet correspond to higher levels of risk taking; healthy financial conditions correspond to low risk taking. Observations of increased vessel losses in recent years of poor performance in some fisheries suggest that fishermen's behavior may indeed conform to MDP model dynamics.

The MDP model of group foraging also offers insight into the efficiency effects of information sharing through code groups in fisheries. Group foraging combined with communal sharing can both increase average food intake and reduce its variance. This result is consistent with observations of the existence of information networks in fisheries characterized by high levels of variability in fish location, and the absence of code groups in fisheries with low levels of locational uncertainty.

These are only two examples of behavioral modelling possibilities in the fishery suggesting that the MDP paradigm is part of a rich research area offering a strong potential for the modelling of human as well as animal behavior. One of the exciting aspects of the MDP model is the history implicit in the state of the actors that is not recognized by other models. A better understanding of the importance of role financial state plays in decision making is one way that MDP models may lead to management policies more attuned to the intricacies of human behavior and therefore more likely to be effective.

Susan S. Hanna

MODELING FOR BIOLOGICAL RESOURCE MANAGEMENT

Wayne M. Getz
Division of Biological Control
University of California
Berkeley, CA 94720

In a management context, analytical models of biological populations can be used to develop a theory of resource conservation and exploitation, or detailed simulation models can be used to evaluate the performance of various management policies with respect to particular resource systems. Both activities can benefit by developing as strong a connection between them as possible. This problem is discussed especially with reference to stage (age, size, etc.) structured resources that encompass fisheries, forests, insect rearing, and wildlife management problems. Other questions dealt with are: evaluating and comparing solutions to narrowly defined problems in more broadly defined contexts; parameter uncertainty in the context of deterministic models; boundary conditions that accompany planning over finite time horizons; the effects of environmental stochasticity in management analyses.

INTRODUCTION

Modeling the growth of biological populations has a long history going back to Gompertz (1825) and Verhulst (1838). Modeling biological resources for the specific purpose of developing management strategies can be traced back almost as far. Faustmann (1849) in the middle of the last century posed and discussed the optimal rotation problem for even-aged forest stand management. Only during the last 30 years, however, has quantitative resource management really come of age with the work of such stalwarts as Gordon (1954), Beverton and Holt (1957), and more recently Clark (1976; 1985). Before the advent of computers, a healthy dose of calculus and linear algebra was required before resources could be modeled for management purposes. Now resources can be modeled with as few quantitative skills as elementary algebra and a course in BASIC programming. Thus resource management analysis can range from the mathematically esoteric to the conceptually inane. Between these extremes, however, models have been developed that enhance our understanding of the biological and economic aspects of resource management. We have also come to appreciate the limitations that exist in applying quantitative methods to biological problems. What are these limitations? Essentially, biological resource systems, like any ecological systems, are too complex and too variable to precisely model as either deterministic processes or as processes that can be characterized by their average behavior.

At the risk of oversimplification, one can view most resource modeling activities as either developing or applying theory. The first of these activities, typified by Clark's 'Mathematical Bioeconomics,' begins with analytically tractable models and then, where possible, appling them to real problems (Clark, 1976; 1985).

The second, typified by the development of simulation models, seeks to examine particular resource problems and then, where possible, to extrapolate to general principles. Both approaches have their strengths and weaknesses, as well as their adherents. The development of a cogent resource management theory obviously enhances our understanding of the field and our ability to translate experiences gained in one area of application to another. At the same time the application of theory to real problems is the bottom line. A theory that is too simple has little value. Thus we must trade what is 'real' with what we are able to 'comprehend.'

We will ultimately be more successful in finding efficient, easily implementable management policies if we look at a problem from both sides; that is, we use both modes of analysis described above. The models we construct, however, will have levels of resolution (complexity) that depend on which modes of analysis we plan to use. The two approaches can be linked by developing a modeling framework that with a little simplification is amenable to mathematical analysis, and with not too much elaboration provides realistic simulation models. One such framework that deals with the problem of harvesting age and stage structured populations, is presented below.

Another essential component to constructing useful models is defining at the outset the exact question to be addressed. If this is not done, then deciding which elements should be included in the model is an arbitrary process. Narrowly defined questions may have analytical solutions. Many hypothetical resource exploitation problems are cast in this context. A good example is finding an optimal harvest policy for a lumped resource (i.e., a resource described by a single variable, usually biomass) that changes according to an elementary growth law (such as the logistic equation). If the problem is more broadly defined and involves maximizing the difference between the profit obtained from directly exploiting a resource and the costs associated with external socioeconomic factors, then analytical tractability invariably disappears. Furthermore, it may be very difficult to quantify these external costs.

One may now ask what value a solution to a narrowly defined question has in the context of a broadly defined problem. The answer obviously depends on the particular problem, but in most cases one would expect that the solution to a narrowly defined problem provides some insight into the solution of a related but more broadly defined problem. Going a step further, the solution to a number of related narrowly defined problems may provide sufficient pieces to a puzzle that we are in a position to make an informed guess as to what would be a good solution to a broad but analytically intractable problem. For example, the analysis of a stochastic marine fisheries harvesting problem could focus on biomass yield and stock dynamics to find a solution that maximizes yield subject to constraints on the variability of annual yield and on the minimum level that the stock is allowed to assume. A second analysis may consider the question of revenue and returns on investment in this fishery. Although the two solutions are not directly comparable, they can be qualitatively evaluated in a context that encompasses both formulations. The insights obtained from both solutions may then lead to the construction of a solution that is suboptimal but good with respect to both narrowly defined problems and is qualitatively satisfactory in the context of the more broadly defined problem.

Very often we address problems in a deterministic framework because the problem is just too hard to solve in a stochastic framework. Whenever we do this, we must remain aware that the structure of our models is a crude approximation to reality and that the values of the parameters used to obtain actual solutions are often rough estimates. We sometimes forget this, and look upon the solution to a particular optimal harvesting problem as sacrosanct. For any particular problem, however, it may be preferable to identify a class of solutions that in some sense are equivalently desirable, and then to decide which of these solutions is best in a broader context.

In the sections that follow, I use a general discrete time model formulation to clarify some of the issues raised here. In particular, I focus on questions relating to uncertainty in model prediction, variability in the state of the resource, and variability in the future stream of profits (or yield). I also deal with the question of choosing the length of the management planning horizon so that ad-hoc boundary conditions can be avoided. Finally, a framework is presented for modeling age and/or stage structured biological resources that can be used to obtain analytical results and used as a basis for constructing simulation models.

DETERMINISM AND UNCERTAINTY

For the sake of clarity, the discussion in this section is presented around resource systems that can be modeled by a general system of nonlinear difference equations. Later, a more structured model is introduced. Consider a resource that has a number of interrelated biological and non-biological components. Suppose the behavior of the resource system is described by the difference equation

$$x(t+1) = f(x(t), u(t), \alpha) \qquad (1)$$

where $x(t) \epsilon R^n$ is the state of the system at time t, $u(t) \epsilon R^m$ is a vector of control variables and $\alpha \epsilon R^s$ is a vector of adjustable model parameters. Very often there is a natural unit of time associated with a particular resource system. For example, many species of animals and plants reproduce on a seasonal basis. Whether this is the unit we select for t, depends on whether it is appropriate to assume that $u(t)$ acting at time t is a reasonable approximation of what occurs in practice, or whether a finer resolution of time is required to characterize the application of $u(t)$. Having chosen a unit for t, let $R(x(t), u(t))$ denote the revenue obtained in the $(t+1)$th time period where the resource is in state $x(t)$ at the beginning of this time period and the level of exploitation is $u(t)$. Let δ denote the revenue discount factor.

Consider the following narrowly defined management problem.

$$\max_{u(0),u(1),u(2),\ldots} J(x_0) = \sum_{t=0}^{\infty} \delta^t R\left(x(t),u(t)\right) \tag{2}$$

subject to equation (1), and the initial condition $x(0) = x_0$. Suppose this problem has a solution denoted by u^α and let the corresponding optimal value be denoted by J^α (for convenience the argument x_0 is suppressed). The parameters α are themselves uncertain but we will assume that both a mean value $\bar{\alpha}_i$ and standard deviation σ_i of the parameter α_i are known for $i=1,\ldots,s$. Define

$$A = \{\alpha \mid \alpha_i \,\epsilon \left[\bar{\alpha}_i - \sigma_i, \bar{\alpha}_i + \sigma_i\right]\} \tag{3}$$

the 1-SD (standard deviation) α parameter set. Define J^A as

$$J^A = \{J^\alpha \mid \alpha \epsilon A\} \tag{4}$$

Since A is a closed subset of R^n, it follows that J^α is continuous in α (consider the augmented control problem in the state space represented by the vector $(x, \alpha)'$, where α satisfies the differential equation $\frac{d\alpha}{dt} = 0$ and $'$ denotes the transpose of a vector). If J^A_{min} and J^A_{max} respectively denote the maximum and minimum values of J^α, $\alpha \epsilon A$, then *a priori* it is not clear where J^α lies in this interval. Although we tend to accept J^α as the maximum value that can be attained in theory, in fact it is only our best estimate. The interval $[J^A_{min}, J^A_{max}]$ provides an estimate of the range of values we can expect to obtain given the uncertainty in the parameters. (Note that this range does not correspond to a 1-SD interval around J^α). To calculate this we would need to generate the density function for J^α as α ranges over its distribution of values (i.e. the convolution of the distribution of α using the functional transformation defined by the solution to the problem is required to evaluate the distribution of J^α). The easiest way to generate this distribution is to use Monte Carlo simulation techniques. In a deterministic setting, however, the interval $[J^A_{min}, J^A_{max}]$ could be generated by evaluating J^α for the values of α at the 'corners' of A if the derivatives $\frac{dJ^\alpha}{d\alpha}$ do not change sign for $\alpha \epsilon A$. If these derivatives do change sign, the situation is more complicated since J^A_{min} and J^A_{max} may correspond to values of α in the interior of A. Here again we can use Monte Carlo simulation to estimate the size of the interval $[J^A_{min}, J^A_{max}]$ by, say, accepting the largest and smallest values of J^α that are obtained after a given number of random simulations (run blocks of 100 simulations until the change in the estimate of J^A_{min} and J^A_{max} is less than some prespecified number). Once we are using Monte Carlo simulation techniques, however, it may then be preferable to estimate the standard deviation associated with J^α as this is statistically more meaningful than estimating the interval $[J^A_{min}, J^A_{max}]$.

Whatever approach we use to generate an estimate for the range of likely values of J^α, including the possibility of selecting A as an ℓ-SD set where ℓ refers to the number of standard deviations (e.g. ℓ = 1.96 -- which corresponds to the 95% confidence interval if the parameters are normally distributed) used to define A in expression (3), it will provide a useful ballpark for assessing the value of suboptimal policies that may be derived from ad hoc considerations. One such consideration is the desirability of implementing a feedback rather than an open loop management policy. For example, we may specify that all management policies must have the form

$$u(t) = Qx(t) \tag{5}$$

in which case the resource exploitation problem reduces to the problem of determining the parameters q_{ij} (elements of the matrix Q) that maximize expression (2) subject to equations (1), (5) and the initial condition $x(0) = x_0$. The advantage of a policy based on equation (5) is that the value of $u(t)$ can always be updated according to the latest estimate of $x(t)$ that may be obtained through direct measurement rather than relying on the value predicted by equation (1). The disadvantage, of course, is the price that must be paid for constraining $u(t)$ to satisfy (5). That is, the value of J in expression (2) will be less when compared with the optimal value obtained under the general formulation, but may be acceptable if its set of 1-SD values sufficiently overlaps with the set J^A defined in equation (4). Furthermore, in each new time interval, as more data associated with the resource becomes available, the optimal values for the elements of Q may be reestimated. That is, the approach can be made adaptive.

The concept of using feedback or adaptive policies is fundamental to managing stochastic resources. Over the past 10-15 years Walters, Hilborn and Ludwig have been active in developing a theory of adaptive resource management (see Walters, 1986, for details). They distinguish between passive and active adaptive prescriptions for management strategies, where active prescriptions include evaluating the amount of additional information that can be gathered when implementing a particular management strategy so that estimates of parameters in the resource model can be improved. In theory, this approach has great merit. In practice, actively adaptive policies may lead to strategies that increase yield or rent variability between years (current rent may be sacrificed to increase estimated future rents) which, in a broader socioeconomic setting, are less acceptable.

The tradeoff between short run and long run gains is an essential part of the management problem. Short run gains are more attractive to individuals in unregulated resources (Gordon, 1954), but also in regulated resources because of the uncertainty associated with the future and the fact that an individual feels more confident having money in his own bank account than being guaranteed a share in a resource over which he has little control. The interests of the society at large are less myopic and, hence, regulatory agencies have been created to help ensure the profitability of the resource over the long run. This raises a number of questions relating to how long is 'long' and how do we deal with uncertainty over the long run. Before the latter can be discussed we need to understand the relationship between planning horizon and strategy in a deterministic systems framework.

PLANNING HORIZON

The problem of choosing an appropriate planning horizon is better understood if we examine the relationship between infinite time horizon problems and problems in which a system is forced into equilibrium at some finite point in time. We start by considering the concept of sustainable rent and how it relates to the dynamic problem defined by equation (1) and expression (2), except now we will notationally suppress the parameter dependent aspect of the problem and rewrite equation (1) as

$$x(t+1) = f\big(x(t),u(t)\big) \tag{6}$$

The maximum sustainable rent problem (MSR) is defined as

$$\max_{u} R(x,u) \tag{7}$$

subject to

$$x=f(x,u). \tag{8}$$

That is we have imposed the equilibrium constraint $x(t+1) = x(t)$ on equation (6) and are looking for the vector $\hat{u}$ and a corresponding equilibrium value $\hat{x}$ that maximizes the rent under equilibrium conditions.

Two important issues arise. The first is the relationship between the MSR solution and the optimal solution to the dynamic problem defined by equation (7) and expression (2). The second is the implementation of deterministic solutions to stochastic problems using an adaptive framework. Consider the first of these issues. The most obvious question relates to the role of the MSR solution pair $(\hat{x},\hat{u})$ in the framework of the dynamic optimization problem. This problem has been examined in the context of harvesting a system of two competing species (Getz, 1979) or, more generally, two interacting species (Haurie, 1982). The analysis was carried out for continuous models and conditions were obtained for the MSR solution, or in the case of discounting for an extremal steady state (ESS) solution associated with the _current-value_ Hamiltonian (see Haurie, 1982) to be stable. In the economics literature, stable ESSs are often referred to as turnpikes. In fact if the problem is linear in the control, the turnpike may be reached in a finite time and become a segment of the optimal solution (Getz, 1979).

Returning to equation (7), consider the set of controls U^e defined by

$$U^e = \{u \text{ s.t. there exists } x_u \text{ for which } (x_u, u) \text{ is a biologically}$$
$$\text{meaningful equilibrium pair for eq. (7)}\} \tag{9}$$

By definition, the turnpike solution must correspond to a stable equilibrium of the dynamic system. Since it is impossible to numerically solve an infinite time horizon optimal control problem, for computational purposes a related finite time horizon problem must be formulated. One approach is to truncate the infinite time horizon problem at t = T either by constraining $x(T)$ to assume some value (or belong to a set of values) or by allowing it to remain free. In the latter case, if no cost is attached to leaving the resource in a particular state, it will often be heavily exploited in the final time intervals before T. This can be avoided by, say, specifying $x(T) = \hat{x}$ (e.g., Getz, 1985). This approach is <u>ad hoc</u> unless, as will be shown below, the discount rate is zero (i.e., $\delta = 1$).

Assume that the optimal solution to the infinite time horizon problem approaches a turnpike as $t \to \infty$. Then one method of approximation is to constrain the problem so that equilibrium is obtained at time T and the corresponding constrained optimal finite time horizon solution provides an approximation to the unconstrained optimal infinite time horizon solution for all T and converges to it as $T \to \infty$. With this in mind, J defined in expression (2) is modified as follows:

$$J_T(x_0) = \sum_{t=0}^{T-1} \delta^t R\big(x(t), u(t)\big) + \frac{\delta^T}{1-\delta} R(x_u, u) \tag{10}$$

where $u(T) = u \in U^e$. Note that the second term in the above equation corresponds to $\Sigma_{t=T}^{\infty} \delta^t R(x_u, u)$. The idea here is to maximize $J_T(x_0)$ for given x_0 subject to equation (7) holding for t = 0,1,...,T-1 and the pair (x_u, u) satisfying the equilibrium constraint (8). If $J_T^*(x_0)$ is the value corresponding to the solution of the optimal T-horizon problem then $J_T^*(x_0)$ approximates the value of the infinite horizon problem and converges to it as $T \to \infty$. The difference between the value of the infinite time horizon and finite time horizon solutions can be regarded as the cost associated with the constraint that the system must be in equilibrium for t > T.

It is worth noting that the optimal equilibrium pair (x_u, u) will depend on x_0, T and δ. However, as $T \to \infty$, if a stable ESS exists, it depends solely on δ, and thus will be written as (x_δ, u_δ). In the context of continuous time systems, Brock and Scheinkman (1976) provide conditions for a unique ESS to be globally asymptotically stable (also see Haurie, 1982). The ESS is constructed using the necessary conditions provided by Pontryagin's Maximum Principle and the associated *current value* Hamiltonian. Specifically, Pontryagin's Maximum Principle (as modified to include discounting --- see Haurie, 1982) states that a solution pair $(\tilde{\mathbf{x}}(t),\tilde{\mathbf{u}}(t))$ is an extremal for the above finite time horizon problem if there exists a *current value* costate variable $\lambda(t)$, multiplier μ and *current value* Hamiltonian (recall that ' is used to denote the transpose of a vector)

$$H\big(\lambda(t+1),\mathbf{x}(t),\mathbf{u}(t),t\big) = \lambda'(t+1)\delta f\big(\mathbf{x}(t),\mathbf{u}(t)\big) + R\big(\mathbf{x}(t),\mathbf{u}(t)\big) \tag{11}$$

such that for $t = 0,\ldots,T-1$

$$\lambda'(t) = \lambda'(t+1)\delta \, \frac{\partial f}{\partial \mathbf{x}} \, \big(\tilde{\mathbf{x}}(t),\tilde{\mathbf{u}}(t)\big) + \frac{\partial R}{\partial \mathbf{x}} \, \big(\tilde{\mathbf{x}}(t),\tilde{\mathbf{u}}(t)\big) \tag{12}$$

$$\lambda'(t+1)\delta \, \frac{\partial f}{\partial \mathbf{u}} \, \big(\tilde{\mathbf{x}}(t),\tilde{\mathbf{u}}(t)\big) + \frac{\partial R}{\partial \mathbf{u}} \, \big(\tilde{\mathbf{x}}(t),\tilde{\mathbf{u}}(t)\big) \tag{13}$$

and the boundary condition

$$(1-\delta)\lambda'(T) = \frac{1-\delta}{\delta^T} \, \mu' \, \frac{\partial f}{\partial \mathbf{x}} \, \big(\tilde{\mathbf{x}}(t),\tilde{\mathbf{u}}(t)\big) + \frac{\partial R}{\partial \mathbf{x}} \, \big(\tilde{\mathbf{x}}(t),\tilde{\mathbf{u}}(t)\big) \tag{14}$$

As $T \to \infty$, the boundary condition is ignored and the equilibrium solution to equations (6) and (12) can be solved for $(\mathbf{x}_\delta,\mathbf{u}_\delta)$. Note that at equilibrium equations (12) and (13) reduce to

$$\lambda_\delta^! \left[1-\delta \, \frac{\partial f}{\partial \mathbf{x}} \, (\mathbf{x}_\delta,\mathbf{u}_\delta) \right] + \frac{\partial R}{\partial \mathbf{x}} \, (\mathbf{x}_\delta,\mathbf{u}_\delta) = 0 \tag{15}$$

$$\lambda_\delta^! \, \delta \, \frac{\partial f}{\partial \mathbf{u}} \, (\mathbf{x}_\delta,\mathbf{u}_\delta) + \frac{\partial R}{\partial \mathbf{u}} \, (\mathbf{x}_\delta,\mathbf{u}_\delta) = 0 \tag{16}$$

where $\lambda_\delta^!$ is the corresponding equilibrium value for the *current value* costate variable (a comparable set of equations to (8), (15) and (16) has been derived by Knapp (1983) using a dynamic programming formulation). From elementary calculus it is follows that these three equations provide a set of necessary condition for $(\mathbf{x}_\delta,\mathbf{u}_\delta)$ to maximize $R(\mathbf{x},\mathbf{u})$ subject to the constraint $\mathbf{x} = \delta f(\mathbf{x},\mathbf{u})$; that is, as $\delta \to 1$ the extremal equilibrium solution $\mathbf{x}_\delta,\mathbf{u}_\delta$ approaches the MSR solution $(\hat{\mathbf{x}},\hat{\mathbf{u}})$.

From the above discussion it is apparent that a second way of approximating solutions to the infinite time horizon problem is by solving the finite time horizon problem as defined above, but subject to the additional constraint that $\mathbf{x}(T) = \mathbf{x}_\delta$. This solution also approaches the infinite time horizon solution as $T \to \infty$, but the additional constraint imposed on $\mathbf{x}(T)$ means that the approximation is not as good as the formulation associated with expression (10).

Note that associated with fixed end-point problems is the question of reachability of the target point or set. Both the reachability problem and the divergence of the value function as $\delta \to 1$ can be avoided by considering the average value associated with $J_T(\mathbf{x}_0)$; that is we maximize

$$J_T(x_0) = \frac{1}{T} \sum_{t=0}^{T-1} R(x(t),u(t)) + R(x_u,u) \tag{17}$$

using the equilibrium rather than the end-point constraint condition.

Once an approximation scheme has been devised, it still remains to be determined what is an appropriate value for T. One approach is to keep increasing T until the improvement in $J_T(x_0)$ is less than a specified value. As discussed in the previous section, however, the most important consideration may not be in finding the exact optimal solution, but in finding a solution that yields a value that is close to optimal and has properties that are desirable in a framework that is broader than the formal definition of the problem.

For example, the problem of designing openloop and feedback harvesting strategies was recently considered for marine fisheries (Getz, 1985), using models in which the elements x_i of the state vector $x(t)$ represent the number of individuals in age class i, i = 1,...,n, the control vector $u(t)$ is a scalar effort variable $u(t)$, and the number of new individuals entering the fishery (recruits) is evaluated as function of the stock level represented by $s(t) = \sum_{i=1}^{n} c_i x_i(t)$, where c_i are constants denoting the relative fecundity of each age class. The problem considered was one of maximizing yield rather than rent and the analysis was carried out using parameters obtained from a South African anchovy fishery. Firstly, the MSY (maximum sustainable yield) solution was calculated. Then three harvesting policies were compared under two different scenarios: one in which the initial condition corresponded to a previously unexploited fishery (virgin stock), and one in which the initial condition corresponded to a fishery at MSY except that the recruitment level was zero at the beginning of the year (failed recruitment). The three policies were:

OPTIMAL 10 YEAR   The yield was maximized subject to the constraint that $x(10) = \hat{x}$ (where the latter denotes the state of the system at MSY).

MSY CONSTANT EFFORT   The MSY effort level $\hat{u}$ was applied even though the fishery was not at equilibrium.

MSY FIXED ESCAPEMENT   The effort level $u(t)$ was chosen in each time interval so that the stock level $s(t)$ was brought back to the MSY stock level $\hat{s} = \sum_{i=1}^{n} c_i \hat{x}_i$ at the beginning of each time step t.

From the results summarized in Table 1 below, it is clear that the suboptimal policies perform almost as well as the optimal 10-year policy. In a broader socioeconomic context, however, the constant effort policy is preferable because effort variation from year-to-year is zero and yield variation is low. The fixed escapement policy also has some desirable qualities; specifically it protects the stock from over-exploitation. These points are discussed more fully in the next section in the context of stochastic resource management.

Table 1: Yields (thousand metric tons) averaged over 10 years for three policies and two scenarios.

Scenario	Harvesting Policy		
	10-year Optimization	Constant Effort	Fixed Escapement
Virgin Stock	131.9	131.3	131.7
Failed Recruitment	102.7	101.4	101.8

STOCHASTIC DYNAMICS

So far, our discussion has focused around deterministic dynamic models. Some concepts associated with deterministic models, such as MSR and ESS are in disrepute (for example, Larkin (1977) has written an epitaph to the concept of maximum sustainable yield (MSY); a concept that still pervades fisheries management). In fact, a central problem in managing such resources as fish stocks is dealing with populations that fluctuate due to stochastic environmental driving variables. Although we may be able to statistically characterize environmental patterns and calculate, using stochastic models (e.g., Beddington and May, 1977; Getz, 1984; Horwood, 1982, 1983; Reed, 1983), what the mean dynamics and associated variance or even what the MESR (maximum expected sustainable rent) will be, most of the time we will be trying to manage a stock that is far from MESR conditions. We need to design management strategies that can adapt to changes in the fishery as the stochastic pendulum swings back and forth between a strong and weak fishery.

The management of variable marine fisheries has received much attention (Hightower and Grossman, 1985; Ruppert, et al., 1985; Walters, 1986; - to mention just a few recent references). This attention has mainly focused on the introduction of environmental variability as a scalar stochastic parameter in a scalar population model (May et al., 1978; Walters, 1975) or on making the stock-recruitment relationship stochastic in an otherwise deterministic age-structured model (Getz and Swartzman, 1981; Hightower and Grossman, 1985). The latter approach was recently applied to an analysis of managing three contrasting stocks (Getz, Francis and Swartzman, 1987): a short-lived highly productive South African anchovy fishery; a long-lived slow growing US West Coast Pacific Ocean perch fishery; and an intermediate Bering Sea pollock fishery. Here I will reiterate some of the questions dealt with in this study and summarize some of the results.

The first question relates to the problem of partitioning the inherent stochasticity in the resource. Although the manager may have a small margin of control over the amount of stochasticity in the fishery (increasing levels of exploitation may result in increasing the coefficient of variation associated with yield - see Beddington and May, 1977) the primary question relates to how much variability should be absorbed by the stock or transmitted into the yield, and how much variability can be tolerated in fishing effort. Stock variation has implications for the long term stability of a resource. Yield variation impacts both the fisherman and the market. Effort variation has implications for capitalization of the fishery and the transfer of effort between related fisheries.

The second issue relates to the design of adaptive management strategies. Ludwig, Walters and colleagues (see Walters, 1986; and reports in this volume) distinguish between passive feedback (or adaptive) strategies and strategies that probe the resource to obtain more information relating to its state and dynamics. Application of such active adaptive strategies requires intangible short term sacrifices for long term gains relating to model improvement while passive adaptive strategies require only tangible short term sacrifices relating to stock recovery. Getz, Francis and Swartzman (1987) considered the performance of three passive feedback policies all designed using MESR as a reference point. The first policy was the application of constant effort at the MESR level. The assumption in the model that yield is an increasing function of stock for given effort levels ensured the stability of this policy. The second was a fixed escapement policy based on returning the stock to the MESR level at the end of each harvesting season. If the stock was in such poor shape that zero fishing effort would still leave the stock below the MESR level at the end of the season then the fishery was closed that season. The third policy was an exploitative combination of the first two: that is, use stock escapement when the stock is above MESR conditions, and apply the MESR effort level when the stock is below MESR conditions.

Monte Carlo simulation studies of these three policies on the three contrasting fisheries mentioned above, resulted in similar yield and stock levels with the greatest difference being in the most variable fishery (anchovy). In this fishery, the difference in average yield was less than 7 between policies but the coefficients of variation associated with the yield, effort and stock differed quite considerably. One of the conclusions of this study was that as long as the policy is designed around MESR conditions, long term yield is relatively unaffected by the structure of the particular policy. Short term yields and the risk associated with collapse of the stock, however, are vastly different and the strategy of choice depends on the elasticity of labor and markets associated with the fishery and an assessment of the vulnerability of the fishery to collapse. The latter is a difficult one, but past experience may shed light on the relative biological robustness of various types of stocks.

STAGE STRUCTURED MODELS

The construction of models can either be approached in an <u>ad hoc</u> manner or a paradigm can be developed for modeling and analysing a class of problems. Recently, for example, Schnute (1985) proposed a general theory for the analysis of fisheries catch and effort data that subsumed almost all previous methods as a special case of his more general approach. In doing so, Schnute made transparent the relationship between these various models and the nature of assumptions that had previously been hidden by the particulars of the approach (for details see Schnute, 1985). As discussed above, the paradigm should provide a framework for both analytical and simulation models. It should also include enough structure (often a drawback of lumped scalar models) so that a number of significant questions relating to a class of problems can be effectively addressed; for example, age related questions. The paradigm, however, should be broad enough to be applicable to a range of resource systems, so that generalizations can be made.

A number of biological resource management problems involve harvesting or controlling individuals that can be classified by age, size or life stage classes. Fish are often harvested with some degree of control on size class, trees can easily be selected by breast-height diameter class, insects reared for biological control purposes can be selected by life stage class (e.g., production of fruitfly pupae is required for sterile insect techniques for controlling fruitfly), wildlife species in game parks can be regulated by culling individuals in a particular age or size class. Life tables based on mortality and fecundity rates can be constructed for these biological populations. Such data leads naturally to the construction of Leslie matrix type models (Leslie, 1945). These have been applied to a number of resource management problems but with limited success due to their linear structure. In fish populations, the most important nonlinearity appears to be in the relationship between spawning stock and the number of individuals entering the youngest age class (i.e., the stock recruitment relationship). In forests, nonlinear ingrowth functions describing the establishment of young trees and nonlinear growth rates describing the movement of trees between stage classes appears to be important in modeling the development of forest stands. In insect populations, the sex of new individuals may depend on population density, which is a rather interesting nonlinear phenomenon that has important application to biological control problems.

The modeling paradigm presented here has application to the above nonlinear age and stage structured resource management problems. It has the form of a discrete time n-dimensional transition model of the state variable $\mathbf{x} = (x_1,...,x_n)'$. It includes nonlinear input function and other nonlinearities that depend on m aggregations of the elements of state variable $\mathbf{x}$; specifically on the elements of a vector $\mathbf{y} = (y_1,...,y_m)'$ given by the transformation

$$\mathbf{y} = A\mathbf{x} \qquad (18)$$

where A is an m × n matrix with elements a_{ij}, i = 1,...,m, j = 1,...,n and m is usually assumed to be much smaller than n.

Consider a biological population in which x_i represents the number of individuals in the i-th stage class. Let the scalar functions $s_i(\mathbf{y})$ and $p_i(\mathbf{y})$, taking values on [0,1], denote the proportion of individuals in stage class i at time t that respectively survive the time interval (t,t+1] and move into the next stage class at time t + 1. Then it follows that the number of individuals in stage class i at time t that remain in stage class i at time t + 1 is $(1-p_i(\mathbf{y}))$. Define the n × n matrices S($\mathbf{y}$) and P($\mathbf{y}$) as follows (note $(S)_{ij}$ denotes the ij-th element of the S, etc.):

$$(S)_{ii}(\mathbf{y}) = s_i(\mathbf{y}) \qquad i = 1,...,n;$$
$$(S)_{ij}(\mathbf{y}) = 0 \qquad i \neq j, \; i,j = 1,...,n, \tag{19}$$

and

$$(P)_{ii}(\mathbf{y}) = (1 - p_i(\mathbf{y})) \qquad i = 1,...,n;$$
$$(P)_{i+1i}(\mathbf{y}) = p_i(\mathbf{y}) \qquad i = 1,...,n-1; \tag{20}$$
$$(P)_{ij}(\mathbf{y}) = 0 \qquad j \neq i,i+1, \; i,j = 1,...,n.$$

Also define an input vector $\mathbf{f}(\mathbf{y}) = (f_1(\mathbf{y}),...,f_n(\mathbf{y}))'$, where $f_i(\mathbf{y})$ is the number of individuals entering the i-th stage class at time t from sources outside the population including births. That is, $f_1(\mathbf{y})$ is usually considered as the number of births, or in fisheries modeling the number of new recruits, while $f_i(\mathbf{y})$, i = 1,...,n, is attributable to migration processes both into and out of the population. Finally, the control vector $\mathbf{u}$ is also n-dimensional since u_i is assumed to be the number of individuals removed from the i-th stage class during the time interval (t,t+1). Without loss of generality we will assume that harvesting actually takes place at the end of this time interval so that the equation for the system dynamics can be written as

$$\mathbf{x}(t+1) = P(\mathbf{y})S(\mathbf{y})\mathbf{x}(t)-\mathbf{u}(t) + \mathbf{f}(\mathbf{y}) \qquad t = 0,1,2,... \; . \tag{21}$$

As long as the elements of P,S and f are independent of $\mathbf{y}$, that is constant, equation (21) is linear and time autonomous. In general this will not be the case for at least some elements, especially $f_1(\mathbf{y})$. Because of the 'linear-like' structure of equation (21), if m (the dimension of $\mathbf{y}$) is low then the model is readily analysed for the existence of equilibria and their stability properties. For example, if m = 1 then finding the equilibrium solution to equation (21) reduces to finding the solution to a single nonlinear algebraic equation (see Getz, 1987).

The model can also be analysed in an equilibrium optimization setting and properties of the MSR and MSY solutions examined. In fact it can be shown that if at most r elements of the vector $\mathbf{f}(\mathbf{y})$ are nonzero and an MSY solution exists, then there exists an MSY solution in which at most r + m stage classes are harvested

(Getz, 1987). This generalizes the MSY bimodal harvesting result obtained by Getz (1980) and Reed (1980) in the context of harvesting fish populations. These fisheries models, as discussed below, are a special case of equation (21) with m = 1 and r = 1. For the nonequilibrium optimization problem, a number of assumptions relating to the structure of the nonlinear functions in equation (21) need to be made before results, such as the stability of ESS solutions etc. (see section entitled Planning Horizon), can be obtained. Assumptions may relate to the signs of the derivatives of the various functions in equation (21) or the convexity and/or positivity of the matrices as functions of **y**. For simulation purposes, however, the elements of these matrices may be made as complicated as we like, but as long as they remain functions of **y** alone, the applicability of the (m+r)-modal harvesting result as well as the general structure of the problem remains intact. That is, the general model links both modes of analysis, as discussed in the Introduction.

There are a number of obvious ways in which equation (21) can be extended into a stochastic setting. Obviously, the elements of the matrices $S(\mathbf{y})$ and $P(\mathbf{y})$ can be made stochastic, but for many systems it may be reasonable to regard the primary source of stochasticity as being associated with the input vector $\mathbf{f(y)}$. In age structured fisheries models, for example, stochasticity is most often introduced by multiplying the function $f_1(\mathbf{y})$, that is the first element of $\mathbf{f(y)}$, by a lognormally distributed random variable (e.g. see Hightower and Grossman, 1985). In general, stochastic optimization problems are much harder to analyze than their deterministic counterparts. However, because equation (21) is easily linearized and this linearized form is so similar to the general form, optimal linear stochastic control formulations, such as Kalman filtering, should provide good approximations to solutions to the more general nonlinear problem.

Equation (21) is broad enough to encompass a number of resource management problems, as is seen by a discussion of the following examples.

<u>Forest Stand Growth Models</u>

A number of simulation models have be developed for the growth of forest stands. Some of this models are quite detailed and keep track of the height and diameter (at breast height) of individual trees within a stand. For management analyses, however, it is convenient to classify trees into diameter classes and ascribe average costs and profits associated with harvesting a tree in a particular diameter class (e.g., see Haight, 1985). Thus in equation (21), the variable x_i would represent the number of individuals in stage class i. It may be adequate set m = 1 and let y (now a scalar variable - hence omission of boldface notation) represent a *basal area index*, that is $y = \sum_{i=1}^{n} w_i x_i$ measures the proportion of the stand covered by trees (w_i are the average cross-sectional area of a tree in the i-th diameter class). One would expect the elements $s_i(y)$ and $p_i(y)$ to be decreasing functions

of y. Only the first element of f(y) would be nonzero and would represent the
ingrowth function. As such, one would expect $f_1(y)$ to initially increase as seed
production increases (with increasing y), but then decrease as crowding begins to
affect the establishment of saplings.

Recent results indicate that equation (21) simulates stand growth as well as
much more detailed simulation models but is much less cumbersome to embed in
numerical algorithms for the purpose of management analyses (Haight and Getz, 1986).

Fisheries Models

Many fisheries models use age rather than size classes (e.g., Getz and
Swartzman, 1981; Hightower and Grossman, 1985; Horwood, 1982; Levin and Goodyear,
1980; Reed, 1983), even though age is indirectly obtained (e.g. by examination of ear
otoliths) and correlated with size, the usual measurement that is made in most
situations. By definition all individuals move up one age class in one time period so
that the transition elements $p_i = 1$, $i = 1,...,n-1$, and only $p_n \neq 1$ because the n-th
class represents all individuals of age n or greater. In most studies, migration is
not considered and the only nonzero element of f(y) is the *stock-recruitment
relationship* $f_1l(y)$. Invariably the vector y is a scalar index y representing the
fecundity, egg potential or biomass of the *spawning stock*. Usually, the survival
parameters s_i are assumed to be independent of y so that $f_1(y)$ is the only
nonlinearity that appears in equation (21). Also the control vector u usually has
the form u = vQx, where v is a scalar effort variable and Q is a diagonal matrix of
catchability coefficients $(Q)_{ii} = q_i$. That is, the control variable is the scalar v.

Mass Rearing Insects

Insect populations are often too volatile and strongly influenced by
stochastic events to model under field conditions. If insects are reared under
controlled laboratory conditions, however, population parameters are deterministic
enough for the construction of predictive models. Insects are mass reared in
laboratories for many purposes including for release as biological control agents of
weeds and other insect species, and for use in sterile insect control programs.
Carey and Vargas (1983) used a Leslie matrix model with a unit time step of two days
to analyse the problem of what proportion of pupae could be harvested from a
sustained fruitfly mass rearing program for irradiation (sterilization) purposes.
More recently, Plant (1986) presented a theoretical analysis of the same problem but
in a context that included the dynamics of both the field and laboratory populations.
Although these studies both used linear models, at least two types of nonlinearities
play an important role in the mass rearing of insects. First, larval mortality rates
may depend on larval density (crowding factor). Second, the sex ratio of eggs laid

by certain parasitic wasps that are used as biological control agents (e.g. trichogramma) depends on the density of fecund females. In the framework of equation (21), this would imply that m = 2 where y_1 is a larval density index (weighted sum of individuals in the age classes corresponding to the larval life stage) and y_2 is a fecund female density index (sum of individual adult females in fecund age classes).

Wildlife Management

Leslie matrix models (Leslie, 1945) have often been applied to wildlife management problems. This includes land and marine mammals in the context of both the conservation and exploitation of species (e.g., Flipse and Veling, 1984; Starfield and Bleloch, 1986; Swartzman, 1984). As for the other types of resource problems, the inclusion of density dependent factors is important. One aspect of equation (21) that may be of particular importance, however, is the question of migration (i.e., some elements f_i of $f(y)$ other than f_1 are not identically zero). Migration is an extremely difficult process to characterize because migration (especially immigration) rates depend to a large extent on the density of surrounding populations which are not included in the model. However, problems can be analysed to see how different bounds on these rates affect the solution to the problems under consideration. For some species it may be possible in practice to use tags to estimate the movements of individuals in and out of a population.

CONCLUSION

The ideas discussed here have been done so in an informal manner. As indicated in the citations, some of the material is developed more fully and formally elsewhere. In particular, comprehensive treatments of stochastic resource management analyses can be found in two recently published texts (Mangel, 1985; Walters, 1986). The utility of applying stage structured models to resource problems, other than the application of the Leslie matrix model with nonlinear recruitment to fisheries problems (e.g. Getz, 1980, Levin and Goodyear, 1980), needs to be fully evaluated; but some recent forest stand management studies appear to be promising (Getz and Haight, 1986; Haight, 1987; Haight and Getz, 1986).

ACKNOWLEDGMENTS

I would like to thank R.C. Francis, R.G. Haight and G.L. Swartzman for discussions that have lead to the formulation of some of the ideas presented here. This work was supported by NSF Grant DMS-8511717 and Grant Number 86-6-18 from the Alfred P. Sloan Foundation.

REFERENCES

Beddington, J.R. and R.M. May. 1977. Harvesting natural populations in a randomly fluctuating environment. Science, Vol. 197, pp. 463-465.

Beverton, R.J.H. and S.J. Holt. 1957. On the Dynamics of Exploited Fish Populations. Ministry of Agriculture, Fisheries and Food (London), Fish. Invest. Ser., Vol. 2, No. 19.

Brock, W.A. and J.S. Scheinkman. 1976. Global asymptotic stability of optimal control systems, with applications to the theory of economic growth. J. Economic Theory, Vol. 12, pp. 164-194.

Carey, J.R. and R.I. Vargas. 1983. Demographic analysis of insect mass rearing: a case study of three tephritids. J. Econ. Entomol., Vol. 78, pp. 523-527.

Clark, C.W. 1976. Mathematical Bioeconomics: the Optimal Management of Renewable Resources, Wiley-Interscience, New York.

Clark, C.W. 1985. Bioeconomic Modelling and Fisheries Management, Wiley-Interscience, New York.

Faustmann, M. 1849. Calculation of the value which forest land and immature stands posses for timber growing. In: Martin Faustmann and the evolution of discounted cash flow, p. 18-34, Commonwealth For. Inst. Paper 42, Oxford, 1968 (translated by W. Linnard).

Flipse, E. and E.J.M. Veling. 1984. An application of the Leslie matrix model to the population dynamics of the hooded seal, *Cystophora cristata* Erxleben. Ecological Modelling, Vol. 24, pp. 43-59.

Getz, W.M. 1979. On harvesting two competing species. J. Optimization Theory & Applications, Vol. 28, pp. 585-602.

Getz, W.M. 1980. Optimal harvesting of structured populations. Math. Biosci., Vol. 48, pp. 279-292.

Getz, W.M. 1984. Production models for nonlinear stochastic age-structured fisheries. Math. Biosci., Vol. 69, pp. 11-30.

Getz, W.M. 1985. Optimal and feedback strategies for managing multicohort populations. J. Optimization Theory and Application, Vol. 46, pp. 505-514.

Getz, W.M. 1987. Harvesting discrete nonlinear age and stage structured populations. J. Optimization Theory and Application, in press

Getz, W.M., R.C. Francis and G.L. Swartzman. 1987. Managing variable marine fisheries. Can. J. Fish. Aquat. Sci., in press.

Getz, W.M., and R.G. Haight. 1986. Finite planning horizon problems and the management of uneven-aged white fir stands. Sloan-Berkeley Working Paper in Population Studies #7, Inst. for Int. Studies, Univ. of Calif., Berkeley.

Getz, W.M. and G.L. Swartzman. 1981. A probability transition matrix model for yield estimation in fisheries. Can. J. Fish. Aquat. Sci., Vol. 38, pp. 847-855.

Gompertz, B. 1825. On the nature of the function expressure of the law of human mortality. Philos. Trans. R. Soc. London, Vol. 115, pp. 513-585.

Gordon, H.S. 1954. The economic theory of a common property resource: the fishery. J. Polit. Econ., Vol. 62, pp. 124-142.

Haight, R.G. 1985. A comparison of dynamic and static economic models of uneven-aged stand management. Forest Sci., Vol. 31, pp. 957-974.

Haight, R.G. 1987. Evaluating the efficiency of even-aged and uneven-aged management. Forest Sci., Vol. 33, in press.

Haight, R.G. and W.M. Getz. 1986. A comparison of stage-structured and single-tree models for stand management. Sloan-Berkeley Working Paper in Population Studies #6, Inst. for Int. Studies, Univ. of Calif., Berkeley.

Haurie, A. 1982. Stability and optimal exploitation over an infinite time horizon of interacting populations. Optimal Control Applications & Methods, Vol. 3, pp. 241-256.

Hightower, J.E. and G.D. Grossman. 1985. Comparison of constant effort harvest policies for fish stocks with variable recruitment. Can. J. Fish. Aquat. Sci., Vol. 42, pp. 982-988.

Horwood, J.W. 1982. The variance of population and yield from an age-structured stock, with application to North Sea herring. J. Cons. Int. Explor. Mer., Vol. 40, pp. 237-244.

Horwood, J.W. 1983. A general linear theory for the variance of yield from fish tocks. Math. Biosci., Vol. 64, pp. 203-225.

Knapp, K.C. 1983. Steady-state solutions to dynamic optimization models with inequality constraints. Land Economics, Vol. 59, pp. 300-304.

Larkin, P.A. 1977. An epitaph for the concept of maximum sustainable yield. Trans. Am. Fish. Soc., Vol. 106, pp. 1-11.

Leslie, P.H. 1945. On the use of matrices in certain population mathematics. Biometrika, Vol. 35, pp. 183-212.

Levin, S.A. and C.P. Goodyear. 1980. Analysis of an age-structured fishery model. J. Math. Biology, Vol. 9, pp. 245-274.

May, R.M., J.R. Beddington, J.W. Horwood and J.G. Shepherd. 1978. Exploiting natural populations in an uncertain world. Math. Biosci., Vol. 42, pp. 219-252.

Mangel, M. 1985. Decision and Control in Uncertain Resource Systems. Academic Press, Orlando, Florida.

Plant, R.E. 1986. The sterile insect technique: a theoretical perspective. In: (M. Mangel, ed.) Systems Analysis in Fruitfly Management. Springer-Verlag, Heidelberg.

Reed, W.J. 1980. Optimum age-specific harvesting in a nonlinear population model. Biometrics, Vol. 36, pp. 579-593.

Reed, W.J. 1983. Recruitment variability and age-structure in harvested populations. Math. Biosci., Vol. 65, pp. 239-268.

Ruppert D., R.L. Reish, R.B. Deriso and R.J. Carroll. 1985. A stochastic population models for managing the Atlantic menhaden (Brevoortia tyrannus) fishery and assessing managerial risks. Can. J. Fish. Aquat. Sci., Vol. 42, pp. 1371-1379.

Schnute, J.A. 1985. A general theory for analysis of catch and effort data. Can. J. Fish. Aquat. Sci., Vol. 42, pp. 414-429.

Starfield, A.M. and A.L. Bleloch. 1986. Building Models for Wildlife Management, Macmillan, New York.

Swartzman, G.L. 1984. Present and future potential models for examining the affects of fisheries on marine mammal populations in the Eastern Bering Sea. In: (B. Melteff, ed.) Proceedings of a Workshop on Biological Interactions Among Marine Mammals and Commercial Fisheries in the South Eastern Bering Sea, Alaska Sea Grant Report 84-1, Univ. of Alaska, Fairbanks, Alaska.

Verhulst, P.F. 1838. Notice sur la loi que la population suit dans son accroissement. Corresp. Math. Phys., Vol. 10, pp. 113-126.

Walters, C.J. 1975. Optimal harvesting strategies for salmon in relation to environmental variability and uncertain production parameters. J. Fish. Res. Board. Can., Vol. 32, pp. 1777-1784.

Walters, C.J. 1986. Adaptive Management of Renewable Resources, Macmillan, New York.

PARTICIPANT'S COMMENTS

Getz's paper addresses some important issues in resource management: dealing with uncertainty, meaningful interpretation and use of the planning horizon in optimization, and the need for better models. I hope that my comments, which are based on a slightly different point of view, will complement his.

In his discussion of uncertainty in parameter values, only the control policies and associated values of J for each alpha are dealt with. It appears to me that there is some value in going beyond this. In situations in which dynamics are uncertain, but can be described probabilistically, it appears that the best policy would be one that maximized J over the distribution of values of alpha.

In the discussion of the planning horizon, the relationships between the solutions to static equilibrium problems and less constrained dynamic problems are clarified for cases in which the "turnpike" is part of the solution to the dynamic problem. It should be kept in mind that for realistic models, with age and size structure, the turnpike is not necessarily part of the general solution to the dynamic problem (see Botsford, 1981b).

With regard to stochastic dynamics, although the notion that we can merely treat erratic population variables as stochastic is widely adhered to, it may discourage further inquiry into the actual causes of the seemingly random behavior. Knowledge of the cause can lead to better management because different causes could imply different policy strategies.

I wholeheartedly agree that there is a need for improved models in renewable resource analysis and management, but would disagree that the stage-structured models proposed are the ultimate answer. At the last conference on control theory and renewable resources, I argued for more realistic models (i.e., models with a closer one-to-one relationship to the essential dynamic mechanism of the modeled populations, Botsford 1981a). Age and size-structured models appeared to hold promise for adequate realism, and their analysis and greater use was recommended.

In the search for better models, one of the requirements on them should be that a model adequately describe the current state of the population as it affects future behavior. For example, the logistic model is not very realistic because it does this poorly [i.e., the current state is represented by the number of (presumably identical) individuals, while behavior in fact depends on age and size specific differences between individuals]. For some renewable resource populations, discrete time age-structured models will be adequate. However, if reproduction or survival depends on size, and growth rate is time varying, then size structure must be included in order to completely specify the state of the population at each time.

The existence of discrete stages in the life histories of some plants and animals has led many researchers to formulate discrete time, stage-specific models in which stage duration is not equal to the fundamental time interval. These models have the drawback outlined above: the current state is not a complete description of future behavior. All individuals in a specific stage are assumed identical, yet biologically they are not. In real populations, those individuals that have just entered a stage do not have the same probability of leaving the stage as those who have been in it for some time. This lack of realism can lead to gross disparities between model behavior and what is biologically reasonable [cf. Bosch (1971) and later comments in Science (1971, Vol. 174, pp. 435-436)]. Lewis (1977) and Nisbet and Gurney (1983) have used stage structured models that do not have this limitation.

References Cited:

Bosch, C.A. 1971. Redwoods: a population model. Science, Vol. 172, pp. 345-349.

Botsford, L.W. 1981a. More realistic fishery models: cycles, collapse, and optimal policy. in T.L. Vincent and J.M. Skowronski (eds.) Renewable Resource Management, Springer-Verlag, New York.

Botsford, L.W. 1981b. Optimal fishery policy for size-specific density dependent population models. J. Math. Biol., Vol. 12, pp. 265-293.

Lewis, E.R. 1977. Network Models in Population Biology, Springer-Verlag, New York. 402 pp.

Nisbet, R.M. and W.S.C. Gurney. 1983. Stage-structured models of uniform larval competition. In: Proc. Res. Symp. on Math. Ecol., Trieste 1982, Springer-Verlag, S. Levin (ed.).

Louis W. Botsford

All too often, proceedings volumes from conferences, even for those like this workshop which had a clear theme, appear to be a motley collection of papers on disparate topics. Somehow, those interactions that flourish in the informal workshop atmosphere become invisible in the cold light of print. This seems particularly to be the case for papers that take rather different approaches to similar types of problems; frequently the spirited discussions that arise from the verbal presentations seem almost inconceivable on later reading. Thus, it is a pleasure to see this paper by Wayne Getz that not only tries to bridge the gap between the analytic modellers and simulators (if I may use those terms), but also sees merit in both approaches! Certainly there are clear examples of each in this volume.

Not surprisingly, Wayne begins by adding his support to the need to properly define the problem. One may suspect that his voice as well will be resoundingly ignored by many, but he really is right; just as there are horses for courses, there are techniques for problems. That is not to say it is easy to resist the lure of always setting one's problem in an uncertain stochastic framework, but frequently that very choice precludes finding an implementable optimal policy. On that point, it is also good to see another plug for sub-optimal policies. Finding the *optimum optimorum* policy certainly has its place - indeed there are examples in this volume of quite counter-intuitive policies that are optimal, the form of which may never have been anticipated without such a search - but it is a recurring theme that there may be a range of policies for a particular problem that, while formally sub-optimal, are only marginally so and are much more practical.

Having found merit in approaching problems from different ends, Wayne attempts to span part of the remaining gap by proposing stage structured models as a paradigm for analyzing a class of problems. With the examples he cites of potential areas for which such models may be appropriate, these models certainly appear to have promise. Indeed one might well argue that so they should, since they are regularly used in those fields by practitioners whose goal is the more modest one of just modelling the dynamics with an acceptable degree of realism. Whether they will prove as successful when the complication of an optimal control setting is superimposed is still an open question. Whatever the outcome, both these models and the other thought provoking views put forward make this paper a valuable contribution.

Geoff Kirkwood

OPTIMAL HARVEST POLICIES FOR FISHERIES WITH
UNCERTAIN STOCK SIZES

G.P. Kirkwood
Division of Fisheries Research
CSIRO Marine Laboratories
GPO Box 1538
Hobart, Tasmania, Australia

The need to estimate stock sizes is a frequently ignored source of uncertainty in the management of renewable resources, such as fisheries. This paper outlines circumstances where this uncertainty is important, and reports the results of a first attempt to model optimal catch quota decisions for a fishery with a fluctuating and uncertain stock abundance.

INTRODUCTION

Managers of renewable resources have perhaps only one certainty: that their decisions will always be based on uncertain information. The appropriate policies to adopt when information is perfect are now well known. But how to modify these policies when knowledge is imperfect is still very much an open question. It is the theme of a number of contributions to this workshop, and it has received considerable attention in the recent literature, where by common consent the techniques of optimal control have been accepted as a major investigative tool.

To set the scene, consider the following simple deterministic model for the dynamics of a renewable resource, which for concreteness I shall take to be a fishery:

$$X_{k+1} = G(S_k) \tag{1}$$

$$S_k = X_k - H_k \, , \tag{2}$$

where X_k is the stock size at the beginning of the k'th period, H_k is the harvest taken during the k'th period and S_k is the escapement remaining after the harvest. Connecting the escapement at the end of the k'th period with the stock size at the beginning of the k+1'th period is a function $G(\cdot)$, which in fisheries jargon is known as the stock-recruitment relationship. Assume that the objective of the fishery manager is to maximize the discounted present value of future harvests:

$$\text{maximize} \sum_{1}^{\infty} \alpha^{k-1} H_k \, , \tag{3}$$

where α is the discount factor, $0 < \alpha < 1$. Then it is well known (e.g., Clark, 1976) that the optimal policy is one of "constant escapement," with optimal harvest H_k^* given by

$$H_k^* = \max(0, X_k - S^*) \ , \tag{4}$$

and the optimal escapement S^* determined by α and the function G.

The assumptions inherent in this formulation are restrictive and unrealistic. Fisheries are stochastic, yet the dynamics are assumed to be deterministic. The stock-recruitment relationship - both its functional form and parameter values - is also assumed to be known. More subtly, it is tacitly assumed that the exact state of the system is known when a management decision is taken. Finally, and this will be the only mention of this point, it is assumed that management objectives can be identified and quantified.

A simple way of converting (1) to a stochastic formulation is to insert a random multiplier to yield

$$X_{k+1} = Z_k \, G(S_k) \ , \tag{5}$$

where the Z_k are independent and identically distributed random variables with mean 1. With the consequential amendment of the objective (3) to include an expectation with respect to the random variables,

$$\text{maximize } E \left\{ \sum_1^\infty \alpha^{k-1} H_k \right\} \ , \tag{6}$$

this is essentially the model studied by Reed (1979), and it will be used during the rest of this paper. Note that in (3) and (6), as opposed to Clark (1976) and Reed (1979), no explicit account is taken of costs. This was done for simplicity. The results of Clark (1976) and Reed (1979) still apply to this special case.

While the incorporation of stochasticity does add an air of realism, the other restrictions remain. In practice, neither the form nor the parameter values of the average stock-recruitment relationship are known. It is possible to proceed using a certainty-equivalent approach, pretending that the current best estimates of the form and parameters of G are in fact correct. However, as outlined succinctly by Walters (1984), much more can be done with actively adaptive policies that contain elements of the original deterministic policy, of caution and of probing (Bar-Shalom, 1976). Whether the uncertainties are in the parameters of the stock-recruitment function (Ludwig and Walters, 1982) or in its functional form (Walters, 1981), it appears that the best policy is occasionally to superimpose substantial probing experiments on an otherwise cautious management plan (Walters, 1984).

The third element of unreality listed above is the assumption that the exact state of the system is known when decisions are made on an appropriate harvest. The potentially serious effect of this can be seen, for example, in the "New Management Procedure" of the International Whaling Commission (IWC). This procedure classifies whale stocks into three categories according to the relationship between the present stock level (PL) and the level at which the maximum sustainable yield (MSY) can be taken (MSYL): "Protection Stock" if PL < 0.9 MSYL; "Sustained Management Stock" if 0.9 MSYL < PL < 1.2 MSYL; "Initial Management Stock" if PL > 1.2 MSYL. If the stock is classified either as Initial Management, or as Sustained Management with PL > MSYL, then the allowable catch limit is 0.9 MSY. For a Sustained Management stock with PL < MSYL, the allowable catch limit is reduced by 10% for every 1% PL is below MSYL. For a Protection stock, the catch limit is zero.

Let us examine the performance of this management procedure (or control rule, given the context of this workshop) when there is uncertainty in PL, ignoring for the purposes of illustration the fact that neither MSY nor MSYL is known exactly. Each year, PL is estimated and compared with the estimated MSYL to determine the catch limit. It is clear that, even if on time-average the stock is in fact being maintained at the optimum level, errors in estimates of PL can easily lead to annual classification changes and major shifts in allowable catch limits. While the size of these shifts may be due to the nature of this particular control rule, problems of this type are almost inevitable in any scheme that does not take proper account of uncertainty in stock sizes. In fact, the management procedure of the IWC is somewhat exceptional in international fisheries management agencies, in that it does specifically include a safety factor. While the nominal goal is to stabilize the stocks at MSYL and to take MSY, the maximum catch allowed is 90% of MSY: a (fixed) 10% safety factor. This goes some way towards alleviating the problem (although for whales not very far, given the prevailing levels of uncertainty), but evidently a much better policy would have both the control rule and safety factor varying with the degree of uncertainty. Such a policy is currently being sought by the IWC.

The IWC example illustrates the problems that can arise if uncertainty in estimates of stock sizes is ignored. In the following sections of this paper, optimal policies that take account of this uncertainty are examined, and the value of gaining better information on the current stock size is investigated. The results presented are based on those reported by Clark and Kirkwood (1986).

POLICIES FOR UNCERTAIN STOCK SIZES

As indicated earlier, the control problem with state equations (5) and (2), and objective function (6), was studied by Reed (1979). He found that, as for the deterministic case, the optimal policy was one of constant escapement. However, the optimal escapement level S* was either equal to or greater than that for the corresponding deterministic model, depending on whether or not that escapement was "self-sustaining." A stock level D was defined as self-sustaining if $Pr(X_{k+1} > D | X_k = D) = 1$. This property would not hold for many plausible distributions of the random variables Z. A higher escapement in the presence of random noise accords with intuition -- caution in the face of uncertainty -- however perfect knowledge about the stock size at the start of each period is still assumed.

To relax this assumption, Clark and Kirkwood (1986) assumed that, at the beginning of the k'th period, there was perfect knowledge of the previous period's escapement S_{k-1}, but not of X_k. In this formulation, the random variables Z play a dual role: representing environmental noise and inducing uncertainty in X_k. This slightly odd characterization of uncertainty in stock size was chosen in order to make solution of the problem feasible. A more realistic portrayal would have had the estimate of the current stock size determined from the past history of the system, including estimates of the earlier stock sizes, but incorporation of such a nonlinear estimation (G is invariably nonlinear) makes the problem quite intractible. In essence, the Clark and Kirkwood (1986) assumption makes the process Markovian, with all the attendant benefits.

Allowing uncertainty in X_k also has implications for the harvest in that period: it is quite possible that poor environmental conditions (low Z_k) could reduce the initial stock size to less than the calculated optimal harvest, in which case the stock would theoretically be exhausted before the full harvest could be taken. Here, if the specified quota is Q_k, the actual harvest taken is

$$H_k = <Q_k, X_k> = \min(Q_k, X_k) . \tag{7}$$

Clark and Kirkwood (1986) acknowledge that this extreme assumption is unrealistic (harvesting would become unprofitable and stop before exhaustion of the stock), but this only affects minor details.

Proceeding with a dynamic programming formulation, let $J_n(S_0)$ be the maximum of objective (6), given escapement in the initial year S_0, but with a finite time horizon n. Then

$$J_{n+1}(S_0) = \max_{Q>0} \{E_\Pi [<Q,X_1> + \alpha J_n(X_1 - <Q,X_1>)|S_0]\} \tag{8}$$

where Π is the Bayesian prior distribution of X_1 given S_0. The value function for an infinite time horizon, $J(S_0)$, satisfies a similar equation.

Not surprisingly, this dynamic programming equation is not susceptible to analytic manipulations; however, comparison with the corresponding equation for the Reed (1979) model does suggest that constant escapement would no longer be the optimal policy, and this was confirmed by the numerical examples cited by Clark and Kirkwood (1986). In these, a stock recruitment function of the form $G(x) = 1-e^{-2x}$ was assumed and two distributions of random noise were examined: uniform and log-normal. The latter distribution was felt to be more typical of fisheries data (Hennemuth et al., 1980).

Results for log-normally distributed noise are shown in Fig. 1, in the form of a graph of optimal expected escapement (i.e., $E_\Pi[X - <X,Q>]$) against expected recruitment, for varying coefficients of variation (CV) of noise. As already noted, the optimal policy is not constant escapement. For moderate coefficients of variation (CV < 0.7), the optimal expected escapement increases with expected recruitment, and for the most part these escapements are higher than those for the deterministic model (CV = 0.0), implying caution. However, for larger CV's (CV > 0.8) and sufficiently high expected recruitments, the optimal policy is to fish out the stock. Essentially this implies removal of any restrictions on catch.

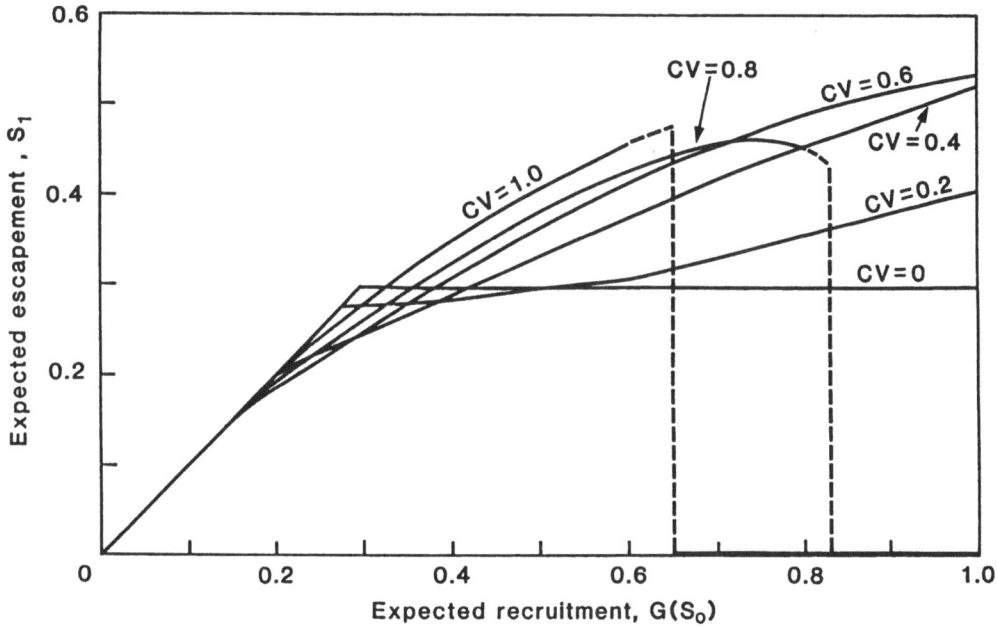

Figure 1.   Optimal expected escapement vs. expected recruitment for example with log-normal noise. The coefficient of variation of the noise ranges from CV = 0 to CV = 1.0. Dotted lines indicate uncertain interpolations.

This result was rather unexpected. However, as Clark and Kirkwood (1986) note, there is a positive probability with a log-normal distribution that the current stock size may exceed any given level, no matter how large. Thus, the higher the level of expected recruitment and the greater the environmental variability, the higher the probability of a very large actual recruitment. Eventually, it turns out that the policy of immediate fishing-out dominates all other quota policies.

Such a radical shift in policy would never be suggested or implemented in practice, of course, and it might be suspected that this optimal policy is an artifact of the particular probability distribution assumed. Clark and Kirkwood (1986) also presented results for noise uniformly distributed in the range $(1-\epsilon, 1+\epsilon)$ for $\epsilon < 1$. In this case, as ϵ increased, again there was a shift in policy away from cautious management, but not to the extent of setting an effectively infinite catch quota. It may be that it is the shift from more to less caution that is the general feature, with the size of the shift being dependent on the assumed distribution of environmental noise, especially on whether the distribution has finite or infinite support.

VALUE OF STOCK SURVEYS

As expected, the numerical examples cited by Clark and Kirkwood (1986) indicated that the value function $J(S_0)$ for fixed S_0 decreased with increasing environmental noise. While the Reed model exhibits similar behavior in circumstances where there is no uncertainty in current stock size, it is reasonable to expect that part of this decrease in the Clark and Kirkwood value function is due to increasing uncertainty in stock size. In a number of fisheries, the pre-season stock is surveyed annually - partly to obtain estimates of the current stock size - and often at considerable expense. It is very apposite, then, to examine under what circumstances stock surveys produce a net benefit through reducing uncertainty in stock size, and to relate the size of that benefit to the intensity of survey.

A partial answer to that question can be obtained by comparison of the Reed model and the Clark and Kirkwood model. In the latter model, uncertainty in stock size was induced into X_k through the random noise Z_k, with S_{k-1} known. If at the start of the k'th period a zero variance stock survey were carried out, yielding perfect knowledge of X_k, then we are back directly to the Reed model. Thus by examining the difference between value functions for the two models with fixed variance for the variables Z_k, it is possible to calculate the maximum expected benefit that could result from a stock survey. If $R(X)$ is the expected present value function for the Reed model with X given, then the maximum expected benefit $V_{max}(S)$ is

$$V_{max}(S) = E_Z [R(Z \, \dot{G}(S))] - J(S) . \qquad (9)$$

For their numerical examples, Clark and Kirkwood (1986) found that high environmental noise can lead to large V_{max}. For instance, when Z is log-normally distributed with coefficient of variation 1.0, the expected return with ideal surveys was approximately 80% greater than when no surveys were done.

To proceed further, the survey must be specifically incorporated as a control variable. Suppose at the beginning of each period the manager can opt to carry out an unbiased survey, resulting in an estimate Y of the stock size with CV σ_Y. In other respects, the model remains as before. Then in a dynamic programming setting, we would seek to solve the following equation for the expected present value function $J^*(S_0)$ of the revised problem:

$$J^*(S_0) = \max_{\sigma_Y > 0} \{E_Y [\max_{Q > 0} \{E_{\Pi_1}[<Q,X_1>+\alpha\, J^*(X_1-<Q,X_1>)]\}]\} , \tag{10}$$

where Π_1 is now the Bayesian prior distribution of X_1 given Y and S_0. Unfortunately, the addition of another control variable drastically increases the complexity of the problem, and certainly brute force solution of (10) is computationally infeasible on realistic computing budgets. Fortunately, with the help of some approximations, preliminary results obtained by the author suggest considerable progress can be made. This will be reported in a forthcoming publication.

DISCUSSION

To those familiar with the most recent literature on optimal management policies for fisheries, the assumption that an annual catch quota is the sole harvest control may seem a little old-fashioned. It is now widely recognized that, in the presence of uncertainty and environmental variability, a catch quota is probably the most unstable of the possible management controls. While I would have little quarrel with that judgement for a fishery contained in a single national jurisdiction, the situation is rather different in international fisheries, which may extend over several national jurisdictions and international waters and be fished by several countries. In many cases, these are managed by international commissions, such as the IWC. In these bodies, it is extremely difficult to get agreement on any control other than catch quotas.

The model considered here actually assumes the conventional objective of maximizing the expected discounted present value of catches. However, if catch quotas are considered old fashioned, then citing MSY as a serious management objective, even in an example, would be positively archaic (see, for example, Larkin 1977). Here again, international fisheries are the exception. While the word "sustainable" may be seen to be inappropriate, certainly "maximum biomass yield" is more appropriate for multi-nation fisheries than "maximum economic yield" (which nation's economy?).

It is the international fishery commissions, with their restricted range of management options, that most keenly feel the need for guidance in dealing with uncertainty. Nonetheless, it is still fair to query the focus on uncertainty in stock size alone, when uncertainty in stock dynamics may well be much greater. To answer this, it is useful to return to the IWC example. For whales, it is quite true that the uncertainty in stock dynamics is at least as great as in stock size. However, policies taking account of uncertainties need some reasonable quantification of the uncertainties, and for the dynamics of whale stocks this is very difficult indeed. Here, stock size stands out; errors in its estimation can more easily be determined, and in multi-stock fisheries those errors can vary quite widely.

For uncertainties in parameters of the stock-recruitment relationship, Ludwig and Walters (1982) found that the optimal management policy called for occasional large probing experiments. Their explanation of this policy is convincing; indeed it provides the very arguments sought by biologists, who have long espoused such experiments, to convince hard-nosed fishery managers of their virtue. When uncertainty lies solely in the stock size, it seems much less likely that probing will be optimal. If so, then one could expect that cautious management would be the order of the day; certainly that is the conventional wisdom. The results presented here suggest, perhaps surprisingly, that this may not always be correct.

For moderate levels of uncertainty, this prescription for caution holds roughly, but the Clark and Kirkwood (1986) analysis suggests it can break down once the degree of uncertainty becomes large enough. To what extent this holds true for other management controls or other characterizations of uncertainty is a matter for further research, although interestingly, similar conclusions were reached by Clark et al. (1985) for a different, but related, problem involving uncertainty in stock sizes. Evidently, the degree of caution exercised must also be a function of the degree of risk-aversion in the management objectives. Here, the standard assumption of a risk-neutral objective has been made. Almost surely, international commissions would be highly risk-averse, especially when faced with the possibility of fishing out a stock.

It is rare (generally because it is often impractical) for models used in optimal control studies to be as realistic as potential users may wish them to be. For example, the model in question (5) for the dynamics of the resource is at best a gross caricature of any model that might be acceptable to biologists, and managers would probably hold the same opinion of objective (6). In such circumstances, it is important to interpret the optimal policies as indicative, rather than prescriptive. When more realism is added to the models of the dynamics, it is not uncommon for the optimum policies to require such a degree of finely tuned management control that they cannot be implemented in practice. In such cases, it is appropriate to look for more easily implemented policies that, while formally sub-optimal, are only marginally worse than the theoretical optimum. This was the approach adopted by Getz (1985).

REFERENCES

Bar-Shalom, Y. 1976. Caution, probing and the value of information in the control of uncertain systems. Ann. Econ. Soc. Meas., Vol. 5, pp. 323-337.

Clark, C.W. 1976. Mathematical Bioeconomics: The Optimal Management of Renewable Resources. New York: Wiley-Interscience.

Clark, C.W., Charles, A.T., Beddington, J.R., and Mangel, M. 1985. Optimal capacity decisions in a developing fishery. Mar. Res. Econ., Vol. 2, pp. 25-53.

Clark, C.W., and Kirkwood, G.P. 1986. On uncertain renewable resource stocks: Optimal harvest policies and the value of stock surveys. J. Environ. Econ. Manag. Vol. 13, pp. 235-244.

Getz, W.M. 1985. Optimal and feedback strategies for managing multicohort populations. J. Opt. Theor. Appl., Vol. 46, pp. 505-514.

Hennemuth, R.C., Palmer, J.E., and Brown, R.B.E. 1980. A statistical description of recruitment in eighteen selected fish stocks. J. Northw. Atl. Fish. Soc., Vol. 1, pp. 101-111.

Larkin, P.A. 1977. An epitaph for the concept of Maximum Sustained Yield. Trans. Am. Fish. Soc., Vol. 106, pp. 1-11.

Ludwig, D. and Walters, C.J. 1982. Optimal harvesting with imprecise parameter estimates. Ecol. Modell., Vol. 14, pp. 273.

Reed, W.J. 1979. Optimum escapement levels in stochastic and deterministic harvesting models. J. Environ. Econ. Manag., Vol. 6, pp. 350-363.

Walters, C.J. 1981. Optimum escapements in the face of alternative recruitment hypotheses. Can. J. Fish. Aquat. Sci., Vol. 38, pp. 704-710.

Walters, C.J. 1984. Managing fisheries under biological uncertainty. In Exploitation of Marine Communities, (R.M. May, ed.), pp. 263-274, Dahlem Konferenzen 1984. Berlin: Springer-Verlag.

PARTICIPANT'S COMMENTS

I was intrigued by a number of issues which were treated only in passing in the paper.

1. The assumptions of this model appear to be satisfied for salmon stocks, where one has a good estimate of the number of spawners, but not of the number of recruits in the next generation.

2. I wonder if the International Whaling Commission might be willing to consider policies which take account of age structure. For mammals, especially under exploitation, population size is not sufficient to characterize the state of the population.

3. I am interested in some of the questions which were raised, but not discussed at any length.

   (a) What are some of the issues which arise in the process of identifying and quantifying management objectives? Do these issues differ between national and international regulatory agencies?
   (b) What are the implications of the inability to get agreement on any control other than catch quotas? Are objectives not being met as a consequence, or is this merely a convenient way for some of the participants to thwart others?

(c) What are the implications for scientists in their attempts to perform useful investigations?

D. Ludwig

Reply

As indicated in the paper, my motivation for studying the effects on optimal catch quota decisions of uncertainty in stock sizes stemmed from my interest in the International Whaling Commission (IWC) and its current search for more effective management regimes. The work itself, however, does not purport to offer such a policy for the IWC. In using the official IWC procedure for determining catch limits to illustrate the problem, it would appear from the points raised by one of the discussants that I have given a misleading impression.

The principal difference between national and international management regulatory agencies is that the international agencies must frame regulations that are in accord with the laws and aspirations of more than one nation. This necessarily restricts the range of available options. Also, other options that may be highly appropriate in single national jurisdictions are simply inappropriate for a multinational body. Often, one such is regulation of fishing effort.

Fixing of annual catch limits is by no means the only form of IWC regulation. Other regulations imposed for different stocks include minimum and/or maximum size limits, closed seasons, restrictions on capture methods, protection of lactating females, and many others. Furthermore, in the period leading up to the IWC decision to set zero catch limits on commercially exploited whale stocks, the procedure for determining catch limits was rarely used. Instead, the scientific advice given to the IWC regularly took the form of estimated replacement yields and projections of the effects on different population components of possible catches, thus capturing the effects of age structure. It remains true, however, that such issues are not directly enshrined in IWC regulations.

I am pleased to discover that the assumptions of this model are satisfied for some salmon stocks. It would be very interesting to apply the model to them, but presumably the concentration on catch quotas would have to be amended.

Geoff Kirkwood

Geoff Kirkwood's paper on "Optimal Harvest Policies for Fisheries with Uncertain Stock Sizes" fits nicely into the framework of a workshop on Renewable Resource Management complementing many other papers in these proceedings, in particular those by Loo Botsford, Carl Walters and Philip Sluczanowski. Geoff considers the stochastic stock recruitment model of Reed (1979) with the added complexity of uncertainty in the previous period's stock size. Although the characterization of uncertainty in the stock size that is adopted is somewhat unusual (so as to make the problem tractable) the results obtained make an interesting comparison with those of Reed. With log-normally distributed noise, the uncertain stock size leads to a more cautious fishing policy except when the coefficient of variation is large, in which case the rather surprising optimal policy is to fish out the stock. Reed's model can be thought of as a special case of that adopted in this paper in which a zero variance survey is carried out yielding perfect knowledge of the stock level. An estimate of the maximum expected benefit that could result from such a survey is then obtained by comparing the optimal value function for the two models.   In certain cases (when the variation is high) this benefit can be considerable.

The author foreshadows further work in which a stock survey is incorporated into the model as a control variable.   However, no results are presented.

Mike Fisher

ANALYSIS OF ENVIRONMENTAL INFLUENCES ON
POPULATION DYNAMICS

Louis W. Botsford
Department of Wildlife and
Fisheries Biology
University of California
Davis, California 95616

Management of animal populations requires an understanding of the
factors that control abundance and how they change in response to
management. There is an increasing awareness that populations are
strongly influenced by random environmental variables and that these
effects should be identified and accounted for in management. Analysis
of environmental influences commonly involves two problems dealt with
here: (1) detection of the environmental variables that affect the
population, and (2) determination of how each environmental variable
affects population dynamics. With regard to the first problem,
"standard" statistical results developed for independent samples are
suspect because of substantial intra-series correlation. However,
various attempts to account for the correlation structure in population
and environmental data do not always yield more accurate rejection
rates. With regard to the second problem, the way in which
environmental variability affects some populations can be determined
from analysis of linearized population models. This leads to a
description of expected effects in terms of the population "frequency
response" which depends on harvest rate and life history characteristics.
Work in both areas underscore the importance of age, size, and sex
structure in populations and management models.

INTRODUCTION

Management of an animal population requires knowledge of how the population
"works", i.e., the factors that control abundance, production, and potential harvest.
Although the dynamic behavior of populations has been historically characterized
primarily as a deterministic process (occasionally with measurement and process
noise), there is an increasing trend in management to both identify and account for
the physical/biological processes responsible for random fluctuations in abundance.
In doing so, two questions inevitably arise: (1) which random environmental factors
have a substantial effect on the population, and (2) exactly how do they affect
population dynamics? I describe here two recent results regarding these two
questions. The first concerns statistical considerations involved in detection of
environmental influences, specifically the evaluation of computed correlations
between environmental and population time series. The second concerns the effect of
environmental forcing on population dynamic, how the environmental time series and
life history characteristics determine the resulting recruitment and catch time
series.

These results were obtained in research to provide a basis for management of
several different animal populations, and they serve as good motivating examples of

the problems involved in analysis of environmental effects. Fig. 1(a) shows estimates of the annual recruitment of young-of-the-year striped bass (*Morone saxatilus*) in the Sacramento/San Joaquin estuary and an associated environmental influence, Sacramento River flow in the spring of each year (the juvenile period). The question of interest here is the relative importance of density-dependence and environmental forcing in the recruitment process (see Stevens et al., 1985 for a recent review). This example is a bit unusual in that information on population dynamics is needed not just to set harvest policy, but also to set the environmental variable, which is also under human control (through dam release schedules, etc.). Fig. 1(b) shows the catch record for the northern California Dungeness crab (*Cancer magister*) fishery and an associated environmental variable, southward wind stress during the late larval period. For this example there are several possible density-dependent recruitment mechanisms that could be causing the observed cyclic fluctuations. Questions of interest here are whether the wind is involved, possibly in conjunction with one of these, and if so, how (see Botsford, 1986 for a recent review). Fig. 1(c) shows recruitment (juveniles per adult) for a population of California quail (*Callipepla californica*) from a semi-arid region of southern California, and an associated environmental variable, total annual rainfall. Questions of interest for this example are whether rainfall influences reproduction (and how), and whether density-dependence is present (see Botsford et al., 1987 for details).

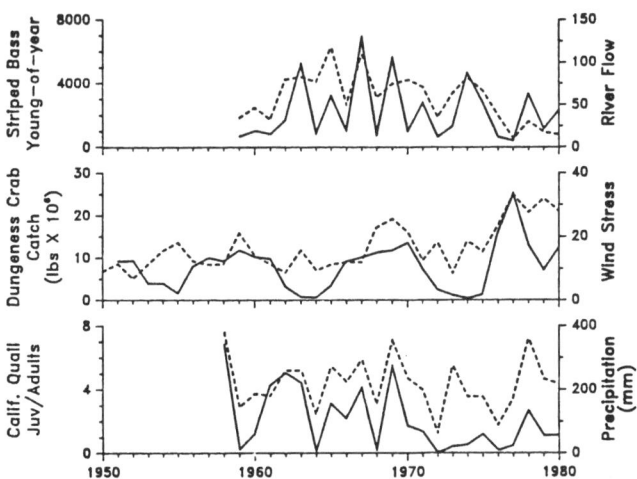

Figure 1.   Three examples of population time series (solid lines) and associated environmental variables (dashed lines):   (a) index of juvenile striped bass in the Sacramento-San Joaquin estuary and flow rate of the Sacramento River averaged over April, May and June, (b) Dungeness crab catch and southward wind stress four years earlier averaged over April, May, and June, and (c) production of young by a population of California quail and total annual precipitation during the previous year (July to June).

DETECTION OF ENVIRONMENTAL EFFECTS

There are several different approaches to detecting an influence of environment on a population's dynamic behavior, but most involve statistical evaluation of the degree to which a population time series and an environmental time series covary. Since this is usually done by computing the correlation between the two series, I will address that approach specifically, but the comments apply generally to other techniques (i.e., regression or spectral techniques).

A relationship established in this fashion is, of course, merely statistical and there is no guarantee that a causal relationship exists. This problem is addressed by requiring the computed correlation coefficient to be greater than a certain minimum value that is associated with a specified probability of concluding that the time series are correlated when they are not. Accurate computation of the value of correlation associated with a specific level of significance (i.e., probability of error) is critical to the effective use of this procedure. Before taking this approach to identifying potential environmental influences, it is important to know the probability of erroneously detecting a causal relationship when it does not exist. In most cases, this probability is much higher than expected.

The usual approach to evaluation of correlation between two series is to compute correlations, then compare them to standard tables of levels of significance for correlation coefficients. These tables are based on assumed independence of samples. Since most environmental population time series are not composed of independent samples, but rather possess some intra-series correlation, this approach is often invalid.

Bartlett (1946) developed an expression that could be used to estimate the variance of an estimated autocorrelation of a stationary Gaussian series. His later expression for the variance of cross correlations (Bartlett, 1966) reduces to the earlier expression. For series of length N

$$\text{var}[R_{xy}(k)] = \frac{1}{N-|k|} \sum_{n=-\infty}^{n=\infty} \rho_{xx}(n)\,\rho_{yy}(n) + \rho_{xy}(n+k)\,\rho_{yx}(n-k)$$

$$+ \rho_{xy}^2(k)\left[\rho_{xy}^2(n) + \frac{1}{2}\rho_{xx}^2(n) + \frac{1}{2}\rho_{yy}^2(n)\right]$$

$$- 2\rho_{xy}(k)\,[\rho_{xx}(n)\,\rho_{xy}(n+k) + \rho_{xy}(-n)\,\rho_{yy}(n+k)] \tag{1}$$

where $r_{xy}(k)$ is the estimated correlation and the ρ's are actual correlations between the two series x and y. Since their true values are not known, estimated values are used in this expression to estimate the variance. The value that corresponds to a specified level of significance is obtained by assuming a distribution of the computed correlations (usually Gaussian). Computation of this expression has been simplified by various researchers by only using parts of it. For example, using only 1/(N-k) to represent the variance corresponds to white processes. Some workers have used only the first term after the summation, while others have used the first two. Since the range of summation is infinite and the series are finite, various means have been developed for setting the limits of summation (e.g., Box and Jenkins, 1976; Fuller, 1976).

Another way of expressing the effect of intra-series correlation on significance levels is in terms of the "effective" number of degrees of freedom. Bayley and Hammersley (1946) developed an estimate for this quantity (N^*) when computing autocorrelations

$$\frac{1}{N^*} = \frac{1}{N} + \frac{2}{N^2} \sum_{j=1}^{N-1} (N-j)\, \rho^2(j) \tag{2}$$

This is similar to using the first term in Bartlett's expression above, except that the $\rho(j)$ terms in the sum are triangularly weighted.

In practice, the most common approach to computing the significance of correlations between environmental variables and population variables is to ignore the above considerations and use the classical result (for independent samples) from a table in a statistics text (e.g., Botsford and Wickham, 1975). However, there are exceptions. For example, Sutcliffe et al. (1976) and Kruse and Huyer (1983), working with oceanographic and fish data, adapted equation (2) for cross correlations (by choosing the smaller of the N^* s). Garret and Toulany (1981) and Koslow (1984) used a different modified version of (2). Peterman and Wong (1984), in their comparison of sockeye salmon stocks used the first term in equation (1). Chelton (1983, 1984) has recommended expressions for N^* based on equation (1) with either the first or the first two terms. Until recently there has, to my knowledge, been no comparison of the various expressions.

Botsford and Wainwright (1987) compared several of these expressions to a newly derived expression that is similar to (1) except that the limits of summation are finite and it contains a triangular weighting of the summed terms. The comparions were made through Monte Carlo simulations of random time series with varying degrees of intra- and inter-series correlation. They showed that generally the new expression gives the best average (averaged over 1000 trials) estimate of the variance, and expressions with fewer terms typically give poorer average estimates. However, expressions with better average variance estimates did not always yield more accurate rejection rates. In some situations (i.e., when the intra-series correlation is low enough) the "text-book" estimate of the variance (i.e., 1/(N-k)), which should apply only to white series, actually yields more accurate rejection rates than the other expressions that allow for intra-series correlation. The "correct" estimates are better for the average, but have such a high variance that one is better off (in terms of average rejection rates) not using them unless they are really needed. In addition to an improved estimate of significance levels in computed correlations, this work provides a guide as to when to use the corrections for intra-series correlations.

EXPECTED EFFECTS OF ENVIRONMENT

Successful identification of environmental influences on populations and development of the means to account for them in management require a general understanding of how random environments influence populations. Mathematical results are available from early concerns regarding the effect of harvest rate on the total variance of the harvest. This problem was first approached by adding white noise to various terms in the logistic model and attempting to relate harvest variance to stability by comparison with characteristic return times (Beddington and May, 1977; May et al., 1978). Others addressed the same question, but accounted for the effects of age structure (e.g., Reed, 1983; Horwood and Shepherd, 1981). This work was based on linearization of models with density-dependent recruitment, about an equilibrium point. Because of the linearization, the latter authors were able to characterize the response to environmental fluctuations in the frequency domain (cf. Gurney and Nisbet, 1979).

A similar linearization approach has been used to determine stability criteria for age and size structured models with density-dependent recruitment (Allen and Basasibwaki, 1974; Levin, 1981; Roughgarden et al., 1985). Botsford and Wickham (1978) and Botsford (1984) used this approach and the following model

$$R_t = B_t \, f(C_t) \tag{3}$$

where R_t is recruitment, B_t is total egg production and $f(\cdot)$ is recruitment survival which depends on C_t, the effective population size (B_t and C_t are weighted sums over the age structure), to determine how harvest rate, cohort size structure (in a fishery with a lower size limit), and the form of the recruitment survival function influence stability. They were primarily concerned with instability that could lead to the cyclic behavior observed in the northern California Dungeness crab population (Fig. 1). Simulation studies showed that results of the analysis of the linearized model accurately predicted actual behavior of the non-linear system. Briefly, increased harvest rate, a more negative slope of the recruitment survival function, and a narrower cohort size distribution lead to a less stable population. Increased harvest rate also increases the frequency of unstable cycles.

Botsford and Brittnacher (1987) have recently analyzed how these three characteristics of a population affect the way in which recruitment is influenced by environmental variability. This was done by adding a second multiplicative survival factor (that depended on a random environmental variable) to the model in equation (3), then analyzing a linearized version. The results were compared to simulations to determine the range of parameter values over which the linear analysis holds.

For the range of parameter values over which the linear model holds, the relationship between the recruitment and the environmental time series can be simply expressed in terms of the frequency response of recruitment (into the population) to environmental "noise" (Fig. 2). This response is easily computed as the discrete Fourier transform of the linearized model. The results in Fig. 2 can be related to and are completely consistent with the results regarding stability and frequency of unstable oscillations. As the population becomes less stable, the frequency response becomes more peaked about a frequency near that of the unstable oscillations. A decrease in harvest rate leads to a less peaked response at a lower frequency. An increase in cohort size width or a decrease in slope of the recruitment survival function also leads to a less peaked response.

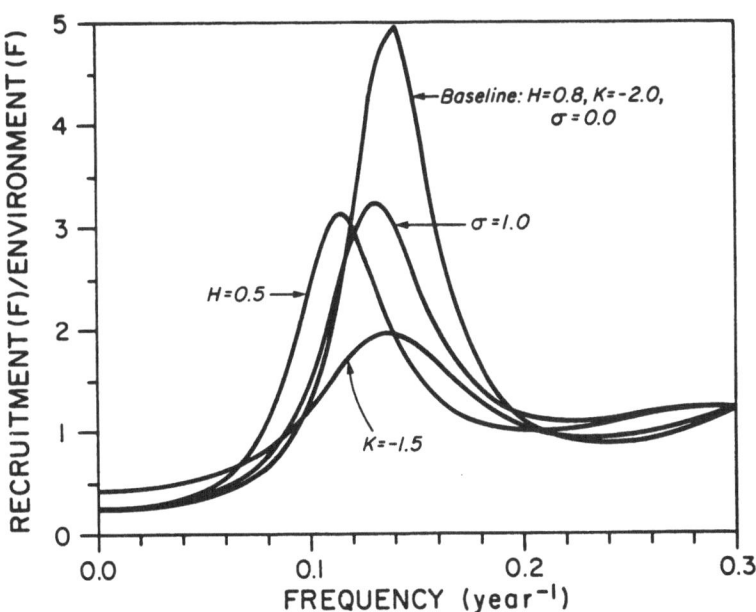

Figure 2.   The ratio of variance in recruitment to variance in an environmental variable at each frequency for a model of the northern California Dungeness crab under various conditions (H = harvest rate, K = normalized slope of the recruitment survival function, σ = width of cohort size distribution in growth-equivalent years).

Comparison of these results to simulations with sinusoidal and white noise inputs showed that they predicted population behavior almost exactly over a wide range of parameter values and "input" amplitudes. They began to fail to accurately predict behavior when the slope of the recruitment survival function became shallow enough that the population as a whole began to behave like a simple, linear population (i.e., there was not enough density dependence to hold it at an equilibrium level).

Another useful result of this work was the importance of the effect of harvest strategy on equilibrium level. As harvest increases, equilibrium typically decreases and the slope of the recruitment survival function becomes shallower, thus leading to a more stable, less variable population. However, this result holds only if reproduction is affected by harvest. If it is not (e.g., in a male-only fishery such as the Dungeness crab, in which males are able to mate before entering the fishery), then equilibrium level is not affected by harvest, and harvest does not have the stabilizing effect on equilibrium.

In addition to determining the effect of variability in the environment on recruitment, we also determined the effect of variability in recruitment on harvest. Since harvest basically consists of summing over year classes, it is a low-pass process (cf. Reed, 1983). Increasing the width of cohort size distribution in a size selective fishery lowers the cutoff frequency of this process. Since this system was linear, a comparison with simulation was not required.

DISCUSSION

Recent results described here indicate that considerably more care should be taken in dealing with environmental influences on harvested populations than has been in the past. The statistical structure (correlation) inherent in these processes precludes simple application of standard techniques based on independent samples. Work presented here regarding statistical significance of computed correlations is one approach to accounting for intra-series correlation when the source of the correlation is not known. The work on responses of populations to environment represents a specific source of correlation. This can possibly be accounted for using well known results regarding correlation between inputs and outputs of linear systems. The fact that the linearized models hold also provides potential interpretation of linear models derived from the population and environmental time series through time series analysis (Box and Jenkins, 1976).

There have been other suggestions for taking advantage of known structure in fisheries data. One example is a scheme to remove noise from recruitment using a filter based on adult age structure (Welch, 1986). There have also been notable attempts to account for other adverse effects of statistical time dependencies that are present in fisheries data (e.g., Walters, 1985; Anderson and Wilen, 1985).

The results also underscore a point I made in my paper at the first of these conferences (Botsford, 1981), the importance of including age and sex structure in analysis of population behavior. The issue of the influence of harvest rate on variability in harvest illustrates the point that analysis of simple models without age and sex structure leads to different, less realistic results than models with age and sex structure. Beddington and May (1977) and May et al. (1978) concluded that since increased harvest rate led to increased characteristic return times, it would lead to greater noise sensitivity. (However, they added the caveat that consideration of age structure could change their conclusions.) Analyses of the effects of age structure by Horwood and Shepherd (1981) and Reed (1983) showed that changes in return time due to changes in age structure bore no consistent relationship to noise sensitivity. As Reed (1983) explained, harvest has two effects, one on the age structure and the other on the equilibrium level. The latter (seen in populations without age structure) changes return time and variability in the same direction, whereas the former changes them in different directions. From the results of Botsford and Brittnacher (1987) described here, it appears that in single sex harvesting, the former is not seen, and that sensitivity to noise depends on the spectral content of the noise as compared to the frequency response of the population (which depends on age structure).

The detection process described here, of course, provides no guarantee that correlations discovered have a causal basis. The probability of detecting a relationship when one does not exist is increased by the fact that researchers typically try several environmental variables until they find one with a significant correlation. This problem is partially compensated for by the fact that environmental variables are typically correlated among themselves (e.g., ocean temperature, upwelling, and tidal height in the California Current system). Rather than demonstrations of actual mechanisms, significant correlations should best be viewed as indicators of potential mechanisms worthy of direct biological/physical investigation.

The fact that the linearized analysis works so well makes identification of causal mechanisms from population-level data even more difficult. The variance at any specific frequency in the recruitment record is the product of variance at that frequency in the environmental time series and the "frequency response" of the population. With only output (recruitment) data and a number of potential inputs (environmental variables) one will often be unable to discern the relative contribution of each. This can lead to a fundamental ambiguity in this type of analysis, of which the Dungeness crab is a good example. The cycles in crab catch can be adequately explained either by one of several density-dependent mechanisms (Botsford, 1986) or an environmental variable, wind stress (Johnson et al., 1986). With aggregate, population-level data, this ambiguity cannot be resolved and direct research on the actual mechanisms is required. Walters (1985b) has noted that this situation is an ideal problem for probing strategies that are designed to differentiate density-dependent effects from density-independent effects by holding density constant.

REFERENCES

Allen, R.L. and P. Basasibwaki. 1974. Properties of age structure models for fish populations. J. Fish. Res. Board Can., Vol. 31, pp. 1119-1125.

Anderson, J.L. and J.E. Wilen. 1985. Estimating the population dynamics of coho salmon (*Oncorhynchus kisutch*) using pooled time series and cross sectional data. Can. J. Fish. Aquat. Sci., Vol. 42, pp. 459-467.

Bayley, G.V. and J.M. Hammersley. 1946. The effective number of independent observations in an autocorrelated series. J. Roy. Statist. Soc., Ser. B, Vol. 8, pp. 184-197.

Bartlett, M.S. 1946. On the theoretical specification and sampling properties of autocorrelated time-series. J. Roy. Stat. Soc. Lond., Suppl. Vol. 8, No. 1, pp. 27-41.

Bartlett, M.S. 1966. An Introduction to Stochastic Processes, 2nd Edition, Cambridge University Press, Cambridge, 362 p.

Beddington, J.R. and R.M. May. 1977. Harvesting natural populations in a randomly fluctuating environment. Science, Vol. 197, pp. 463-465.

Botsford, L.W. 1981. More realistic fishery models: Cycles, collapse, and optimal policy. In T. Vincent and J. Skowronski (eds.), Renewable Resource Management, Proc., 1980, pp. 6-20.

Botsford, L.W. 1984. Effect of individual growth rate on expected behavior of the northern California Dungeness crab (*Cancer magister*) fishery. Can. J. Fish. Aquat. Sci., Vol. 41, pp. 99-107.

Botsford, L.W. 1986. Population dynamics of the Dungeness crab (*Cancer magister*), pp. 140-153, in G.S. Jamieson and N. Bourne (eds.) North Pacific Workshop on Stock Assessment and Management of Invertebrates, Can. Spec. Pub. Fish. Aquat. Sci. 92.

Botsford, L.W. and J.G. Brittnacher. 1987. Effects of environment on harvested populations with density-dependent recruitment and age and size structure. (in preparation).

Botsford, L.W. and T.C. Wainwright. 1987. Analysis of correlation in population and environmental time series. (in preparation).

Botsford, L.W. and D.E. Wickham. 1975. Correlation of upwelling index and Dungeness crab catch. U.S. Fish. Bull., Vol. 73, pp. 901-907.

Botsford, L.W. and D.E. Wickham. 1978. Behavior of age-specific, density-dependent models and the northern California Dungeness crab (*Cancer magister*) fishery. J. Fish. Res. Board Can., Vol. 35, pp. 833-843.

Botsford, L.W., T.C. Wainwright, J. Smith, S. Mastrup, and D. Lott. 1987. Influence of rainfall on a population of California quail (*Callipepla californica*) from a semi-arid region of California. (submitted.)

Box, G.E.P. and G.M. Jenkins. 1976. Time Series Analysis: Forecasting and Control. (Rev. Ed.), Holden-Day, Oakland, 575 p.

Chelton, D.B. 1984. Commentary: Short-term climatic variability in the northeast Pacific Ocean. pp. 87-99, in W.G. Pearcy (ed.) The Influence of Ocean Conditions on the Production of Salmonids in the North Pacific. Oregon State University Press, Corvallis, Oregon.

Chelton, D.B. 1983. Effects of sampling errors in statistical estimation. Deep-Sea Res., Vol. 30, pp. 1083-1103.

Fuller, W.A. 1976. Introduction to Statistical Time Series, John Wiley and Sons, New York, 470 p.

Garret, C.J.R. and B. Toulany. 1981. Variability of the flow through the Strait of Belle Isle. J. Mar. Res., Vol. 39, pp. 163-189.

Gurney, W.S.C. and R.M. Nisbet. 1980. Age- and density-dependent population dynamics in static and variable environments. Theoret. Pop. Biol., Vol. 17, pp. 321-344.

Horwood, J.W. and J.G. Shepherd. 1981. The sensitivity of age structured populations to environmental variability. Math. Biosci., Vol. 57, pp. 59-82.

Johnson, D.F., L.W. Botsford, R.D. Methot, and T.C. Wainwright. 1986. Wind stress and cycles in the Dungeness crab (Cancer magister) catch off California, Oregon and Washington. Can. J. Fish. Aquat. Sci., Vol. 43, pp. 838-845.

Koslow, J.A. 1984. Recruitment patterns in northwest Atlantic fish stocks. Can. J. Fish. Aquat. Sci., Vol. 41, pp. 1722-1729.

Kruse, G.H. and A. Huyer. 1983. Relationships among shelf temperatures, coastal sea level, and the coastal upwelling index off Newport, Oregon. Can. J. Fish. Aquat. Sci., Vol. 40, pp. 238-242.

Levin, S. 1981. Age structure and stability in multiple-age spawning populations. In T. Vincent and J. Skowronski (eds.), Renewable Resource Management, Proc., 1980, pp. 21-45.

May, R.M., J.R. Beddington, J.W. Horwood, and J.G. Shepherd. 1978. Exploiting natural populations in an uncertain world. Math Biosci., Vol. 42, pp. 219-252.

Peterman, R.M. and F.Y.C. Wong. 1984. Cross correlations between reconstructed ocean abundances of Bristol Bay and British Columbia Sockeye salmon (Oncorhynchus nerka). Can. J. Fish. Aquat. Sci., Vol. 41, pp. 1814-1824.

Reed, W.J. 1983. Recruitment variability and age structure in harvested animal populations. Math Biosci., Vol. 65, pp. 239-268.

Roughgarden, J., Y. Iwasa, and C. Baxter. 1985. Demographic theory for an open marine population with space-limited recruitment. Ecology, Vol. 66, pp. 54-67.

Stevens, D.E., D.W. Kohlhorst, L.W. Miller, and D.W. Kelly. 1985. The decline of striped bass in the Sacramento-San Joaquin Estuary, California. Trans. Amer. Fish. Soc., Vol. 114, pp. 12-30.

Sutcliffe, W.H., Jr., R.H. Loucks, and K.F. Drinkwater. 1976. Coastal circulation and physical oceanography of the Scotian Shelf and the Gulf of Maine. J. Fish. Res. Board Can., Vol. 33, pp. 98-115.

Walters, C.J. 1985a. Bias in the estimation of functional relationships from time series data. Can. J. Fish. Aquat. Sci., Vol. 42, pp. 147-149.

Walters, C.J. 1985b. Pathological behavior of managed populations when production relationships are assessed from natural experiments. In M. Mangel (ed.), Resource Management, Springer-Verlag, New York, pp. 89-104.

Welch, D.W. 1986. Identifying the stock-recruitment relationship for age-structured populations using time-invariant matched linear filters. Can. J. Fish. Aquat. Sci., Vol. 43, pp. 108-123.

PARTICIPANT'S COMMENTS

I shall take advantage of my unrefereed status as a commentator to draw stronger conclusions from this work than the author.

The first part of the paper is concerned with attempts to use cross-correlations to determine environmental influences. The conclusion from Botsford and Wainwright is that statistics which test for significant cross-correlations are themselves badly behaved: they suffer from a horrible trade-off between bias and efficiency.

Beyond such technical difficulties, there are more essential problems, which are raised by the author at the end of the paper. The statistical literature is full of warnings about inferences of causation from correlation. A very powerful argument is given by Mosteller and Tukey (1977, pp. 320-331). It begins as follows

George Box has [almost] said (1966): "The only way to find out what will happen when a complex system is disturbed is to disturb the system, not merely observe it passively." These words of caution about "natural experiments" are uncomfortably strong. Yet in today's world, we see no alternative to accepting them as, if anything, too weak."

Mosteller and Tukey go on to illustrate a number of common pitfalls, which demonstrate their point.

We may conclude from the present work that successful identification of environmental influences will require:

1. Consideration of density-dependence.

2. Consideration of age or size structure, if that is affected by the environment or by human actions.

3. Consideration of other plausible biological and physical effects and influences.

4. Qualitative analysis of linearized models, and

5. Examination of their validity by means of simulations of the full nonlinear model in plausible circumstances.

6. Experiments, to differentiate various potential mechanisms of environmental influence.

References Cited

Box, G.E.P. 1966. Use and abuse of regression. Technometrics, Vol. 8, pp. 625-629.

Mosteller, F.W. and J.W. Tukey. 1977. Data Analysis and Regression. Addison Wesley, Reading, Mass.

Donald Ludwig

The search for environmental correlates and explanations of fish stock variation has become a major research focus in recent years. Doubtless some interesting causal relationships will be found, and a few of these may even be of some value in management (at least to help improve short term forecasting). The analysis of linearized models is producing some surprising predictions about the conditions leading to cyclic behavior, and a synthesis of ideas that used to be fiercely debated under the headings environment versus density dependence or "Thompson-Burkenroad" debates.

As Botsford notes, there have been two main criticisms of environmental analysis. First, management requires a statistical description of how populations "work", but this requirement is not the same as knowing what factors cause the variation. Knowledge of causality would help to sort out conflicting management recommendations for cyclic populations (fish hard if density dependence is strong, fish less if environment is important), but that knowledge cannot be obtained by just watching correlations over time (effects of environment and density will remain confounded). Second, fisheries scientists have lots of environmental data series to draw upon, and any diligent search is bound to uncover some significant (but likely spurious) correlations; Barlett's and other corrections do not really deal with this problem, beyond suggesting how diligent the scientist will need to be.

Carl Walters

IDENTIFICATION AND CONTROL OF STOCHASTIC LINEAR MULTISPECIES ECOSYSTEM MODELS

Yosef Cohen
Department of Fisheries and Wildlife
University of Minnesota
St. Paul, MN 55108

Identification, verification and validation of linear state-space models are discussed. A model with three states: the densities of waterfowl species (mallard, *Anas platyrhynchos*, blue-winged teal, *Anas discors*, and pintail, *Anas acuta*) and one input, the density of ponds, was identified based on data from annual surveys done by the U.S. Fish and Wildlife Service in north-central North America for the last 30 years. An optimal stochastic set-point controller was designed for the system with the following goals in mind: (1) keeping populations as close as possible to their 30-year averages; (2) penalties for deviations from the goal were equal for the three species; and (3) penalty on applying the control (i.e., regulating the density of ponds) is to be considered. The density of each species separately depends on the density of ponds. However, because of interspecific interactions, the optimal controller dictates that to increase the population of pintail, the density of ponds should be decreased. Other stochastic controllers: tracking a desired goal (e.g., density of waterfowl), where the goal itself changes over time; and the effect of unmodeled disturbances and modeling errors on the optimal controller are discussed.

IDENTIFICATION, VERIFICATION AND VALIDATION OF STATE SPACE MODELS

To design a control for an ecosystem, one needs first to identify a mathematical model and then estimate its parameters. Despite extensive theoretical work on the subject, model identification remains a difficult task for two main reasons. First, a delineation of sufficient model complexity is usually subjective; one simply cannot test all possible system mechanisms and eliminate those that prove unimportant in affecting system dynamics. Second, much of the theory of identification (Walter, 1980) is difficult to apply, especially in ecosystem studies, where the feasible number of perturbation experiments, which are required for proper identification, is limited. One way to overcome this problem is to adopt a local, ad hoc approach. A particular structure of convenience is chosen arbitrarily and the parameters are estimated. Next, the model is examined for conformity to assumptions. This stage is termed model verification (Box and Jenkins, 1976), and consists primarily of examining residuals and their auto-correlations. Finally, the model is validated with independent data.

The particular model of interest is:

$$x(k+1) = Ax(k) + Bu(k) + Gw(k) \tag{1}$$

where $x(k)$ is an n-dimensional state vector and $u(k)$ is the r-dimensional control input to be applied. A and B are n by n and n by r transition and input matrices. G is an n by S noise input matrix. w(k) is s-dimensional zero-mean white Gaussian discrete-time noise with

$$E\{w(i)w^T(j)\} = Q\delta_{ij} \qquad (2)$$

and assumed to be independent of x(0). x(0) itself is modeled as a Gaussian random vector with mean x_0 and covariance P_0. In addition, sampled-data measurements are assumed available from the system in the form

$$z(i) = Hx(k) + v(k) \qquad (3)$$

where v($\cdot$) is m-dimensional zero-mean white Gaussian discrete-time noise with

$$E\{v(i)v^T(j)\} = R\delta_{ij} \qquad (4)$$

and assumed independent of both x(0) and w(j,$\cdot$). With some modifications, the models herein can be extended to include time dependencies; i.e., A(t), B(t), G(t), Q(t), H(t), and R(t).

Identification

To identify a model of the form (1), define the augmented state vector

$$x'(k) = \begin{bmatrix} u(k) \\ x(k) \end{bmatrix} \qquad (5)$$

where x'(k) represents a realization of a stationary multivariate time series of dimension n+r. If the series are not stationary, differencing, as well as other transformations, may be applied to the data (Tsay and Tiao, 1984). Furthermore, if the data contains a seasonal component, which is reflected in frequencies with dominant amplitudes, this seasonality can easily be removed by differencing. Akaike (1976, 1980) proposed a method of identifying state-space models of the form:

$$q(k) = Fq(k-1) + Ce(k) \qquad (6)$$

where q(k) is a vector process of dimension s, whose first r+n components comprise x'(k). The remaining components contain the additional information needed to forecast q(k); i.e., the past history of the process. F is an s by s transition matrix and C is an s by r+n input matrix. e(k) is a sequence of independent, r-dimensional random vectors called innovations or shocks with a common covariance matrix S.

To convert (6) to (1)--if input/output relationships need to be verified--one must ensure that for a model which identifies s = r+n, the parameters $F_{ij}(i=1,...,r, j=r+1,...,r+n)$ are all zero, or equivalently insignificant. If input/output relationships are apparent from knowledge of the system (e.g., it is unlikely that the density of ducks will affect the density of ponds, whereas the density of ponds may affect the density of ducts), then these F_{ij} are specified as zero during model identification. Pham-Dihn-Tuan (1978) proved that auto-regressive moving-average and state-space models of the form (6) are equivalent.

Verification and Validation

Once a model is identified, the parameters are estimated, and the model may be verified by examination of the residuals' covariance matrices and auto-correlations. If, after model fitting, (2) is satisfied, then the the model is essentially verified. Verification of (4) is more difficult and requires independent data set for the construction of (3). If the residuals do not conform to the model's underlying assumptions, then the model may be re-estimated with initial parameter estimates as derived from first identification and with some parameters--particularly those that are insignificant--set to zero. The residuals are then analyzed. If after a number of trials, residual analysis still indicates that the model assumptions are not satisfied--for example the residuals are auto- and cross-correlated with time delays--then the modeler is faced with the choice of either identifying a different generic model; e.g., a nonlinear model, or designing a linear controller with unmodeled disturbances and admitted modeling errors, to be discussed below.

Models may be validated in two ways: First, the model is applied to independent data set, and its closeness of fit and residuals are examined as above. Second, the model may be identified from time series from which the last few observations were excluded, and then model-based forecasts are compared to actual data. If possible, both methods should be applied. Clearly, adoption of a particular model depends on the purpose of modeling. As trivial as it sounds, this point is often overlooked in the selection of a modeling approach. For example, in many management problems, the manager is not overly concerned with biological mechanisms operating within the system, but rather with forecasting the effect of relatively orthodox management actions. In such cases, a simple black-box linear model (see IDENTIFICATION, VALIDATION AND VERIFICATION OF TRANSFER-FUNCTION MODELS) can produce results superior to more sophisticated models for which data are unavailable.

Design of Control

Set-point Controller. In ecosystem management, one may often be interested in keeping the system at some desired reference point using appropriate controls. Such controllers are termed nonzero set-point controllers (Kwakernaak and Sivan, 1972). The optimal control of linear stochastic systems is similar to deterministic systems (Fleming and Rishel, 1975). To derive a set-point controller for (1), consider the system

$$x(k+1) = Ax(k) + Bu(k); \qquad x(0) = x_0 \tag{7}$$

$$y(k) = Cx(k) + Du(k) \tag{8}$$

with $x(\cdot)$, $u(\cdot)$, and $y(\cdot)$ of dimensions n, r, and p, denoting the state, input and output, respectively. A model of this form can be identified as in the section titled Identification, Validation and Verification of Transfer-Function Models. Suppose it is desired to maintain $y(k)$ at a nonzero equilibrium value y^0 with zero steady state error vector. Furthermore, assume, for the moment, that the measurements are not corrupted. Both the nominal control u^0 and x^0 that will hold the system at the desired equilibrium may be determined from

$$x^0 = Ax^0 + Bu^0 \tag{9}$$

$$y^0 = Cx^0 + Du^0 .$$

When $p \neq r$, a solution to (9) cannot be found by conventional matrix inversion. Yet, a unique minimum norm solution exists (Takahashi et al., 1970).

To derive the minimization criterion, consider the perturbation variables

$$\delta x(k) = x(k) - x^0$$

$$\delta u(k) = u(k) - u^0$$

$$\delta y(k) = y(k) - y^0$$

These variables satisfy (7) and (8) with perturbation variables replacing the respective variables. The cost criterion to be minimized is then

$$J = \sum_{j=0}^{N} \frac{1}{2} \begin{bmatrix} \delta x(i) \\ \delta u(i) \end{bmatrix}^T \begin{bmatrix} X(i) & S(i) \\ S^T(i) & U(i) \end{bmatrix} \begin{bmatrix} \delta x(i) \\ \delta u(i) \end{bmatrix} + \frac{1}{2} \delta x^T (N+1) X_f \delta x(N+1) \tag{10}$$

where N+1 is the final time of interest. If X_f is diagonal, then the diagonal terms are chosen to reflect the importance of maintaining each component of $x(N+1)$ close to the chosen set-point. The entries in the weighing matrices reflect the relative importance of maintaining individual state and control components deviation at small values over each sample period; e.g., if these matrices are diagonal, then the larger the diagonal term, the closer the corresponding variable will be kept to the desired steady-state value. $S(i)$ can be chosen to allow cross terms between $u(i)$ and $x(i)$ and discount terms can be included in $X(i)$, $U(i)$ and $S(i)$. The solutions for the optimal controller $u^*(k)$ is well known (e.g., Maybeck, 1982):

$$u^*(k) = u^0 - G^*(k) [x(k) - x^0] \tag{11}$$

with

$$G^*(k) = [U(k) + B^T K(k+1)B]^{-1} [B^T K(k+1)A + S^T(k)] \qquad (12)$$

where $K(k)$ satisfies the backward Riccati recursion

$$K(k) = X(k) + A^T K(k+1)A - [B^T K(k+1)A + S^T(k)]^T G^*(k) \qquad (13)$$

solved backwards from $K(N+1) = X_f$ (Lainiotis, 1975; Silverman, 1976, Pappas et al., 1980). When only noise corrupted measurements are available from the system as in (3), the controller can be easily designed by defining a perturbation state variable

$$\delta z(k) = z(k) - z^0$$

and a Kalman filter may be applied to derive an estimate of $x(k)$, which then replaces $x(k)$ in (11) (Maybeck, 1982).

Trackers. Another objective of ecosystem management may be to achieve a control such that the output y from (8) tracks a reference output, y_r. This situation may arise in cases where the system output is subject to some other ecosystem processes of overriding importance. For example, one may want to manage wildlife populations in a large forest where the forest itself undergoes a slow, statistically predictable succession. In such cases, in addition to feedback terms, such as in (11), the controller should include feed-forward terms that would account for the forthcoming changes. The reference output may be modeled as:

$$x_r(k+1) = A x_r(k) + G_r w_r(k) \qquad (14)$$

$$y_r(k) = C_r x_r(k)$$

with $w_r(\cdot,\cdot)$ a discrete-time, zero-mean, white Gaussian noise with auto-correlation

$$E\{w_r(i)w^T(j)\} = Q_r \delta_{ij} \qquad (15)$$

where w and w_r assumed independent, $x_r(0)$ independent of $x(0)$, and the dynamics driving noises are independent of the state initial conditions.

In this case, the tracking errors

$$e(k) = y(k) - y_r(k)$$

are to be minimized, while restricting excessive amount of control with appropriate cost functions. The objective is then to minimize

$$J = E\{ \frac{1}{2} \sum_{i=0}^{N} [e^T(i)Y(i)e(i) + u^T(i)U(i)u(i) + 2e^T(i)S(i)u(i)]$$

$$+ \frac{1}{2} e^T(N+1)Y_i e(N+1)\}$$

with the penalty matrices Y and U interpreted as before, and S allowed to be nonzero for reasons such as: (1) the control is applied over an interval, e.g., harvesting season; (2) the addition of feed-forward terms in D in (8); and (3) the fact that a sampled data control is applied to a continuous-time system. (1)-(3) are further discussed in Kalman and Koepcke (1958), Athans (1971, 1972), Dorato and Levis (1971), Bertsekas and Shreve (1979).

The controller for this system is well known (Kwakernaak and Sivan, 1972) and will not be repeated here, except to point out that is of the form:

$$u^*(k) = -[G_1^*(k) \ G_2^*(k)] \ [x(k) \ x_r(k)]^T \ , \tag{16}$$

where G_1^* is independent of the reference variable to be tracked and G_2^* includes feed-forward term. When time invariant systems are modeled with constant cost matrices, then steady state, constant G_1^* and G_2^* will be achieved. However, since the system is not controllable with respect to $u(k)$, asymptotic stability is required; i.e., the eigenvalues of A must all lie strictly in the unit circle on the complex plane.

An important case for applications arises when only noise-corrupted measurements are available from both the system and the reference. For example, it is well known that observers tend to increasingly underestimate the number of birds in a waterfowl flock as the flock's size increases. Then, both (3) and

$$z_r(k) = y_r(k) + v_r(k)$$

should be added to the model with $x(k)$ and $x_r(k)$ in (16) replaced by their appropriate estimates, which can be derived by application of the Kalman filter (Maybeck, 1982).

Controllers for Unmodeled Disturbances and Modeling Errors

Suppose that (7) and (8) are modified to include unmodeled constant disturbances (Sandell and Athans, 1973):

$$x(k+1) = Ax(k) + Bu(k) + d \tag{17}$$

and modeling errors are admitted. The set-point controller (11) in this case may be inadequate, for now y is not necessarily equal y^0 in steady state. A new controller needs to be designed such that the error between an updated y^0 and $y(k)$ is compensated for. To achieve such control, $x(k)$ is augmented by the difference equations

$$q(k) = q(k-1) + [y(k-1) - y^0] \tag{18}$$

and a new set-point controller can be synthesized (Lee and Athans, 1977). The new controller then will use $y^0(k)$, instead of y^0, since the management goal itself needs to be updated. There are a number of technical problems, such as proper specification of q(0) and invertability of matrices, which need to be addressed before applications are undertaken. It can be shown that for this controller y(k) converges to y^0 in the limit (as t→∞), despite the effect of the unmodeled noise d.

IDENTIFICATION, VALIDATION AND VERIFICATION OF TRANSFER-FUNCTION MODELS

The basic idea here is to identify an l-order auto-regressive model of the form

$$y(k+1) = A_1 y(k) + \cdots + A_l y(k-l) + B_1 u(k) + \cdots + B_p u(k-p) + Gw(k) \qquad (19)$$

where y is an n-dimensional vector of outputs, u is a m-dimensional vector of inputs (controls), A_i and B_i are parameter matrices of appropriate dimensions (with some elements zero), and w(k) is an n-dimensional zero-mean Gaussian vector (with property (2)) with innovation matrix G. The delay periods are l and p, with l > p. Identification of such models is well known (Akaike, 1976, 1980). A disadvantage of purely auto-regressive models is that the number of parameters is larger than that of mixed, auto-regressive moving-average models (Box and Jenkins, 1976). An advantage of models such as (19) is their simplicity, and the ubiquity of standard identification algorithms (e.g., SAS, 1985).

Once an l-order difference equations model is identified, it can be transformed to the form (7) and (8). This can be accomplished by methods such as parallel programming (Cadzow, 1975). The method consists of taking the z-transform of (19), and using partial fraction expansions, with appropriate substitutions, to derive (7) and (8) (see, for example, Cohen, 1986).

EXAMPLE

The Data

Many aspects of data collection, and the data itself, are described in Martin et al. (1979). Briefly, every spring, since 1955, the major breeding grounds of waterfowl species in North America are surveyed and the number of ponds and abundance of 10 waterfowl species are recorded. The sampling units are organized in 49 strata (Fig. 1). Strata 26-49 include most of the breeding grounds in North America, and because of its geological history, this area is characterized by numerous ponds. I chose strata 26-49 for the analysis, and considered the density of waterfowl species as state variables. The density of ponds was defined as input. At the time of writing, I did not have exact data for the area of each stratum, so that the density figures are based on reported population sizes (and number of ponds) and on roughly estimated areas of the various strata.

Figure 1.   Strata locations for censusing waterfowl populations (from Pospahala et al., 1979).

Identification

The first problem was to decide now many species should be included in the model.   Correlations and partial correlations between the densities of the 10 waterfowl species, with lags of up to three years, revealed that the fluctuations in densities of three species: mallard, blue-winged teal and pintail were most highly synchronized.   Thus, a three-component state vector was chosen with $x_1(k)$, $x_2(k)$, and $x_3(k)$ denoting the mean of the ln-density of mallard, blue-winged teal and pintail at year k, for 1955 through 1984.   Yearly means were calculated for strata 26-40.   For the purpose of model identification, the data for 1981-1984 were excluded.

The model was identified for strata 26-40 using the statespace procedure (SAS, 1985), with the transfer coefficients--in the matrix F in (6)--between the state variables and the input variable specified as zero. Since the time series are relatively short, and since the number of parameters to be estimated needs to be as small as possible, lags of no more than one year were included in the initial model identification.   Thus, a total of 12 parameters were estimated for ensemble mean of

15 time series each 27 years long. These means, for the density of mallard, ±2 SD, are shown in Fig. 2, as well as the model fit, its two SD and forecasts of up to three years with their 2 SD. Similar results were obtained for the blue-winged teal and the pintail.

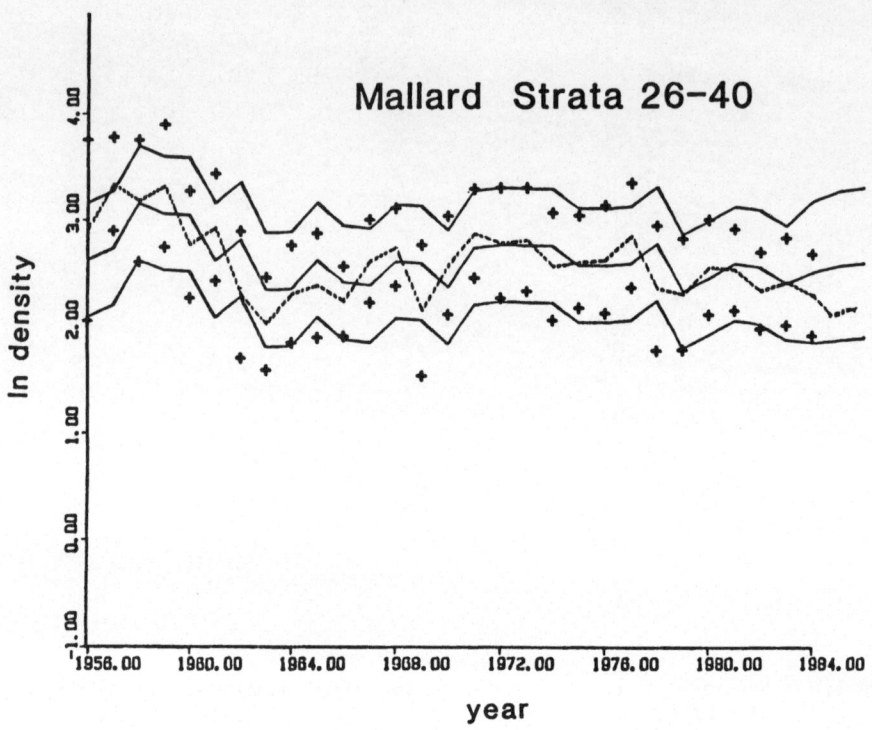

Figure 2. Model fit and data ±2 SD for strata 26-40 for mallard. The three solid lines are the model (center solid line) and its 2 SD; broken line shows data based means; and +'s show 2 SD around the data based means. Data up to 1980 were included in fitting the model. Thus, from 1981 on, the data are actual forecasts. r^2 for model fit = 0.4

Verification and Validation

The model forecasts were judged to be adequate (Fig. 2). Further verification was provided by using the identified parameters with the data for strata 41-49 (Fig. 3). Note that in this case, the results for 1955-1981 are in fact one year forecasts, and for 1980-1984 are 2-, 3-, and 4-year forecasts respectively. Overall, the model reflects the data adequately.

Set-point Controller. To develop a set-point linear controller, I fitted a model to stratum 26 only (Fig. 4). This was done simply because application of a controller requires the best model possible. Since the process of model identification is neither tedious nor expensive, one should apply a controller to each stratum separately after identifying specific models to each stratum.

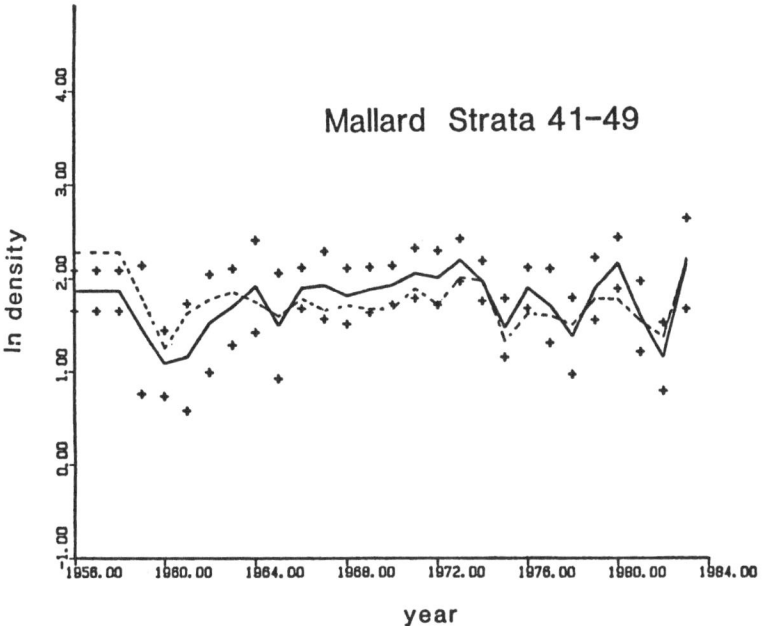

Figure 3.   Model fit (broken line) and data based means ±2 SD (solid line and +'s) for strata 41-49.   Note that model parameters were derived from fitting to strata 29-40.   Thus, the model represents (up to 1980) real one year forecasts.   From 1980 on, it represents 2-, 3-, and 4-year forecasts. r^2 for model fit = 0.42.

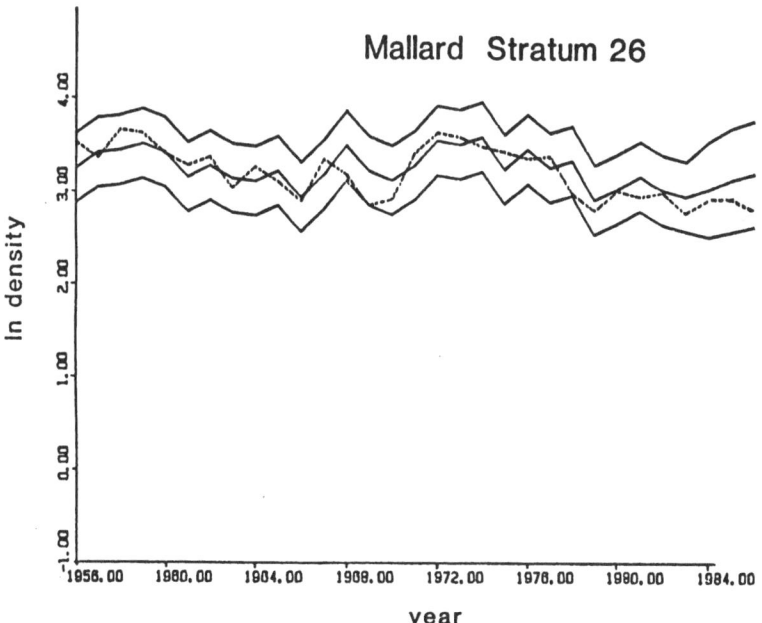

Figure 4.   Model fit, its 2 SD (solid lines) and data (broken line) for mallard density in stratum 26. r^2 = 0.62.

After fitting the model to stratum 26, I developed the optimal controller with the following objectives/steps in mind: [1] the management goal is to regulate the density of ponds such that the population density of the mallard, blue-winged teal and pintail would remain as close as possible to their 30-year average; i.e., $x_1^0 = 3.250$, $x_2^0 = 2.617$, and $x_3^0 = 2.244$ (all quantities are in ln). [2] The penalty for deviations from this goal is $1/[0.2(x_i^0)^2]$ for $i = 1,2,3$, with no interactions among these penalties. [3] Similar penalty was accepted for X_f. Thus, X and X_f in (10) are diagonal. [4] The value of u^0 which satisfied these requirements was determined. Since there is only one input and three states, (9) was solved in a least square sense to give $u^0 = 1.234$. [5] The penalty on the amount of control applied was calculated as $1/(u^0)^2$. [6] The planning horizon is for 100 years.

Based on [1]-[6], (12) and (13) were solved backward. After about 10 iterations, G^* stabilized to $G^* = [0.204, 0.031, -0.085]$. In other words, with the system and control penalties as described (i.e., species are equal in terms of their influence on the amount of control necessary), the optimal control stipulates that when mallard or blue-winged teal densities are below average, the density of ponds should be increased. On the other hand, when pintail is below average, and perhaps because of interspecific interactions, the density of ponds should be decreased. Furthermore, the density of mallard is more sensitive to changes in pond density than those of blue-winged teal or pintail.

In particular applications, the penalties on deviations from the goal and on the amount of applied control--as expressed in X, X_f, and U--should be weighed accordingly. If the density of ponds cannot be easily regulated, then the penalty term U should be larger. This will then result in less control applied, and therefore less tight regulation. If regulating the density of a particular species is to be prefered over the others, then the diagonal terms in X and X_f should be adjusted accordingly. The advantage of using such an approach is that the manager is then forced to clarify to herself the important issues in management; e.g., are species equally important? is the amount of control to be applied feasible? etc.

A typical control trajectory is shown in Fig. 5. Here, I generated a series of values for the densities of the three species as follows: starting from $x(0) = x^0$, a random noise was added, drawn from a multivariable normal distribution with mean zero and covariance matrix as determined from the model fit (eq. (6)). Then, the feedback control from (11)-(13) was applied to produce a typical 50-year trajectory for u^* and x. Note that the values thus produced are far from what may be accepted by managers. Acceptable values may be achieved by appropriate penalties in the cost matrices U, X, and X_f. These derived penalties can then be examined by the manager to determine whether control is worthwhile in the first place, and what may be accomplished by feasible controls.

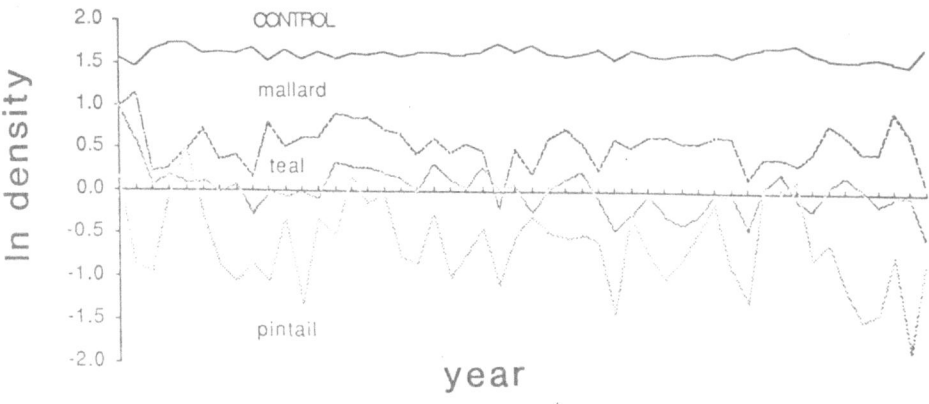

Figure 5. Typical control trajectory for set-point controller for 50 years.

CONCLUSION

The approach outlined is useful when a manager is interested in short term forecasts, and in derivation of short term control policies, which may be periodically updated. Since the model fit was adequate, and since the verification with a somewhat independent data set and forecasts were sufficiently close, there is no reason to undertake more sophisticated approaches such as adaptive control (Walters, this volume) or to assume nonlinearities in either model or parameters. Such approaches are data intensive, and more often than not sufficient data are unavailable. Finally, it should be realized that models such as (1) may be biologically "unreasonable". For example, the identified model may indicate that two species which are biologically unrelated appear as if they compete. One needs to keep in mind that the models are data based, and no Cartesianism (identification of cause and effect) is attempted.

No attempt was made to fit a nonlinear model, such as the Lotka-Voltera competition model. Even though interspecific competition among these three species has not been proven, a nonlinear model may fit the data more closely and result in more accurate forecasts. Yet, the basic approach of identification, verification, and validation would not change; the identification algorithms would. Even if the system examined is linear, the parameters themselves may be nonlinear, which again makes the whole identification procedure nonlinear. Finally, the costs and penalties as applied here are fictitious.

ACKNOWLEDGMENTS

Jeff Stone provided assistance in programming and reviewed the manuscript. Financial support was provided by the University of Minnesota Agricultural Experiment Station. Contribution Number 14804 of The University of Minnesota Agricultural Experiment Station. Comments by Dr. Ed Mansfield improved the manuscript.

REFERENCES

Akaike, H. 1976. Canonical correlation analysis of time series and the use of an information criterion. Pp. 27-98 in R.K. Mehrns, and D.G. Lainiotis (eds.), System Identification: Advances and Case Studies. Academic Press, New York.

Akaike, H. 1980. On the identification of state space models and their use in control. Pp. 175-187 in D.R. Brillinger, and G.C. Tiao (eds.). Directions in Time Series. Institute of Mathematical Statistics, University of Wisconsin, Madison, Wisconsin.

Athans, M. 1971. The role and use of the Linear-Quadratic-Gaussian problem in control system design. IEEE Trans. Automat. Control AC-16, No. 6, pp. 529-552.

Athans, M. 1972. The discrete-time Linear-Quadratic-Gaussian stochastic control problem. Ann. Econ. Soc. Measurement, Vol. 1, No. 4, pp. 449-491.

Box, G.E.P., and G.M. Jenkins. 1976. Time Series Analysis Forecasting and Control. Holden-Day, San Francisco.

Bertsekas, D., and S. Shreve. 1979. Stochastic Optimal Control: The Discrete-Time Case. Academic Press, New York.

Cadzow, J.A. 1975. Discrete-Time Systems. Prentice-Hall, Englewood Cliffs, New Jersey.

Cohen, Y. 1986. Discrete state models for analysis and management of ecosystems. In J. Verner, M.L. Morrison, and C.J. Ralph (eds.). Modeling Habitat Relationships of Terrestrial Vertebrates. University of Wisconsin Press, Madison, Wisconsin. In press.

Dorato, P., and A.H. Levis. 1971. Optimal linear regulators: the discrete-time case. IEEE Trans. Automat. Control AC-16, No. 6, pp. 613-620.

Fleming, W.H., and R.W. Rishel. 1975. Deterministic and stochastic optimal control. J. Inst. Math. Appl., Vol. 15, pp. 319-342.

Kalman, R.E., and R.W. Koepcke. 1958. Optimal synthesis of linear sampling control systems using generalized performance indices. Trans. ASME, Vol. 80, pp. 1800-1826.

Kwakernaak, H., and R. Sivan. 1972. The maximally achievable accuracy of linear optimal regulator and linear optimal filters. IEEE Trans. Automat. Control AC-17, No. 1, pp. 79-86.

Lainiotis, D.G. 1975. Discrete Riccati equation solutions: partitioned algorithms. IEEE Trans. Automat. Control AC-20, pp. 555-556.

Lee, W., and M. Athans. 1977. The discrete-time compensated Kalman filter. Rep. ESL-P-791. MIT Electronic Systems Lab., Cambridge, Massachusetts.

Martin, F.W., R.S. Pospahala, and J.D. Nichols. 1979. Assessment and population management of North American migratory birds. Statistical Ecology, Vol. 10, pp. 187-239.

Maybeck, P.S. 1982. Stochastic Estimation, and Control, Academic Press, New York.

Pappas, T., A.J. Laub, and N.R. Sandell, Jr. 1980. On the numerical solution of the discrete-time algebraic Riccati equation. IEEE Trans. Automat. Control AC-25, No. 4, pp. 631-641.

Pham-Dihn-Tuan. 1978. On the fitting of multivariate processes of the autoregressive-moving average type. Biometrika, Vol. 65, pp. 99-107.

Sandell, N.R., Jr., and M. Athans. 1973. On type-L multivariable linear systems. Automatica, Vol. 9, No. 1, pp. 131-136.

SAS. 1985. SAS Institute, Inc., Cary, North Carolina.

Silverman, L.M. 1976. Discrete Riccati equations: alternative algorithms, asymptotic properties, and system theory interpretations. Pp. 313-386 in C.T. Leondes (ed.). Control and Dynamic Systems: Advances in Theory and Applications. Vol. 12, Academic Press, New York.

Takahashi, Y., M.J. Rabins, and D.M. Auslander. 1970. Control and Dynamic Systems. Addison-Wesley, Reading, Massachusetts.

Tsay, R.S. and G.C. Tiao. 1984. Consistent estimates of autoregressive parameters and extended sample autocorrelation functions for stationary and nonstationary ARMA models. J. Amer. Stat. Assoc., Vol. 79, pp. 84-96.

Walter, E. 1982. Identifiability of state space models. Lect. Notes Biomath, No. 46.

Widnall, W.S. 1968. Applications of Optimal Control Theory to Computer Controller Design. MIT Press, Cambridge, Massachusetts.

PARTICIPANT'S COMMENTS

The methods of linear time series analysis and control are a fairly recent development (Box & Jenkins, 1976), and are probably not yet too familiar to many resource managers. According to my colleague, Ed Mansfield, there seems to be considerable scope for practical application of these techniques to resource management. Ed was been obtaining interesting results in the forecasting of forest fire frequencies and fish catches. Cohen's paper provides another interesting application.

No justification is offered for the chosen objective of keeping duck populations as close as possible to their historical average. No doubt Ducks Unlimited will have something to say about this, especially since Cohen finds that this would require reduction in pond density. But at least this objective is a refreshing change from the usual dogma, that more of everything is always better. Clearly Cohen's method would work equally well on any other set-point objective short of the infinitude of ducks implicit in DU's very name.

Colin W. Clark

AN EVOLUTIONARY RESPONSE TO HARVESTING

Thomas L. Vincent
Aerospace and Mechanical Engineering

Joel S. Brown
Ecology and Evolutionary Biology

University of Arizona
Tucson, Arizona 85721

Evolutionary game theory uses a generating function to define the fitness of any individual bounded by the same evolutionary constraints. Using a single generating function, we investigate the effect of harvesting with a model which has a single evolutionarily stable strategy (ESS). We show that harvesting will promote an evolutionary response from the community which will result in a decrease in yield. Of particular interest are the circumstances under which harvesting promotes the coexistence of several strategies in a community that would otherwise have a single-strategy ESS.

INTRODUCTION

The goals of managing an ecosystem are diverse and include maximizing sustainable yield, controlling pests and diseases, maintaining or promoting diversity, and enhancing recreational or aesthetic values. The impact of a management program on an ecosystem may be broadly classified as (1) ecological and (2) evolutionary. The ecological impact involves changes in the dynamics of existing or introduced populations, as evidenced by ecosystem stability, persistence, and extinction. Evolutionary impact is associated with changes in evolutionary dynamics, as evidenced by ecosystem traits or characteristics which do evolve.

Evolutionary impact includes both directional and disruptive selection. In the last 100 years, intense hunting pressures have resulted in dramatic reductions in the body size of Nile crocodiles and in the tusk size of African elephants. Drug and pesticide resistence among disease agents and insects is a frequent occurrence and is of growing concern. The over 125 recognized breeds of dog are testimony to the ability of a management program to exercise disruptive selection on a species.

Most discourse on environmental management considers ecological rather than evolutionary impact (see, for example, Goh, 1980; Vincent and Skowronski, 1981; or several papers in this volume). The inadequacy of ecological models for management purposes is recognized (Beddington, et al., 1981); however, there have been very few attempts (Pulliam, 1981; Rosenzweig, this volume) to incorporate evolutionary effects into management models. Ecological considerations often appear more tangible and immediate; evolutionary effects seem slow and ambiguous in comparison. Modeling the evolutionary implications of management practices is complex since it requires simultaneous considerations of ecological processes (population dynamics) and evolutionary processes (changes in the frequency of traits and phenotypes). Here we present an approach for considering both of these processes in a model that should prove useful in addressing management questions.

Elsewhere (Vincent and Brown, 1984a, 1984b; Vincent, 1985; Brown and Vincent, 1986), we have developed a continuous evolutionary game theory utilizing the evolutionarily stable strategies (ESS) concept of Maynard Smith and Price (1973). Strategies are defined as heritable traits or phenotypes of individuals. The theory is used to find the ESS for models including both ecological and evolutionary processes. The ESS may be either a single strategy or a coalition of coexisting strategies.

In this paper, we briefly sketch the theory and apply it to a harvesting model. We examine the conditions under which harvesting will exert either directional or disruptive selection. We also compare the effect of evolution on yield.

A REVIEW OF EVOLUTIONARY GAME THEORY

Consider a community of individuals in which the population is in the process of coevolving. We will identify the individual by the strategy (heritable phenotype) it possesses. Strategies are assumed to be points in an n-dimensional Euclidean space, E^n, and may be represented by the row vector

$$u = [u_1, \ldots, u_n] . \tag{1}$$

In general, these strategies may not take on any value in E^n, but must lie in some subset of E^n. It is assumed that all strategies are drawn from the same subset $U \subseteq E^n$. Let $r(t)$ be the total number of distinct strategies in the community at time t, let $\tilde{u}$ be the row vector of all such strategies present,

$$\tilde{u} = [u^1, \ldots, u^r], \tag{2}$$

let N_i be the population density of individuals using strategy u^i, and let N be the row vector of the population densities,

$$N = [N_1, \ldots, N_r] . \tag{3}$$

Assume that the population dynamics of the community can be expressed in terms of a discrete time model of the form

$$N_i(t + 1) = N_i(t)H_i(\tilde{u}, N) \qquad i = 1, \ldots, r . \tag{4}$$

By definition, H_i is the fitness of individuals using strategy u^i.

For what follows, we need to identify certain strategies in the population as being separate from the rest. Let the vector

$$u^0 = [u^1, \ldots, u^\sigma] \tag{5}$$

denote the first $\sigma < r$ strategies of $\tilde{u}$, and let

$$u^m = [u^{\sigma+1}, \ldots, u^r] \tag{6}$$

denote the remaining (mutant) strategies. It follows that

$$\tilde{u} = [u^0, u^m] . \tag{7}$$

Let us now define the following ordering for the population vector N:

$$N = 0 \qquad \text{if and only if } N_i = 0 \qquad i = 1, \ldots, r$$

$$N > 0 \qquad \text{if and only if } N_i > 0 \qquad i = 1, \ldots, r \tag{8}$$

$$N \underset{\sim}{>} 0 \qquad \text{if and only if } N_i > 0 \qquad i = 1, \ldots, \sigma$$

$$N_i = 0 \qquad i = \sigma + 1, \ldots, r$$

<u>Definition</u>: A vector u^0, $u^i \in U$, $i = 1, \ldots, \sigma$, is said to be a <u>coalition vector</u> if there exists a population vector $N^* \underset{\sim}{>} 0$ such that for all $u^i \in U$, $i = q + 1, \ldots, r$, and for all $N(0) \underset{\sim}{>} 0$, equation (4) yields

$$\lim_{t \to \infty} N(t) = N^* . \tag{9}$$

At the equilibrium point, all strategies in the coalition vector must have the same fitness

$$H_i(\tilde{u}, N^*) = 1 . \tag{10}$$

The fraction of the total population using strategies in the coalition vector is called the <u>coalition frequency</u>, P_0.

$$P_0 = \sum_{i=1}^{\sigma} \frac{N_i(t)}{N(t)} . \tag{11}$$

<u>Definition</u>: A coalition vector u^0 is said to be an ESS if there exists a generation time $t_m > 0$ such that for all strategy vectors $u^i \in U$, $i = q + 1, \ldots, r$, and all $N(0) > 0$, equations (4) and (11) yield a monotone increasing sequence for the coalition frequency P_0 when $t = t_m, t_{m+1}, \ldots$, and

$$\lim_{t \to \infty} N(t) = N^* \to \lim_{t \to \infty} P_0(t) = 1 . \tag{12}$$

We have previously shown (Brown and Vincent, 1986, lemma 2.1) that if u^0 is an ESS with $r > q$, then as $t \to \infty$, the fitness associated with any strategy in the coalition u^0 must be greater than the average fitness of the remaining (mutant) strategies. That is,

$$H_j(\tilde{u}, N) > \bar{H}_m \qquad \text{for } j = 1, \ldots, \sigma \tag{13}$$

where

$$\bar{H}_m = \frac{\displaystyle\sum_{i=\sigma+1}^{r} N_i(t) N_i(\tilde{u}, N)}{\displaystyle\sum_{i=\sigma+1}^{r} N_i(t)} \tag{14}$$

is the average fitness of the mutants. Since this condition must hold for any number of mutant strategies, it must hold in particular for one mutant. In this case, (13) simplifies and we obtain

$$H_j(\tilde{u}, N) > H_{\sigma+1}(\tilde{u}, N) \qquad \text{for } j = 1, \ldots, \sigma \tag{15}$$

since in the limit $\underset{t \to \infty}{N(t)} \to N^*$ and $H_j(\tilde{u}, N^*) = 1$ we have

$$H_{\sigma+1}(\tilde{u}, N^*) < H_j(\tilde{u}, N^*) = 1 \qquad j = 1, \ldots, \sigma . \tag{16}$$

This result is made more useful with the introduction of a generating function.

Definition: A function $G(u, \tilde{u}, N)$ is said to be a generating function for the population if

$$G(u^i, \tilde{u}, N) \equiv H_i(\tilde{u}, N) \qquad i = 1, \ldots, r . \tag{17}$$

In other words, G is a generating function if the fitness of individuals using a particular strategy u^i is obtained by replacing u in G by that strategy. In terms of the generating function, condition (16) becomes

$$G(u^{\sigma+1}, \tilde{u}, N^*) < G(u^j, \tilde{u}, N^*) = 1 \qquad j = 1, \ldots, \sigma. \tag{18}$$

This result may be summarized in the following theorem.

Theorem: Let u^0 be a coalition vector with the equilibrium density vector $N^* \geq 0$. Let $G(u, \tilde{u}, N)$ be a generating function for the population. If u^0 is an ESS under the dynamics given by (4), then the function

$$G(u, \tilde{u}, N^*)$$

must take on a global maximum with respect to $u \in U$ at $u^1, \ldots, u^\sigma$, with the maximum value of G at each of these points equal to 1.

NECESSARY CONDITIONS FOR AN ESS WITH HARVESTING

Our objective is to examine the possible evolutionary outcomes from subjecting a population to "harvesting" by "predators" external to the system (e.g., man). We will assume that the harvesting strategy of the predator is not influenced by the prey. Our specific aim is to examine the evolution of the prey as influenced by the predator.

The methods discussed may be applied to a wide class of models. We choose to work with the discrete form of the Lokta-Volterra model here because of its wide use and familiarity. We will modify the model, however, by adding a harvesting term.

Assume that the number of prey individuals in the population using strategy u^i in the next generation is given by

$$N_i(t + 1) = N_i(t) \left\{ 1 + \frac{R}{k(u^i)} \left[k(u^i) - \sum_{j=1}^{r} N_j(t)\alpha(u^i, u^j) \right] - h(u^i) \right\} \tag{19}$$

where $N_i(t)$ is the number of individuals using strategy u^i in the current generation, R is a common intrinsic growth rate, $k(u^i)$ is the carrying capacity of an individual using strategy u^i, $\alpha(u^i, u^j)$ is the competitive effect of individuals using strategy u^j on the fitness of individuals using strategy u^i, and $h(u^i)$ is the effect of harvesting by the predator. A generating function exists for this system and is given by

$$G(u, \tilde{u}, N) = 1 + \frac{R}{k(u)} \left[k(u) - \sum_{j=1}^{r} N_j\alpha(u, u_j) \right] - h(u) . \tag{20}$$

In seeking to find an ESS, the necessary conditions as obtained from the theorem will depend upon the number of strategies in the coalition vector. Since we do not know a priori the number of strategies in the coalition vector, we begin by seeking an ESS with a coalition of one and see if we obtain multiple solutions. If an ESS coalition of one does not exist, we can then seek a coalition of two, etc.

In this section, we develop for the above model the necessary conditions for a coalition of one and two strategies. Necessary conditions for a coalition of greater than two involve similar procedures.

We first assume a coalition of one ($\sigma = 1$). In this case, the appropriate G function for use with the theorem is given by

$$G(u, \tilde{u}, N^*) = 1 + \frac{R}{k(u)} \left[k(u) - N_1^*\alpha(u, u^1) \right] - h(u) . \tag{21}$$

Since the strategy set is unbounded, a necessary condition to maximize G at u^1 is given by

$$\frac{\partial G(u^1, \tilde{u}, N^*)}{\partial u} = 0 \tag{22}$$

with the equilibrium condition

$$G(u^1, \tilde{u}, N^*) = 1 . \tag{23}$$

In terms of the generating function (20), we obtain

$$\frac{R}{k(u^1)} \left[\frac{\partial k(u^1)}{\partial u} - N_1^* \frac{\partial \alpha(u^1, u^1)}{\partial u} \right] - \frac{R}{[k(u^1)]^2}$$

$$\left[k(u^1) - N_1^*\alpha(u^1, u^1) \right] \frac{\partial k(u^1)}{\partial u} - \frac{\partial h(u^1)}{\partial u} = 0 \tag{24}$$

and

$$\frac{R}{k(u^1)} \left[k(u^1) - N_1^*\alpha(u^1, u^1) \right] = h(u^1). \tag{25}$$

Note that (24) may be used to simplify (23) to obtain

$$\frac{R}{k(u^1)} \left[\frac{\partial k(u^1)}{\partial u} - N_1^* \frac{\partial \alpha(u^1, u^1)}{\partial u} \right] - \frac{h(u^1)}{k(u^1)} \frac{\partial k(u^1)}{\partial u} - \frac{\partial h(u^1)}{\partial u} = 0 . \tag{26}$$

For a coalition of two ($\sigma = 2$), the appropriate G function to use with the theorem is given by

$$G(u, \tilde{u}, N^*) = 1 + \frac{R}{k(u)} \left[k(u) - N_1^*\alpha(u, u^1) - N_2^*\alpha(u, u^2) \right] - h(u) . \tag{27}$$

Again, since the strategy set is unbounded, necessary conditions to maximize G at u^1 and u^2 are given by

$$\frac{\partial G(u^1, \tilde{u}, N^*)}{\partial u} = 0 \tag{28}$$

$$\frac{\partial G(u^2, \tilde{u}, N^*)}{\partial u} = 0 \tag{29}$$

with the equilibrium conditions

$$G(u^1, \tilde{u}, N^*) = 1 \tag{30}$$

$$G(u^2, \tilde{u}, N^*) = 1 . \tag{31}$$

Using the same procedures to obtain (25) and (26), the equilibrium conditions may be written as

$$\frac{R}{k(u^1)} \left[k(u^1) - N_1^* \alpha(u^1, u^1) - N_2^* \alpha(u^1, u^2) \right] = h(u^1) \tag{32}$$

$$\frac{R}{k(u^2)} \left[k(u^2) - N_1^* \alpha(u^2, u^1) - N_2^* \alpha(u^2, u^2) \right] = h(u^2) \tag{33}$$

and the optimizing conditions may be written as

$$\frac{R}{k(u^1)} \left[\frac{\partial k(u^1)}{\partial u} - N_1^* \frac{\partial \alpha(u^1, u^1)}{\partial u} - N_2^* \frac{\partial \alpha(u^1, u^2)}{\partial u} \right]$$

$$- \frac{h(u^1)}{k(u^1)} \frac{\partial k(u^1)}{\partial u} - \frac{\partial h(u^1)}{\partial u} = 0 \tag{34}$$

$$\frac{R}{k(u^2)} \left[\frac{\partial k(u^2)}{\partial u} - N_1^* \frac{\partial \alpha(u^2, u^1)}{\partial u} - N_2^* \frac{\partial \alpha(u^2, u^2)}{\partial u} \right]$$

$$- \frac{h(u^2)}{k(u^2)} \frac{\partial k(u^2)}{\partial u} - \frac{\partial h(u^2)}{\partial u} = 0 . \tag{35}$$

If the interaction coefficient α has the following symmetry properties

$$\alpha(u, u^j) = 1 \qquad \text{if } u = u^j \tag{36}$$

$$\frac{\partial \alpha(u, u^j)}{\partial u} = 0 \qquad \text{if } u = u^j , \tag{37}$$

then the necessary conditions for a coalition of one strategy as given by (25) and (26) reduce to

$$N_1^* = k(u^1) \left[1 - \frac{h(u^1)}{R} \right] \tag{38}$$

$$\frac{R - h(u^1)}{k(u^1)} \frac{\partial k(u^1)}{\partial u} = \frac{\partial h(u^1)}{\partial u} \tag{39}$$

and the necessary conditions for a coalition of two as given by (32)-(35) reduce to

$$N_1^* + N_2^* \alpha(u^1, u^2) = k(u^1)\left[1 - \frac{h(u^1)}{R}\right] \tag{40}$$

$$N_1^* \alpha(u^2, u^1) + N_2^* = k(u^2)\left[1 - \frac{h(u^2)}{R}\right] \tag{41}$$

$$\frac{R - h(u^1)}{k(u^1)} \frac{\partial k(u^1)}{\partial u} = \frac{\partial h(u^1)}{\partial u} + \frac{R}{k(u^1)} N_2^* \frac{\partial \alpha(u^1, u^2)}{\partial u} \tag{42}$$

$$\frac{R - h(u^2)}{k(u^2)} \frac{\partial k(u^2)}{\partial u} = \frac{\partial h(u^2)}{\partial u} + \frac{R}{k(u^2)} N_1^* \frac{\partial \alpha(u^2, u^1)}{\partial u} \; . \tag{43}$$

In order to illustrate the use of these results in what follows, we will consider a specific example where $R = 0.25$ and

$$k(u) = 100 \; e^{-\frac{(u-1)^2}{2}} \tag{44}$$

$$\alpha(u, u^j) = 1 - \frac{(u - u^j)^2}{16} \qquad \text{(for any j) .} \tag{45}$$

THE ESS UNDER NO HARVESTING

In order to examine the effect of harvest, we will first examine the case with no harvesting,

$$h(u) \equiv 0 \; . \tag{46}$$

The necessary conditions for a coalition of one are

$$N_1^* = k(u^1) \tag{47}$$

$$\frac{\partial k(u^1)}{\partial u} = 0 \; . \tag{48}$$

With $k(u^1)$ given by (44), we obtain

$$u^1 = 1 \tag{49}$$

$$N_1^* = 100 \; . \tag{50}$$

A geometric interpretation of this solution in relation to the theorem is obtained by plotting the generating function (21) minus one (G-1) as a function of u with N_1^* and u^1 given by (49) and (50). Figure 1 illustrates this plot. As required by the theorem, G as defined by (21) takes on a maximum value of 1 when $u = 1$. We deduce from Figure 1 that this is a global maximum and hence $u^* = 1$, $N_1^* = 100$ represents a true ESS. Since this ESS for the population is unique, we need not investigate the possibility of a coalition of two or higher.

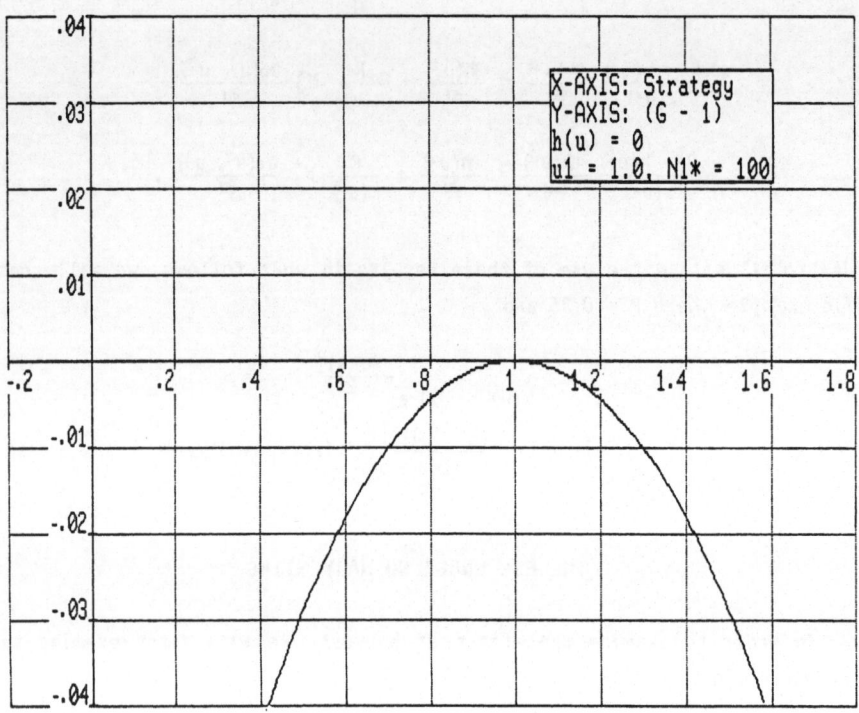

Figure 1. The ESS Under No Harvesting.

THE ESS UNDER LINEAR HARVEST

Consider now a linear harvest function of the form

$$h(u) = 0.1u . \qquad (51)$$

If u were body size, then (51) could correspond to a situation where a premium is paid for a catch of larger body size or where nets or traps select linearly for larger body size. Again, we start by examining the necessary conditions for a coalition of one. From (38) and (39) we obtain

$$(0.25 - 0.1u^1)(1 - u^1) = 0.1 \qquad (52)$$

$$N_1^* = 100 \ e^{-\frac{(u^1-1)^2}{2}} \left[1 - \frac{0.1u}{0.25}\right] . \tag{53}$$

From (52) we obtain $u^1 = 0.5$ and $u^1 = 3$. With $u^1 = 0.5$ we obtain $N_1^* = 70.599$, and with $u^1 = 3$ we obtain $N_1^* = -2.707$. It follows that the second solution is not valid, and

$$u^1 = 0.5 \tag{54}$$

$$N_1^* = 70.599 \tag{55}$$

remains as the ESS candidate. Figure 2 illustrates a plot of the generating function (21) minus one (G - 1) as a function of u with u^1 and N_1^* given by (54) and (55). We conclude from this figure that (54) and (55) are indeed an ESS. It is of interest to examine the plot of the generating function minus one (with the harvest term included) with u^1 and N_1^* given by the no-harvest ESS (49) and (50), as shown in Figure 3. Note how the maximum value of this function shifts to the left with a value of G > 1 and encourages the evolution of u in this direction. The strategy corresponding to the maximum value of G in Figure 3 is actually to the left of the ESS obtained in Figure 2.

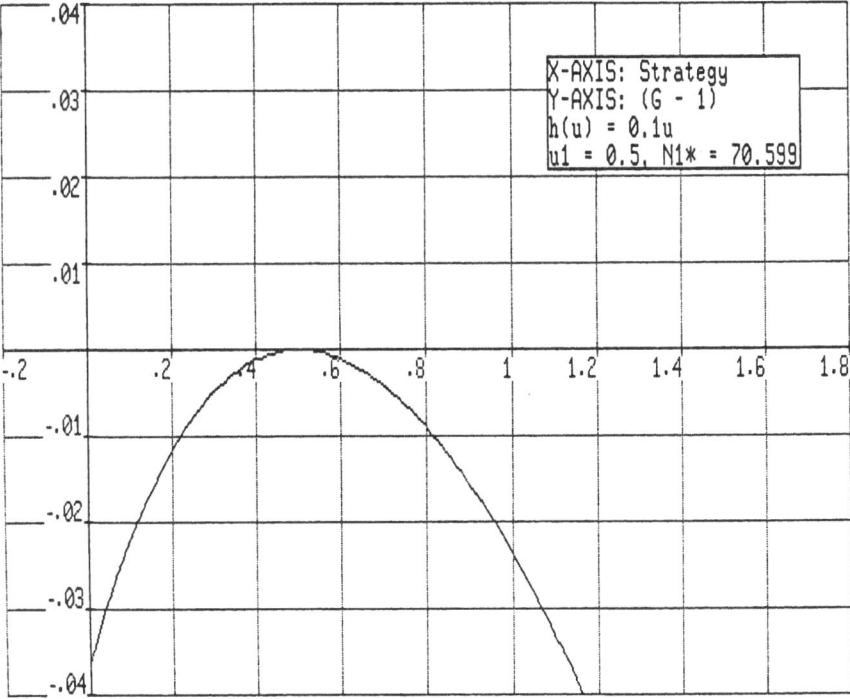

Figure 2. The ESS Under Linear Harvest.

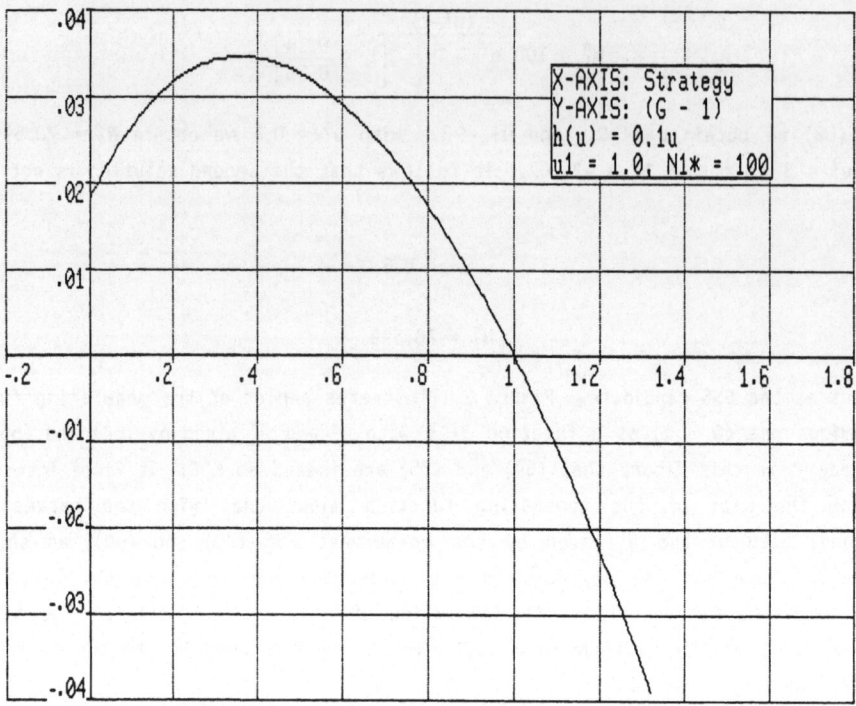

Figure 3. Harvesting the No-Harvest ESS.

Thus, the effect of this linear harvest will be to cause the population to evolve from the phenotype $u = 1$ to $u = 0.5$ (e.g., one-half the body size). We may now calculate the equilibrium population and steady-state yield before and after evolution takes place. Before the population evolves ($u^1 = 1$), the equilibrium population under harvesting is

$$N_1 = 100\left[1 - \frac{0.1}{0.25}\right] = 60 \tag{56}$$

and the steady-state yield is

$$\text{Yield} = N_1 h(u^1) = 6 . \tag{57}$$

After the population evolves to the ESS ($u^1 = 0.5$), the equilibrium population is

$$N_1 = 100\, e^{-\frac{(0.5)^2}{2}}\left[1 - \frac{0.05}{0.25}\right] = 70.6 \tag{58}$$

and the steady-state yield is

$$\text{Yield} = N_1h(u^1) = 3.53 \ . \tag{59}$$

We have the interesting before-and-after evolutionary effect that the yield will decrease while the equilibrium population increases with individuals at one-half the value of their original evolutionary parameter.

THE ESS UNDER A HUMP-SHAPED HARVESTING FUNCTION

As a final harvesting function we use

$$h(u) = 0.15e^{-\frac{(u-1)^2}{2}} \ . \tag{60}$$

This could correspond to a situation where the evolutionary parameter is again body size, but a premium is paid for the body size $u = 1$. Using the methods of the previous section, we again first seek a coalition of one. We obtain

$$u^1 = 1.6038 \quad \text{with } N_1^* = 41.667 \tag{61}$$

$$u = 0.3961 \quad \text{with } N^* = 41.667 \tag{62}$$

However, neither of these solutions represents an ESS, as can be verified by plotting the generating function (21) minus one $(G - 1)$ using (61), as shown in Figure 4, and the generating function (21) minus one $(G - 1)$ with (62), as shown in Figure 5. In each case, the generating function has a global maximum greater than one at a point other than the proposed ESS. Hence, neither solution is an ESS. We thus seek an ESS with a coalition of two strategies. Using conditions (40)-(43) for this problem, we obtain

$$u^1 = 0.275 \quad \text{with } N_1^* = 22.16 \tag{63}$$

$$u^2 = 1.725 \quad \text{with } N_2^* = 22.16 \ . \tag{64}$$

The coalition $u^1 = 0.275$, $u^2 = 1.725$ does indeed represent an ESS, as can be verified by plotting the generating function (27) minus one $(G - 1)$ using (63) and (64), as shown in Figure 6.

The effect of harvesting in this case is to produce the coevolution of two phenotypes. The effect on yield is similar to the previous case. Before the population evolves ($u^1 = 1$), the equilibrium population under harvesting as obtained from (38) is

$$N_1 = 40 \tag{65}$$

with a corresponding

$$\text{Yield} = N_1h(u^1) = 6 \ . \tag{66}$$

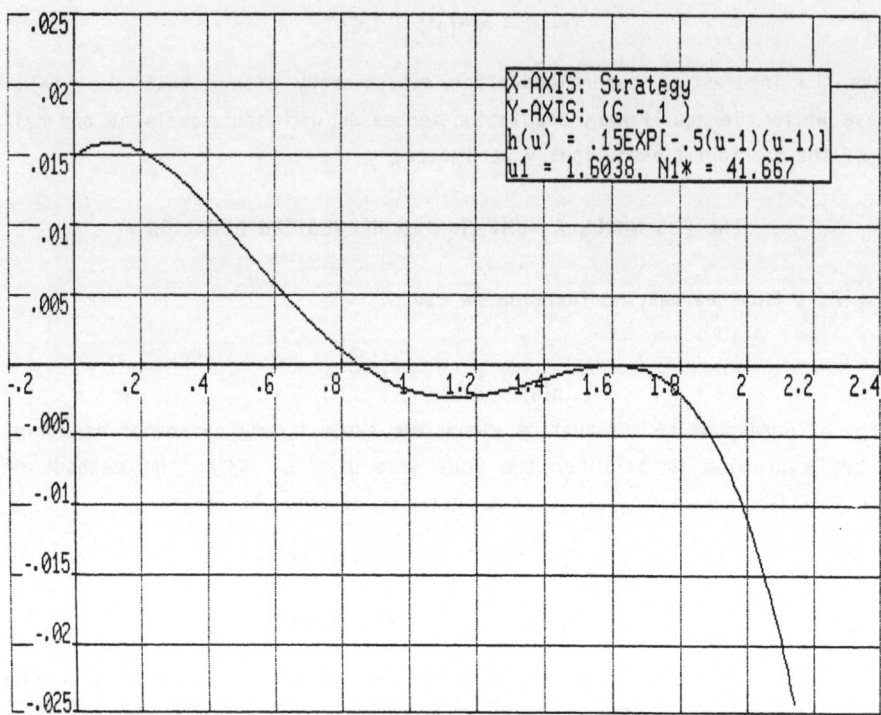

Figure 4. The First Coalition of One Solution Under Nonlinear Harvest.

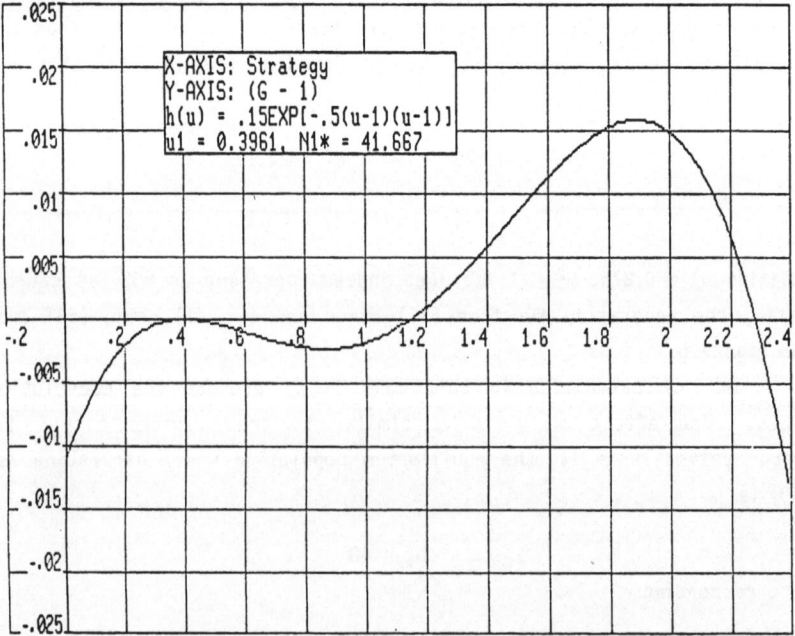

Figure 5. The Second Coalition of One Solution Under Nonliner Harvest.

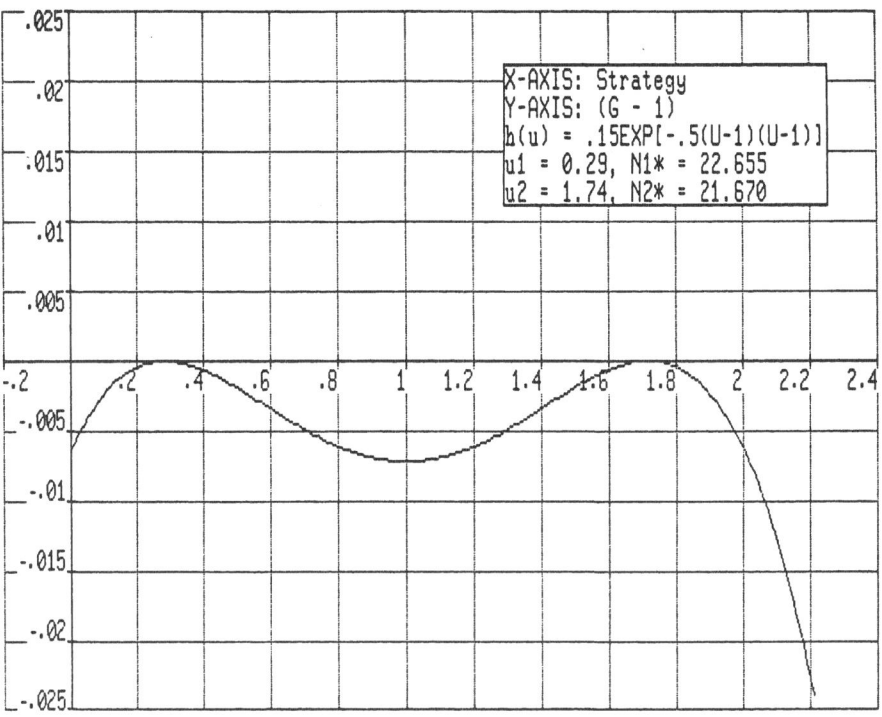

Figure 6. The ESS Under Nonlinear Harvest.

After the population evolves to the ESS coalition (u^1 = 0.275, u^2 = 1.725), the equilibrium population under harvesting as obtained from (40) and (41) is

$$N_1 = 22.16 , \qquad N_2 = 22.16 \tag{67}$$

with a corresponding

$$\text{Yield} = N_1 h(u^1) + N_2 h(u^2) = 5.11 . \tag{68}$$

Note that both phenotypes are harvested. Again, the before-and-after evolutionary effect is that the yield will decrease while the equilibrium population increases [i.e., compare (65) with the sum of $N_1 + N_2$ in (67)].

DISCUSSION

The examples illustrate the effect of harvesting on the ESS of a previously unexploited population. The models are both frequency and density dependent. The fitness of an individual is simultaneously a function of population density, the frequencies of strategies among the population, and the intensity of harvesting.

The shift in the ESS is shown graphically in Figures 3-5. These plots are frequency-dependent adaptive landscapes. The number of peaks in the landscape provides a clue as to the number of strategies in the ESS coalition.

Whether intentional or accidental, harvesting and cropping of prey species by humans introduces a new selective pressure on the population. Body size, for example, is one such evolutionary trait which may be brought under selection. The inclination of humans to harvest the largest individuals (it is often illegal to harvest crabs, shellfish, or fish below a threshold size) or the use of traps and nets that selectively collect larger individuals corresponds to a harvest function qualitatively similar to the linear harvest function. This results in directional selection and, at the new ESS, a decline in yield is obtained.

Often, harvest techniques are designed to be most efficient at collecting individuals using the strategy which predominates in the population. Insecticides are an example of this. They are designed to be effective against insect traits that actually occur, not those that may occur. The humped-shaped harvest function is most effective at collecting individuals using the existing ESS. For this example, harvesting results in disruptive selection, and the new ESS is a coalition of two strategies. In fact, for this case, harvesting mimics models of predator-mediated coexistence of prey. While developmental or genetic constraints may prevent evolution toward the new ESS (Maynard Smith, 1981; Eshel, 1982; Bomze, et al., 1983), continued harvesting will constantly encourage the two distinct strategies.

The game theoretic approach to evolution is not restricted to the specific Lokta-Volterra model used for the examples. Potentially, any management model of population dynamics can be used. An evolutionary model is generated by incorporating into the management model an evolutionary variable. The resultant model gives the ecological implications of an individual possessing a particular strategy. The modeling challenge is to identify evolutionarily flexible traits and to understand the effects of those traits on fitness and population dynamics.

We would like to conclude with an anecdote on the selective influence of human harvesting of prey (related to us by P. Smallwood). Researchers bred laboratory mice for feeding experiments with raptors. As is frequently the case, the docile mice were handled by gently lifting them by their tails. Soon, to the amusement of the researchers, the colony of mice consisted of individuals that held their tails flat to the ground! The objective of this paper is to provide a tool for understanding the evolutionary implications of management practices. The selective pressures of management programs on the organisms we exploit will often be less humorous and more profound than in the mice example above.

REFERENCES

Beddington, J., D. Botkin, and S.A. Levin. 1981. Mathematical models and resource management. Lecture Notes in Biomathematics, Vol. 40, pp. 1-5. Springer-Verlag, Berlin.

Bomze, I.M., P. Schuster, and K. Sigmund. 1983. The role of Mendelian genetics in strategic models of animal behaviour. J. Theor. Biol., Vol. 101, pp. 19-38.

Brown, J.S. and T.L. Vincent. 1986. A theory for the evolutionary game. Theor. Pop. Biol., Vol. 30, No. 3, in press.

Eshel, I. 1982. Evolutionarily stable strategies and viability selection in Mendelian populations. Theor. Pop. Biol., Vol. 22, pp. 204-217.

Goh, B.S. 1980. Management and Analysis of Biological Populations. Elsevier, Amsterdam.

Maynard Smith, J. 1981. Will a sexual population evolve to an ESS? Amer. Nat., Vol. 117, pp. 1015-1018.

Maynard Smith, J. and G.R. Price. 1973. The logic of animal conflicts. Nature, Vol. 246, pp. 15-18.

Pulliam, H.R. 1981. Optimal management of optimal foragers. Lecture Notes in Biomathematics, Vol. 40, pp. 46-53. Springer-Verlag, Berlin.

Vincent, T.L. 1985. Evolutionary games. J. Opt. Theory Applic., Vol. 46, pp. 605-612.

Vincent, T.L. and J. S.Brown. 1984a. The effects of competition on flowering time of annual plants. In S.A. Levin and T.G. Hallam (eds.), Lecture Notes in Biomathematics, Vol. 54, pp. 42-54. Springer-Verlag, Berlin.

Vincent, T.L. and J.S. Brown. 1984b. Stability in an evolutionary game. Theor. Pop. Biol., Vol. 26, pp. 408-427.

Vincent, T.L. and J.M. Skowronski (eds.). 1981. Renewable resource management. Lecture Notes in Biomathematics, Vol. 40. Springer-Verlag, Berlin.

PARTICIPANT'S COMMENTS

Vincent and Brown make an effective case for the consideration of evolutionary change in managed populations. Historically, the evolutionary time scale has been viewed as being much longer than the ecological; but there is increasing evidence from a variety of situations that substantial evolutionary change can occur in a few generations, especially under strong management stress. Examples range from the evolution of pesticide and antibiotic resistance, to the rapid evolutionary changes in plant-pathogen and host-parasite systems, to the evolution of early maturation and spawning in fish populations.

Vincent and Brown's observations that harvesting can promote coexistence has a variety of well-known parallels in the ecological and evolutionary literature. Gause, in his classic studies of coexistence among protozoans, sampled without replacement in sufficient amounts to alter the competitive outcome. Even indiscriminate removals ("rarefaction"), such as might occur from flushing or localized disturbance, can shift the competitive balance in the direction of more opportunistic species. Preferential harvesting of a competitive dominant, or even indiscriminate localized disturbance, can lead to a stable balance among competing

types that thereby subdivide a successional gradient. Such mechanisms are not fine-
tuning, but are thought to provide the major pathways to maintaining diversity in
many forest, grassland, intertidal, and other communities.

S.A. Levin

The paper and its predecessors (Vincent and Brown, 1984; Vincent, 1985) are
motivated by the needs of biology and inspired by the theory of differential games.
The purpose of this comment is to point out some possible connections leading to
fields as far apart as economics and laser theory. It appears that at least some of
these connections could be useful by providing alternative ways of attacking the
problem and corroborating the results already obtained.

In the Introduction, the authors observed that modelling the evolutionary
implications of management practice is complicated by the existence of two vastly
different time scales (evolutionary and managerial). Large dynamic systems of this
type crop up in many fields: in economics ('nested dynamics' - see Weidlich and Haag,
1983), physics (e.g., flow instabilities, phase transitions, theory of lasers), and
chemistry (complicated reactive kinetics including the well known Belousov-
Zhabotinsky reaction - see Haken, 1977, 1983, 1984; Nicolis and Priogogine, 1977;
Güttinger and Eikermeier, 1979).

The general approach for investigating such systems consists in isolating the
slow, stable modes, and in their elimination using 'adiabatic approximation'. This
leads to a drastic reduction in the number of variables, whose dynamics is then
investigated. For a detailed explanation of adiabatic approximation, slaving
principle, etc., see Haken (1984).

The following example should be sufficient to establish the relevance of
this procedure to the Vincent-Brown problem. Consider a simple form of Lotka-
Volterra dynamics (Luenberger, 1979)

$$\dot{N}_i = \left[\beta_i - \gamma_i \sum_{j=i}^{r} \alpha_j N_j \right] N_i, \qquad i = 1, \ldots, r: \qquad (1)$$

where N_i is interpreted as number of individuals using certain survival strategy u_i
at a time t and $\beta_i - \beta_i^0 - h_i$, with $h_i > 0$ being the harvesting intensity. The
parameters α, β, γ can be viewed as functions of the strategy vector u so that the
particular values α_i, β_i, γ_i are fixed when the individuals choose their survival
strategies.

It is easy to show that this system has $r + 1$ possible equilibrium points:

$$(0,0,\ldots0): (N_1^*,0,\ldots0): \ldots(0,\ldots N_i^*,0,\ldots0): (0,\ldots N_r^*),$$

where $N_i^* = \beta_i/\alpha_i\gamma_i$. We observe that in any of these equilibrium points, there is at
most one type of individuals surviving. This principle of competitive exclusion
(Luenberger, 1979) means that in a system governed by (1), there will be no
coalitions in the sense of Vincent-Brown and harvesting can only induce switching
from one surviving phenotype to another.

Assuming (without loss of generality) that the surviving individuals are N_1,
then the equilibrium $(N_1^*,0,\ldots0)$ is stable if

$$\beta_1 > 0 \qquad \text{and} \qquad \beta_1/\gamma_1 > \beta_i/\gamma_i \qquad (2)$$

for all $i \neq 1$. The reader can easily verify the following proposition: If the parameters of system (1) satisfy the conditions (2) or, alternatively, if the survival strategy u_1^* can be selected in such a way that (2) is satisfied, then there exists in the vicinity of equilibrium vector $N^* = (N^*,...0)$ an initial vector $(N_1(0),...N_r(0))$ for which $N \rightarrow N^*$ and $\dot{N}_1/N_1 > \dot{N}/N_i$ for all $t > 0$.

The strategy u_1^* can be identified with a (local) evolutionary stable strategy (Vincent and Brown, 1984).

More complicated problems can be formulated and analyzed. For example, see the results obtained by Denebourg and Allen presented in Nicolis and Prigogine (1977). There are a few interesting possibilities open for exploration.

1. It should be possible to analyze the phenotype survival in the more complicated Lotka-Volterra systems using the ecological Liapunov function (Luenberger, 1979). This could make it possible to introduce the methods for control of uncertain systems pioneered by Leitmann into management of systems with evolutionary response (see Lee and Leitmann, this volume; Leitmann, 1980; Skowronski, 1984).

2. While in relatively simple systems the evolutionary response to harvesting is limited to switching from one phenotype to another, or to changing the phenotype mix, it seems possible that harvesting can transform an equilibrium into a limit cycle or chaotic attractor and vice versa. Similar transformations have been observed elsewhere (Haken, 1984) where the paper by Diener and Poston, in Haken (1984) deserves special attention.

References

Güttinger and H. Eikermeier. 1979. Structural Stability in Physics, Springer, Berlin

Haken, H. 1983. Synergetics, 3rd Edition, Springer, Berlin.

Haken, H. 1983. Advanced Synergetics, Springer, Berlin.

Haken, H. (ed.). 1984. Chaos and Order in Nature, Springer, Berlin.

Leitmann, G. 1980. Deterministic control of uncertain systems, Acta Astronautica, Vol. 7, pp. 1457-1461.

Luenberger. 1979. Introduction to Dynamic Systems, Wiley, New York.

Nicolis, G., I. Prigogine. 1977. Self-Organization in Nonequilibrium Systems, Wiley-Interscience, New York.

Skowronski, J.M. 1984. Applied Liapunov Dynamics, Systems and Control Engineering Consultants, Brisbane.

Vincent, T.L. 1985. Evolutionary games, Journal of Optimization Theory and Applications, Vol. 46, pp. 605-612.

Vincent, T.L., J.S. Brown. 1984. Stability in an evolutionary game, Theoretical Population Biology, Vol. 26, pp. 408-427.

Weidlich, W., G. Haag. 1983. Quantitative Sociology, Springer, Berlin.

P.F. Lesse

DENSITY-DEPENDENT HABITAT SELECTION:
A TOOL FOR MORE EFFECTIVE POPULATION MANAGEMENT

Michael L. Rosenzweig
Department of Ecology and Evolutionary Biology
The University of Arizona
Tucson, Arizona 85721

Habitats vary in quality and individuals choose them accordingly. Some species distribute themselves non-randomly among habitats so as to equalize the net reproductive rates of individuals in different habitats (Ideal Free Distribution). Habitat selectivity is greatest when population sizes are small.

Determining habitat preferences may be done in the laboratory or the field. The latter may be accomplished by studying the covariance of census and habitat qualities, or by studying the non-randomness of population distribution among habitats. All methods have confirmed the density-dependence of habitat selectivity.

Density-sensitive habitat distributions may be used to reduce the cost of regular censuses. In some cases, they may make it practical to replace censuses and yield limits with habitat reserves.

When populations decline greatly, they are supported by the most favored habitats. Substantial fractions of some populations may depend on such cradle habitats. Pest populations may be controlled most cheaply by concentrating on their cradle habitats, although natural selection might interfere.

Managers will have to take cognizance of habitat selection in order to achieve maximum sustainable yields. Sometimes, harvesting should be concentrated in habitats where death rates are naturally high or birth rates naturally low. But long-term considerations in cases where some individuals dominate others may complicate achieving maximum sustainable yields.

INTRODUCTION

No organism is insensitive to the variety of habitats which the Earth offers. Some habitats convey only death; others, life and reproductive success. On a less trivial plane, those that bring success can be ranked according to their profitability, and organisms should evolve to respond physiologically and/or behaviorally to such a hierarchy.

The point of this chapter is that it may be useful for managers of natural populations to take cognizance of the fact that organisms have adapted to a hierarchy of habitat qualities, and to apply to management practices what is discovered about such adaptation. To make this clear, I shall begin by explaining the concept of density-dependent habitat selection. Then I shall examine three schemes for detecting habitat selection. Finally, a few suggestions for using what is discovered are introduced. These include possible improvements in population monitoring, pest control, resource species protection and maximum sustainable yields. The suggestions are not meant as an exhaustive list, but merely as a hint of the usefulness of the concpet.

INTRASPECIFIC DENSITY-DEPENDENT HABITAT SELECTION

When a species is faced with a variety of profitable habitats in space and/or time, it may contrive to limit its activity to a subset. This is habitat selection in its broadest sense. The limitation must serve to increase the fitness of the limited individuals, or natural selection cannot establish it as an enduring strategy. Fitness -- which may be defined as the instantaneous per capita rate of change or as the per capita replacement rate per finite period -- fitness, therefore, is the "good" which must anchor any utility function in the sea of possible objects of optimization.

The ecologist knows that eventually, as population size grows, the resources of a finite space are fully utilized and the population ceases its growth. Thus, it is the fate of growing populations to reduce their average fitnesses to unity, another way of saying one-for-one replacement. Fretwell and Lucas (1970) combined this insight with the concept of habitat selection, and derived the following elegant, robust framework for studying intraspecific habitat selection.

Let there be an array of habitat types, H_i. The population size of a species in each habitat type is n_i; its total population is $N = \Sigma n_i$. Although other assumptions have been suggested and studied, assume that each individual need not move around, but can remain as long as it wants to in any habitat patch; if it lives its whole life there, it leaves behind W_i descendants. W_i is thus its fitness in habitat type i.

Assume further that individuals interact with each other only by depleting each other's resources. Therefore, we can expect fitness to be a monotonically decreasing function of population size, $W_i(n_i)$.

Now we assume that at least two of the habitat types can support the species. In other words, $W_i(n_i) > 1$ for some values of n_i evaluated for i = 1 and i = 2. In Figure 1, four habitat types can support the species; the fifth cannot.

When the species is uncommon, say $N = n_i = A$ in Fig. 1, it ought to occupy only one habitat type. But its habitat selectivity wanes as it grows. Past A, some individuals will occupy habitat 2, enough in fact to equalize the fitnesses of individuals in habitats 1 and 2. An example is represented by the triangles over the densities B_2 and B_1. As N grows, habitats 3 and 4 join the list of acceptable types. At carrying capacity, represented by the dots in the figure, the population is subdivided among all four according to the values K_1 in 1, K_2 in 2, etc. Such a distribution equalizes fitnesses among individuals in all habitats by adjustment of their population densities. Fretwell termed it an ideal free distribution.

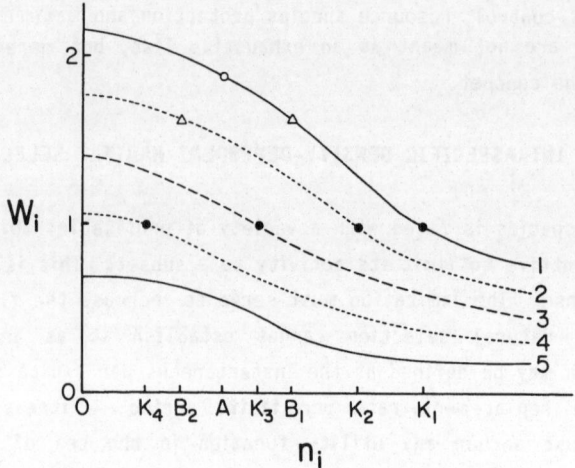

Figure 1. Fretwell's theory of habitat selection leading to ideal free distributions. The variable n_i is the population size in a particular habitat type, i. W_i is fitness in i. The five curves show fitness in five different habitat types. At any total population size, individuals should distribute themselves among habitats so that their fitnesses fall on a horizontal line, achieving equality. The triangles or dots offer an example.

It is possible for the functions $W_i(n_i)$ to cross (Holt, 1985). It is also possible for other interactions such as territoriality to interfere with the ideal free distribution (Fretwell, 1972). The ideal free distribution will not result in equalization of fitnesses if individuals must travel between patches instead of occupying only one (Rosenzweig, 1974). Moreover, others (Svardson, 1949; Morisita, 1950) had dealt with the problem before Fretwell and modelled it somewhat differently. Nevertheless, an overall principle emerges: habitat selectivity is greatest when population size is small, and successively less favorable habitats are occupied as population grows.

SCHEMES FOR ELUCIDATING HABITAT SELECTION

Laboratory choice experiments. Sidestepping entirely the difficulties of field observation, investigators who choose this method are hoping that organisms are programmed with somewhat rigid preferences which can be revealed in an artificial situation. These investigators are often rewarded.

A sophisticated series of laboratory trials descended directly from the work of Morisita. Several Japanese ecologists tested whether various species' habitat selectivity was density dependent (Morisita, 1952; Ito, 1952; Kosaka, 1956). A good example is the work of Kosaka (1956). Flatfish of two species were let into a pool with a varying number of others of their species. Half the pool's bottom was sand, half gravel. Both species strongly preferred the sandy half if they were introduced as individuals (e.g., 98% of *Limanda yokohamae* settled on the sand.) But their preferences decayed as they were forced to share the pool (e.g., 60% of *L. yokohamae* used the sand when twelve fish shared the pool.)

Laboratory experiments such as these are suggestive. But they have traded relevance to natural circumstances for ease of experimental control and interpretability. That makes them difficult to use by a population manager. He or she can probably use only the general ideas which emerge from laboratory studies. Recognizing that, both other strategies tackle the imprecise world of real habitats.

Covariance studies. If a species exhibits habitat selectivity in nature, it stands to reason that its census will covary nonrandomly with that of one or more habitat properties which index the habitat's quality for the species. For example, a forest-loving warbler might have a population size positively correlated with the biomass of trees over 20 cm dbh (diameter at breast height).

Hopeful ecologists traipse into the field censusing their target species and measuring simultaneously as large an array of environmental variables as they can. The data are then subjected to some form of covariance analysis (multiple regression is popular). A few even repeat their work to guard against statistical accidents. Despite the crudity of this approach, it often meets with good fortune, perhaps because only correlation and not mechanism is required for success.

Sometimes covariance studies arrive at baroque formulations of proposed habitat selectivities. For instance, my colleagues and I (Rosenzweig, Abramsky and Brand, 1984) needed as precise a statistical fit to some gerbil population data as we could achieve. We measured over twenty habitat variables and crafted a regression equation for each species of gerbil which we censused. Figure 2 shows a sample result for *Gerbillus allenbyi*. The ordinate (ORD) is percent (by weight) soil particles smaller than 0.25 mm diameter; the abscissa (ABC) is a measure of the total amount of vegetation in a standard sample. Contour lines are plotted by solving the regression equation for five different values (1,5,10,20,40) of N_A, the predicted census value of *G. allenbyi*. The equation is

$$N_A = -5.21 + 7.62 \quad 10^- w \tag{1}$$

where $\underline{w}$ is a complicated implicit function of the habitat variables:

$$w = ORD + 5ABC^{2.5}/ABC^{.023w} \tag{2}$$

The methods used to arrive at this equation were <u>ad hoc</u>, inelegant, and best explained when the reader has a free sabbatical year. But the point is that this sort of thing can be, and is being, done. Eq. (1) "explains" 92% of the variance in *G. allenbyi*'s census with only two independent variables. Repeated work (Abramsky, Rosenzweig and Brand, 1985) indicated that we were surely after the right sort of variables for this species.

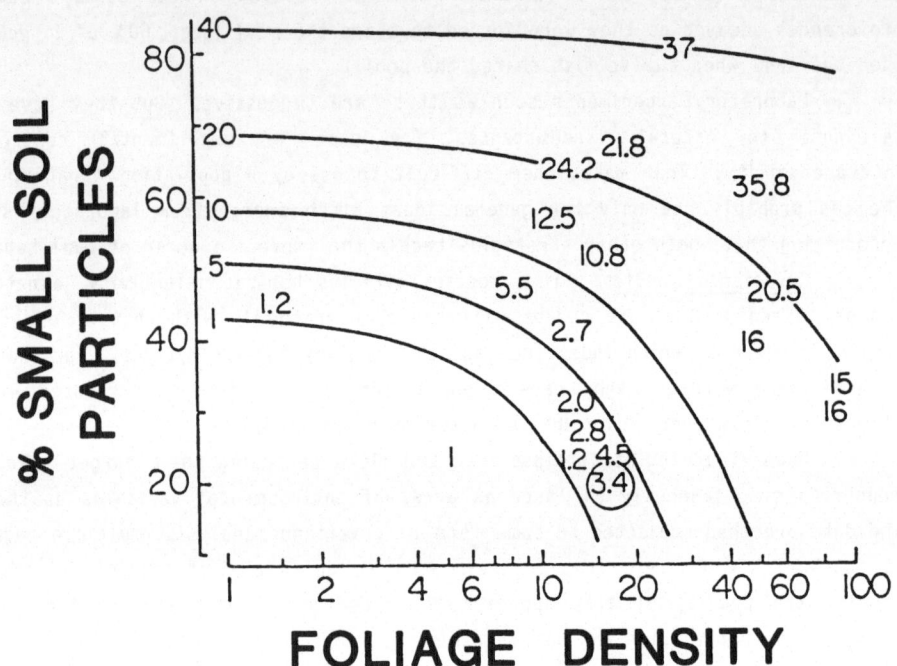

Figure 2. Population sizes of Allenby's gerbil fit to two habitat variables. Small soil particles are those smaller than 0.25 mm diameter. The numbers arranged along the edge are the values of the contour lines (1,5,10,20,40). Other numbers are actual censuses. (From Rosenzweig et al., copyright 1984 by *Oikos*.)

Nevertheless, covariance studies of habitat selection have problems if selectivity is density-dependent (as it ought to be). These have been explored to a limited extent (Crowell and Pimm, 1976; Van Horne, 1983; Abramsky et al., 1985). Basically, the problem is that when it is rare, a species selects its best habitat(s). Thus, low censuses may be associated often with the best habitat types. The flaw is not fatal to the method, but does inspire more caution than is usually the rule in covariance studies. To date, only covariance studies provide a means to describe the habitat hierarchy of a species in nature.

Distribution analysis. The most recently developed scheme for studying habitat selection is distribution analysis (Rosenzweig and Abramsky, 1985). It is specifically directed at uncovering patterns associated with density-dependent habitat selection.

The fundamental idea is that species' selectivities should be reflected in the number of habitats they use. If their individuals stick to one habitat type, selectivity is high. If they are found in all, more or less evenly, then selectivity is absent.

Distributional evenness is assessed in the following general manner. A hypothesis about the nature of habitat types for the species is erected (perhaps from a covariance analysis). Sites are found with fairly homogeneous extents of each of the proposed habitat types. Populations are sampled in a set of grids at each site. Each grid is a homogeneous habitat type. Habitat types may be replicated at a site (i.e., two or more grids may constitute the same habitat), but whatever is done at one site must be done at all of them. Population size is either varied experimentally or allowed to vary naturally so that its effect upon distributional evenness can be determined.

The favored index for measuring evenness is Fisher's y.

$$ y = \sum_{j=1}^{m} \frac{n_j^2}{N^2} \tag{3} $$

where $\underline{m}$ is the number of grids per site. (Note that if each habitat is represented by only one grid per site, $\underline{m}$ equals the number of hypothesized habitat types; otherwise it is larger.) The problem with y is that it is sample-size dependent: small samples can occupy only small numbers of grids. Thus, as N grows, so will y, even if the species is not habitat selecting at all.

It is possible to transform y so as to take advantage of its sample-size dependence (Rosenzweig and Abramsky, 1985). Define:

$$ y' = nMy - m - N + 1 \tag{4} $$

Now if we let pj be the probability that a single individual will reside in grid j, and we also transform N to x':

$$ x' = N - 1 \tag{5} $$

then Rosenzweig and Abramsky (1985) have shown that

$$ E\{y'\} = \left[m \sum_{j}^{m} p_j^2 - 1 \right] x' \tag{6} $$

Eq. (6) affords us a diagnostic tool for detecting habitat selection and assessing its dependence on density. Each set of pj values may be thought of as a strategy of habitat use. From (6), we see that if this strategy is constant, we expect the values of y' to fall on a straight line of slope M:

$$M = m \sum_{}^{m} p_j^2 - 1 \qquad (7)$$

emanating from the origin in (x',y') space. In fact, if $p_j = 1/m$ for all j, then there is no selectivity and the straight line will have zero slope. The more selective the species, the more disparate the p_j values. The more disparate they are, the higher is (7), the slope of the line.

Because all strategies emanate from the origin, every single point yields an estimate of the selectivity. Hence, if

$$\frac{dM}{dx'} \neq 0$$

then the selectivity is density-dependent. Fretwell's hypothesis is equivalent to saying that dM/dx' should be negative.

Rosenzweig and Abramsky (1985) discuss various hypotheses of density-dependence and how they should appear in (x',y') space. It will be sufficient here to examine just one of their data sets, the one for *G. allenbyi*. Figure 3 is quite straightforward: *G. allenbyi*'s distribution is far from random. Over low densities, it appears to exhibit a substantial, constant selectivity. Then, at about 90 gerbils, the selectivity decays sharply until it differs little from a random distribution.

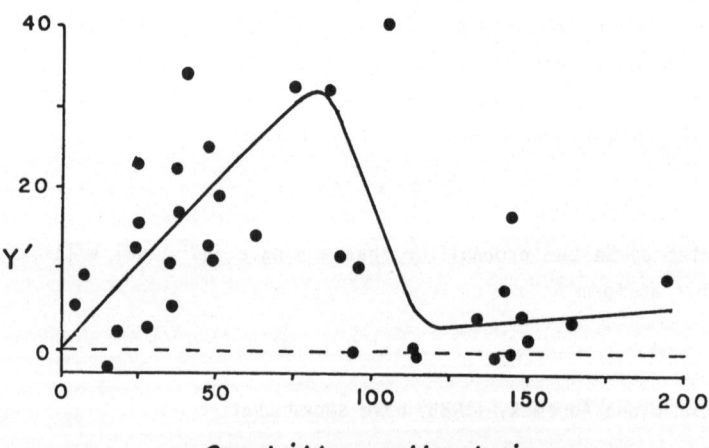

Gerbillus allenbyi - 1

Figure 3.   Population size of Allenby's gerbil sharply affects its habitat selection. Population data are the same as those in Fig. 2. The abscissa is the census value minus one individual. The ordinate is the transformed Fisher index of evenness. The dashed line is the expected regression for a species which is distributed randomly. See text for explanation. (From Rosenzweig and Abramsky, copyright 1985 by the University of Chicago.)

The distributional method promises to be a powerful ally of covariance analysis. It has even been applied to multispecies circumstances (Rosenzweig and Abramsky, 1986).

APPLYING HABITAT SELECTION TO MANAGEMENT

No one doubts that optimal harvesting of resources by humans should depend upon knowledge of habitat selection. At the very least, those actually doing the harvesting must adopt a time-energy-locale budget which reflects the predilections of their quarry. But there are more subtle points, too.

Censusing. Any practical program of population management must begin with a routine census technique to monitor the condition of the population. Often this aspect of management is difficult and expensive. Censusing must take place as close to harvest time as possible in order to insure minimum error in setting catch limits. Yet, the closer to harvest time, the smaller the sample which can be obtained before decisions need to be made. Taking advantage of our knowledge of density-dependent habitat selection can help the population more accurately reveal its condition, as we shall now see.

Recall from Fig. 1 and its associated text that each population size, N, is distributed among the habitat types in a unique fashion. In particular, any use of a habitat occurs only after a certain size N has been reached. For example, in Fig. 1, habitat 2 is used if and only if $N > A$. And habitat 3 is used if and only if $N > B_1 + B_2$.

A manager with a nomogram for his population, such as Fig. 1, could determine its approximate size by determining the poorest habitat type presently in use. This streamlines the idea of obtaining presence/absence data (Mangel, this volume) even further. We need obtain no data from the best habitats. It is the poorer ones which reveal the presence of populations large enough to harvest. One can even imagine a sampling protocol in which the habitats were searched in inverse order, poorest first. Because it is true that the poorer the habitat with a significant presence, the more there is to harvest, it must also be true that such a protocol would determine the presence of large populations most quickly, and thus allow proportionately more time for proportionately larger harvests.

Habitat Reserves. The ultimate simplification in setting thresholds beyond which one may exploit a population may be possible if habitats occur in large patches and are well marked (e.g., near shore vs. far shore, or different fishing depths). This is clearly the case for certain eastern Pacific fishes (Hanna, this volume). Here, one may protect the habitat-set within which will be housed the threshold population. Then the human exploiters may be let loose in the rest to harvest at will. The commercial exploiters will soon find it in their own interest to assume the job of censusing; governments will no longer need to do it, although they may elect to assist in the census so as to optimize the efficiency of their fisheries.

In a real sense, the habitat protection technique is not a new idea and has long been in use. Its most common application may be seen by recognizing that habitats have a time dimension too. Therefore, whenever a government establishes a closed season, it is protecting a temporal habitat.

But the idea has also been used spatially. Witness the many wildlife reserves set up to protect various endangered terrestrial species around the globe. There is no reason why a species has to flirt with extinction before we find it profitable to protect it from exploitation in a set of habitats.

Cradle Habitats. We have learned that small populations exercising density-dependent habitat selection should contract their habitat use to a subset of the profitable habitat types. These habitats nurture the population and prepare it for another expansion phase by encouraging its maximal growth. Let us therefore call them the cradle habitat(s). Recognition of cradle habitats may offer an opportunity not only to the resource manager, but also to the manager of a pest population.

Resources may depend upon space-time cradle habitats to a surprising extent. A chilling example is the population of Sitka black-tailed deer on Alaska's southeastern islands (Fagen, ms; Kirchhoff, Schoen and Thomas, ms.; Schoen and Kirchhoff, 1985; Kirchhoff, Schoen and Wallmo, 1983). Unlike other deer, Sitka deer are not necessarily helped by forest-clearing practices which encourage the growth of deer browse. The problem is that in some years, there is so much snow that the browse is buried and unavailable. During such periods, the deer depend upon and seek out the thickest parts of the old-growth spruce forest. Here, the trees bear much of the snow overhead, leaving the forest floor's browse accessible. Such old-growth patches constitute less than 10% of the area of the islands, and severely snowy years come once in 5 or so. When they do, the old, thick forest succors the deer for only a few months, say one-third of the year. Thus, the space-time extent of the cradle habitat is less than 1%. Yet it has been estimated that over 90% of the deer herd depends upon it for life itself. Managers must know about such cradle habitats or risk wild fluctuations in populations and harvests.

Faced with the problem of minimizing the population size, pest managers may also find it useful to manage cradle habitats. By constantly draining them of their pests, managers may turn cradle habitats into pest vacuum cleaners. A pest population, controlled only in its cradle, should flow in from lesser habitat types, only to be caught or destroyed. As long as it is possible to keep the population in the cradle habitat below the threshold for spilling over into secondary habitat types, the entire pest population should move into the cradle and subject itself to control. Because control can be achieved in a fraction of the space available to the pest, and with permanent installations, it should be conducted relatively cheaply and efficiently.

There is a long-term difficulty associated with pest management which does not bother the resource manager. Pest-control devices in the cradle habitats change them into not very good places for pests to live. In fact, what were the best habitats, have been rendered by the pest manager into very, very poor habitat choices. They will continue to be chosen first only by dint of misguided instinct. There is no reason to believe that natural selection is not capable of deciphering this change and acting accordingly. (See in particular Vincent and Brown, this volume.) A complete pest management scheme must therefore take natural selection into account and find a way to thwart it. I doubt there will be a general method for doing this; ad hoc solutions may be required for each pest.

Maximum sustainable yields. A population distributed ideally among a hierarchy of habitat types, takes on a degree of complexity unknown in simpler systems. The manager may need to be as conscious of the distribution as of the inequality of ages, sizes and physiological conditions among individuals. One issue likely to surface is how to achieve maximum sustainable yields.

In an ideal free distribution, all individuals have the same per capita reproductive rate. So, at first, one might imagine that it makes no difference to the yield which habitat(s) is worked. Probably, nothing could be more misleading.

First, the cost of harvesting may vary in different habitats. But this cost does not enter the ideal free distribution. Surely optimizing net yields will require avoiding costly harvesting methods when possible.

Second, equality of net reproductive rates may be achieved despite vast disparities in turnover. Birth and death rates in one habitat may be very high; in another, very low. Exploiting the individuals with low death rates should depress their subpopulation's net reproduction much more than exploiting individuals likely to die anyway. Hence, the prudent predator (Slobodkin, 1968) must take account of habitat-to-habitat variation in death rates in order to approach the ideal of imposing only compensatory mortality (Rosenzweig, 1978) on the exploited population. The mathematical treatment of this problem may prove very interesting.

Finally, not all populations achieve ideal free distributions. Some, although strong density-dependent habitat selectors, conform to a distribution which Fretwell (1972) calls the Ideal Despotic Distribution.

IDD's result when some individuals are able to occupy and defend patches of habitat which are considerably better than average. Challengers have the problem, not of determining whether their fitness outside the better patch is less than it would be inside, but of determining whether their fitness would be better if they persisted with their challenge or if they accepted their subordinate status. In a very large number of cases, challengers do acquiesce and remain subordinates at least for a time.

Managers faced with IDD's certainly do not want to harvest uniformly in all habitats. The despots are doing most of the producing and have low death rates.

The subordinates are doing little or no producing, and may have higher death rates too. It looks like they should be saddled with much or all of the burden of exploitation.

Yet, two problems remain for the manager (who, it is becoming obvious, most urgently needs the gift of prophecy). First, subordinates are often younger individuals. Their current production may be low, but they embody the future of the population. One does not want to remove them all. Moreover, they are ready to move into the preferred habitat as soon as they get the chance. Many field studies have confirmed this fact. Thus, the optimal strategy for the ideal manager is to take the despots just before their productivity declines with age, and when they are between productive events, and when their place can be quickly taken by a subordinate. A tall order undoubtedly, but there may be a way to approximate it in some cases. Meanwhile, at least we have a qualitative idea of the goal.

The second problem may be considerably less tractable. If one tries to manage an IDD by harvesting subordinates, natural selection is bound to rear its creative head. The subordinate's acquiescence depends upon its not suffering so much in the poorer habitat. Otherwise it should evolve to pay the price of persisting in its challenge. The manager who thrusts most of the burden of exploitation upon the subordinates is raising the level of suffering among them. It is not too far-fetched to imagine that the burden can get so great that "have-nots" will evolve who continually disturb the peace of the "haves." The result may be a breakdown of the social structure of the exploited species, and its replacement by who knows what. In any case, "haves" will pay more to reproduce, and sustainable yields ought to fall. I cannot suggest how this may be avoided.

ACKNOWLDEGMENTS

The concept of cradle habitats and relationship to pest control emerged from extensive conversations with Dr. Z. Abramsky. The author was supported by NSF Grant BSR-8103487.

REFERENCES

Abramsky, Z., M.L. Rosenzweig and S. Brand. 1985. Habitat selection in Israel desert rodents: comparison of a traditional and a new method of analysis. Oikos, Vol. 45, pp. 79-88.

Crowell, K.L. and S.L. Pimm. 1976. Competition and niche shifts of mice introduced into small islands. Oikos, Vol. 27, pp. 251-258.

Fagen, R. An ideal free distribution method for assessing productivity changes from habitat use statistics. Ms.

Fretwell, S.D. 1972. Populations in a Seasonal Environment. Princeton University Press, Princeton, NJ.

Fretwell, S.D. and H.L. Lucas, Jr. 1970. On territorial behavior and other factors influencing habitat distribution in birds. I. Theoretical development. Acta Biotheor., Vol. 19, pp. 16-36.

Hanna, S. The structure of fishing systems and the implementation of management policy. (This volume).

Holt, R.D. 1985. Population dynamics in two-patch environments: some anomalous consequences of an optimal habitat distribution. Theor. Pop. Biol., Vol. 28, pp. 181-208.

Ito, Y. 1952. The growth form of populations in some aphids, with special reference to the relation between population density and movements (in Japanese with English summary). Res. Pop. Ecol., Vol. 1, pp. 36-48.

Kirchhoff, M.D., J.W. Schoen and M. Thomas. Seasonal distribution and habitat use by Sitka black-tailed deer in southeastern Alaska. Ms.

Kirchhoff, M.D., J.W. Schoen and O.C. Wallmo. 1983. Black-tailed deer use in relation to forest clear-cut edges in southeastern Alaska. J. Wildl. Manage., Vol. 47, pp. 497-501.

Kosaka, M. 1956. Experimental studies on the habitat preference and evaluation of environment by flatfishes, *Limanda yokohamae* (Gunthek) and *Kareius bicoloratus* (Basilewsky) (in Japanese with English summary). Bull. Jap. Soc. Sci. Fish., Vol. 22, pp. 284-288.

Mangel, M. Sampling surveys for a highly organized population. (This volume).

Morisita, M. 1950. Dispersal and population density of a water-strider, *Gerris lacustris* L (in Japanese). Contr. Physiol. Ecol. Kyoto Univ. No. 65.

Morisita, M. 1952. Habitat preference and evaluation of environment of an animal. Experimental studies on the population density of an ant-lion, *Glenuroides japonicus* M'L. (I) (in Japanese with English summary). Phys. and Ecol., Vol. 5, pp. 1-16.

Rosenzweig, M.L. 1974. On the evolution of habitat selection. Proc. First Intern. Congr. Ecol., pp. 401-404.

Rosenzweig, M.L. 1977. Aspects of biological exploitation. Quart. Rev. Biol., Vol. 52, pp. 371-380.

Rosenzweig, M.L. and Z. Abramsky. 1985. Detecting density-dependent habitat selection. Amer. Natur., Vol. 126, pp. 405-417.

Rosenzweig, M.L. and Z. Abramsky. 1986. Centrifugal community organization. Oikos, Vol. 46, pp. 339-348.

Rosenzweig, M.L., Z. Abramsky and S. Brand. 1984. Estimating species interactions in heterogeneous environments. Oikos, Vol. 43, pp. 329-340.

Schoen, J.W. and M.D. Kirchhoff. 1985. Seasonal distribution and home-range patterns of Sitka black-tailed deer on Admiralty Island, southeast Alaska. J. Wildl. Manage., Vol. 49, pp. 96-103.

Slobodkin, L.B. 1968. How to be a predator. Amer. Zool., Vol. 8, pp. 43-51.

Svardson, G. 1949. Competition and habitat selection in birds. Oikos, Vol. 1, pp. 157-174.

Van Horne, B. 1983. Density as a misleading indicator of habitat quality. J. Wildl. Manage., Vol. 47, pp. 893-901.

Vincent, T. and Joel Brown. Havesting an evolutionary model. (This volume.)

PARTICIPANT'S COMMENTS

My discussion will be directed to some issues involved in the use of density-dependent habitat selection in population management, in particular to issues that arise when the management problem involves the control of human exploiters.

Assuming that the objective of the animal population is to maximize fitness in the long run, management as a process must contend with the challenge of melding this objective with the exploiters' objective of maximizing the economic yield. Maximum economic yield, as distinguished from maximum physical yield, is measured as the returns earned from yield minus the costs incurred in harvesting. The fact that different habitats may differ in revenues and costs of harvest combined with the simultaneous process of different groups optimizing over different objectives functions makes the management process a difficult one and the economics of various management strategies an important consideration.

As Rosenzweig points out, management by either temporal or spatial habitat is a well-tried technique. Rosenzweig's point is that making use of the knowledge of density-dependent habitat selection will have two major benefits: 1) more efficient sampling of populations; 2) area- or time-specific management regulations that capitalize on relative differences in habitat quality. This type of regulation must also contend with several economic factors which influence the costs and benefits of habitat management and its appropriateness as a management tool for any given population. The fishery provides good examples of these factors.

Costs of harvesting will be affected by habitat management if closed areas redirect fishing effort to areas associated with different travel times or search times. Temporal closures may also affect harvesting costs if seasonality is an important determinant of market conditions, weather conditions, or animal quality. Either time or area closures may cause short run changes in supply (e.g., a glut of fish on the market during a shortened season) that will affect the economic yield to fishermen.

The market interaction between protected species and other species is a further consideration for management. The long run impact of redirected fishing effort may be an increased level of effort on substitute species. The size distribution of vessels in a fleet is a factor in the cost of redirected fishing effort since changes in exploitation patterns will be required. Temporal closures may benefit one vessel size class over another if specialization in fishing patterns exists by vessel size. Area closures may benefit vessels in one geographic area over another through the proximity of ports to open areas and may also lead to crowding of vessels and gear in limited fishing areas.

Enforcement costs will vary with the type of habitat protection; it is likely to be costlier to monitor and enforce spatial closures than temporal closures. Data acquisition costs may decrease, as Rosenzweig argues, if habitat selectivity implies that poorer habitats may be sampled first. Habitat management has the potential for alleviating the problem of discarded fish, a significant problem in multispecies fisheries where limits are placed on single species.

The point of this discussion is that there are many economic factors complicating the management problem. In addition to the requirement that the populstion exhibit density-dependent habitat selection, successful habitat management will depend on the economic conditions affecting the exploiters of those populations. A resource's stage of exploitation may well determine the ease with which habitat management techniques blend with the economic system. A fully exploited resource is likely to exhibit more extreme responses to habitat management through changes in both exploitation costs and strategies that affect related resources.

As a final note, the selection implications discussed by Rosenzweig of the Ideal Despotic Distribution present the possibility of inadvertent population changes that should be kept in the forefront of the discussion. This possibility points to the advantages of using adaptive management techniques that experiment with different alternatives and concurrently generate information about a population's response to management.

Susan Hanna

There is an interesting parallel between the ecological theory of habitat selection and the bioeconomic theory of a spatially distributed fishery. Imagine that the fishing banks consist of a sizeable number of identifiable discrete aggregations, or "pools," containing multiple species of fish, some perhaps of much greater market value than others. The fishing boats distribute themselves over the banks, concentrating at first on the richer pools, but in so doing gradually degrading those pools toward parity with the originally marginal ones. As the season progresses, fishing effort spreads more widely, in such a way that net return to effort is always constant among the exploited pools.

Now, the dynamics of response of the "prey" fish populations depend in a complicated way on their growth, mortality, and migration rates. Nevertheless, there are general principles, that operate at the level of the fishery, that control the "habitat selection" of the population of "predator" fishing boats. Of overriding influence is the "common property" character of fishery exploitation, the usual situation where no one can own the fish, so no one has an incentive for conservation in their harvest. Characteristically, fishery effort will be too heavy overall, and concentrated too exclusively on the richer pools. These offer the larger immediate profit, but also may be the ones most vital as cradle habitat and most susceptible to over-fishing.

Focusing once more on the lower trophic level, it seems clear than an ecological theory of habitat selection by the fish ought to help to predict their response to harvest. Conversely, their habitat selection patterns will certainly be affected by the exploitation practices of the fisherman, and hence ultimately by the institutional and management regime under which the fishery operates.

Clark, Colin, 1982. Concentration profiles and the production and management of marine fisheries in W. Eichhorn (ed.), Economic Theory of Natural Resources. Physica-Verlag, Wurtzburg-Wien.

McKelvey, Robert, 1986. The economic regulation of targeting behavior in a multispecies fishery. Natural Resource Modeling. Vol. 1, No. 1.

Robert McKelvey

PART II

CONTROLS/TECHNIQUES

APPROACHES TO ADAPTIVE POLICY DESIGN FOR HARVEST MANAGEMENT

Carl J. Walters
Institute of Animal Resource Ecology
The University of British Columbia
Vancouver, B.C., V6T 1W5

In the face of uncertainty about production relationships, it may be worthwhile to deliberately vary harvest rates so as to produce informative contrasts in stock size. Such dual effects of control (on immediate performance and on information available to future decision makers) create very challenging modelling and optimization problems even for simple systems. This paper reviews recent literature on dual control and suggests directions for future research related to renewable resource problems.

INTRODUCTION

> "... the optimal controller has to know how to use what it knows as well as what it knows about what it shall know."
> - Bar-Shalom (1981)

Control theory has provided a variety of useful insights about renewable resource management, on issues ranging from optimal investment trajectories during resource development to the structure of feedback policies for coping with unpredictable environmental variation. While it is difficult to find instances where optimal control calculations have been applied precisely and quantitatively to actual field situations, it is common to see qualitative results restated as guidelines or targets in management plans. In particular, it has become fashionable to talk about dealing with uncertainty in terms of "adaptive management", which has been taken to mean everything from regular review of monitoring results and production parameter estimates through to the development of elaborate "probing" experiments to measure system responses outside the range of recent historical experience. The concept of management as experimentation has been particularly appealing to resource managers with biological backgrounds, who tend to mistrust simple dynamic models and to see all sorts of mechanisms that might produce unpredictable and surprising ecological responses to management actions.

This paper reviews recent developments in the control system and renewable resource literature concerning the design of optimal adaptive policies and suggests directions for future research. The general literature on simultaneous identification and control has grown explosively in the last decade, so I will focus specifically on studies dealing with the so-called "dual effect of control", where it is recognized that management actions may affect future uncertainty as well as immediate system performance. These studies indicate that it can indeed be worthwhile to engage in probing experiments, but there are formidable mathematical and computational difficulties in finding the optimum balance between probing and cautious behavior.

A PROTOTYPICAL PROBLEM

As motivation for the review and introduction to some terminology, this section develops a simple harvesting model and shows how it leads to a difficult stochastic optimization problem. Consider a harvested population for which the stock dynamics can be adequately represented by the stochastic production function

$$N_{t+1} = \alpha S_t^b e^{w_t} \tag{1}$$

where N_{t+1} is stock size available for harvest in year t+1, S_t is the stock after harvest in year t, w_t is a normally distributed "environmental effect" with mean 0 and $cov(v_t, v_{t+k}) = 0$ for k > 0, and α and b are parameters. Suppose that management can control the "instantaneous fishing rate" F_t so as to regulate catch C_t through the catch equation

$$C_t = N_t(1-e^{-F_t}) \tag{2}$$

to leave $S_t = N_t-C_t$ each year. Finally, suppose that the management objective is to maximize a risk averse, discounted utility function V of the catches:

$$V = \sum_{t=0}^{\infty} \lambda^t \log(C_t) \tag{3}$$

where $\lambda < 1$ is a discount factor. Letting $x_t = \log(N_t)$, eq. (1) can obviously be rewritten as the linear dynamics,

$$x_{t+1} = a + b(x_t - F_t) + w_t \tag{4}$$

where $a = \log(\alpha)$, and the utility function becomes

$$V = \sum_{t=0}^{\infty} \lambda^t \left[x_t + \log(1-e^{-F_t}) \right] \tag{5}$$

Eqs. (4) and (5) appear to represent a trivial problem in optimal control.

If the parameters a and b are assumed known, it is easily shown (Deriso, 1986; Walters and Ludwig, 1986) that the "certainty-equivalent" optimal policy is

$$F_{CE} = -\log(\lambda b) \tag{6}$$

i.e., a constant instantaneous fishing rate for all times t > 0, independent of x_t and of the statistical properties of w_t. If b is recognized to be uncertain and is characterized by a prior probability distribution $P_0(b)$ with mean $\hat{b}_0$ and variance σ_0^2, and if it is further assumed that no further information will be gathered about b in the future, then the "cautious" optimal policy which maximizes the expected value of

V over the distribution $P_0(b)$ is given approximately by (Walters and Ludwig, 1986)

$$F_{CA} = -\log \left[\frac{\lambda \hat{b}_0 (1-\lambda \hat{b}_0)^2 + \lambda^2 \sigma_0^2}{(1-\lambda \hat{b}_0)^2 + \lambda^2 \sigma_0^2} \right] \tag{7}$$

which is again a constant fishing rate for all $t > 0$. F_{CA} is cautious in that it is a decreasing function of the uncertainty measure σ_0^2 for b, and approaches F_{CE} as this measure approaches zero.

A "passively adaptive" management strategy would be to apply either F_{CE} or F_{CA}, then use resulting stock responses to reestimate a and b, for example by recursive linear regression in the format of eq. (4) or by Bayesian updating of $P_t(b)$. Provided F_{CA} or F_{CE} then induces informative changes in $x_t - F_t$ relative to historical experience, there will be more or less rapid learning depending on the variance of w_t. Such informative changes would be expected, for example, in the early development of a fishery, and then even F_{CE} or F_{CA} would be seen as experimental policies. Unfortunately, good monitoring programs cannot be economically justified or practically implemented in the early development of most resources, so good data are usually only available for $x_t - F_t$ after most of the informative transient responses have already occurred.

Even if the passively adaptive strategy does not immediately provide informative contrasts in $x_t - F_t$, it is "asymptotically optimal" in the sense that the regression estimates $\hat{b}_t$ will eventually converge to b and σ_t^2 will approach zero (so F_t will approach the optimum defined by eq. 6), due to "free" information provided by stochastic variation in $x_t - F_t$. This is easily seen by noting that the linear regression estimator $\hat{\underline{\beta}} = (\hat{a}\ \hat{b})'$ can be written as

$$\hat{\underline{\beta}}_t = \underline{\beta} + (X'X)^{-1} X'\underline{w} \tag{8}$$

where X is the $t \times 2$ matrix $(\underline{1}\ \underline{d})$ and $d_i = x_{i-1} - F_{i-1}$. The error term (and bias; see Walters, 1985) $(X'X)^{-1}X'\underline{w}$ decreases with increasing t, provided w_t has nonzero variance. Unfortunately, it is easily shown by Monte Carlo simulation that the estimates converge very slowly for biologically realistic variances in w; 50-100 years data may be needed in some cases to insure that F_{CE} is within 10% of the correct optimum. The b parameter represents density dependent interactions between organisms and their ecological environment, and since this environment will change b is not likely to remain constant on such long time scales even if the discount rate is low enough to justify concern about what will happen. In short, it is practically meaningless to point out that the passive strategy is asymptotically optimal.

Consider now the possibility of taking an "actively adaptive", informative action at time zero, for example a very high F_0 that sets $x_0 - F_0$ far from the historical average and induces a transient in $x_t - F_t$ that will also be informative even if $F_t = F_{CE}$ for $t > 0$. This action has a cost (reduced x_t, $t > 0$ in eq. 5), but also a benefit through its complicated effects on $\hat{\beta}_t$, σ_t^2 and hence on the ability of future decision makers to make better choices F_t. To model this "dual effect of control", so as to seek an F_0 that optimizes the cost/benefit tradeoff, we are forced to model not only the state dynamics x_t but also the learning dynamics for $P_t(b)$. See the introductory quote by Bar-Shalom (1981). What initially appeared to be a trivial control problem with a simple solution (eq. 6) is in fact a messy nonlinear problem that is of high or infinite dimensionality (depending on how the dynamics of $P_t(b)$ are approximated). Not only is it conceptually difficult to decide what learning model to use (i.e., what statistical tools will future decision makers use?), but also there are formidable numerical problems due to the fact that the tails of $P_t(b)$ become more important than its expected value $\hat{b}_0$: the potential benefits of learning are greatest for the unlikely opportunities and dangers represented by b values far smaller or larger than $\hat{b}_0$.

APPROACHES TO ADAPTIVE CONTROL: A REVIEW

In the previous section, I hinted that there are two basic strategies for sequential decision making in the face of uncertainty: passively and actively adaptive. Both of these strategies assume than an information-updating feedback strategy of some sort will be optimal in the first place. For more precise discussions and proofs of this point, see Aström (1976), Bertsekas (1976), and Bar-Shalom (1981). The general conclusion is that the optimal decision at time t will be some function of the "information state" available for decision making at that time. At most, this information state will be all parameters (possibly infinite) needed to define a priori understanding ($P_0(b)$ in the example), plus all data gathered subsequently (x_t, F_t time series in the example). Under various statistical assumptions, the information state can be represented by a fixed-dimension vector of "sufficient statistics" for the uncertain states and parameters (Streibel, 1965, 1975; Bar-Shalom 1981; Gessing and Jacobs, 1984). A relatively simple way to approximate the information state for complex systems is by the means and covariances of uncertain states and parameters ($\hat{b}_t$, σ_t^2 of $P_t(b)$ in the example); this is known as the "wide sense" information state (Tse, et al., 1973; Bar-Shalom and Tse, 1974).

Passively adaptive policies are actually globally optimal for some stochastic control problems, such as the linear-quadratic regulator with known coefficient matrices. General conditions for the certainty-equivalent control choice to be optimum are presented in several texts, and particularly clear presentations are in papers by Wonham (1968), Bar-Shalom and Tse (1974), and Gessing and Jacobs (1984). Basically these conditions involve the control being "neutral" in the sense that it does not influence future probability distributions for uncertain states and parameters. Nonlinear systems and linear systems with unknown parameters generally do not meet these conditions.

Much research on adaptive control has been developed under the assumption that passive policies are "close enough" to optimum so that explicit consideration of dual effects (see above) is not worthwhile. The large literature on "self-tuning regulators" falls mainly in this category (see review in Aström, 1983), as does most of the literature on asymptotically optimal policies (see for examples Kumar, 1983; Hernandez-Lerma, 1985). Passive but cautious policies analogous to eq. (7), derived by assuming no future learning, have often been suggested as alternatives to certainty-equivalent passive policies (Jacobs and Patchell, 1972; Chow, 1975; Wittenmark, 1975).

The tradeoff between short term system performance versus longer term learning was first formalized by Fel'dbaum (1960-61; 1965), who coined the phrase "dual effect of control" and defined the terms "caution" and "probing" to represent departures from certainty-equivalent behavior. His ideas were pursued during the 1960's by Soviet scientists, who found them to be computationally intractible and to be of limited practical significance in the design of automatic control systems (Prof. Ya. Z. Tsypkin, Moscow, pers. comm., 1983). During the same period, there was growing interest in related sequential decision problems such as the "two arm bandit" (when to sample from alternative reward processes) and partially observed Markov decision processes (Aström, 1965, 1969; Bellman, 1961; review in Kumar and Seidman, 1981). By the mid 1970's, the general dynamic programming functional equation for adaptive control was well known (Wittenmark, 1975; Bertsekas, 1976), and a few numerical algorithms had been suggested for approximating the dual effects (Tse, et al. 1973; Norman, 1976; Walters and Hilborn, 1976). Analytical solutions have been obtained for a few simple problems (Athans, et al., 1977; Kumar and Seidman, 1981; Kolonko and Benzing, 1985).

Over the past ten years, research on the dual control problem has resulted in a variety of approaches for trying to avoid the dimensionality problem imposed by the need to represent future information states. These approaches can be roughly classified into five groups of descending computational complexity.

(1) Numerical Dynamic Programming - here the future information state is represented by Bayes posterior probabilities on alternative models or by the mean and covariance of the Bayes posterior density for parameter estimates; the information state measures are discretized on a grid for approximation of the dynamic programming value function, and an optimal stationary policy is sought for all information state combinations in the grid (Wenk and Bar-Shalom, 1980; Walters, 1981; Ludwig and Walters, 1982; Bayard and Eslami, 1985).

(2) Wide Sense Dual Control - here for each policy choice at t = 0 given a particular initial state, perturbation analysis is carried out to approximate uncertainties around a nominal trajectory of future wide sense information states (mean and covariance of states and parameters); the optimum choice is

found by numerical search (the original algorithm is in Tse, et al., 1973; it is reviewed and interpreted in terms of caution and information value terms in Bar-Shalom and Tse, 1976, and Bar-Shalom, 1981; for renewable resource applications, see Smith and Walters, 1981 and Walters, 1986).

(3) <u>Simplifications of the Wide Sense Algorithm</u> - the original algorithm can be simplified by ignoring some second order perturbation terms; for an example see Norman (1976).

(4) <u>Direct Representation of Information Value in the Value Function</u> - here the idea is to set up the value function as though there were going to be no future learning, then add explicit terms that represent assumed future value from immediate probing (Padilla, et al., 1980; Chen and Zarrop, 1985; Sato et al., 1985).

(5) <u>Optimization in Policy Space</u> - here one begins by assuming a functional form for the relationship between the (stationary) optimum policy choice and some information state measures (usually the wide sense measures), then optimum parameter values for this form are sought by general optimization techniques such as stochastic approximation (Walters, 1986); this approach can be combined with dynamic programming (Ludwig and Walters, 1982; Bayard and Eslami, 1985).

Beyond these approaches, there has been some progress in recasting problems in terms of other frameworks such as the multiarm bandit (Glazebrook, 1983), but so far as I am aware these efforts have not led to promising numerical algorithms.

Numerical exercises with dynamic programming and the wide sense algorithm have consistently resulted in two striking predictions about the optimum control in the presence of dual effects. First, this control is either very close to the cautious (or certainty-equivalent) optimum, or else is a quite drastic probe away from that optimum. In other words, modest experiments or supposedly informative "dithering" are apparently not good policy choices in general; such choices degrade immediate system performance without substantially reducing uncertainty. Second, the optimum probing choice has only modestly (5-20%) higher expected value than the best cautious choice. Part of the reason for this second prediction was noted earlier: high rewards from probing are assigned low prior probabilities, while losses (if the certainty-equivalent choice is correct) are assigned high prior probabilities. Another reason is evident in the dynamic programming results of Ludwig and Walters (1982), and I call this the "procrastination effect": if the current decision maker anticipates that his successors will make optimum adaptive choices, including probing experiments if they face high uncertainty, then he will see it as less important to initiate such experiments immediately.

RESEARCH DIRECTIONS IN ADAPTIVE MANAGEMENT

From the standpoint of applied mathematics and control theory, there are at least four research areas where valuable insights might be gained about adaptive policies for renewable resource management. First, analytical solutions for simple prototype problems could tell us a lot about the functional form of optimal policies in relation to information state measures, and provide a baseline for testing various numerical approximations; see for example the analysis by Dersin et al. (1981) on the scalar linear-quadratic regulator with stochastic coefficients. It is particularly important to learn more from simple cases about the relative performance of passive versus active adaptive strategies.

Second, there is a need for systematic comparison of alternative algorithms for approximating the dual effects of control, over the range of approaches suggested in the previous section. For example, comparison of dynamic programming versus wide sense dual control algorithms for a harvesting problem has indicated that the wide sense algorithm may systematically overestimate the importance of probing (Walters, 1986).

Third, it may be profitable to develop procedures for recasting problems involving uncertainty about continuous variables (unknown states, parameters) into discrete frameworks (finite collections of alternative hypotheses, partially observed Markov decision processes). The work of Wenk and Bar-Shalom (1980) suggests that there are exciting possibilities for simplified dynamic programming solutions in such settings.

Finally, it is important to recognize that most managed populations and ecosystems are not unique and indivisible entities. Instead, they are divided into a collection of more or less discrete spatial "replicates" (Walters, 1986) that can often be managed differently. Thus, there are opportunities to set up dynamic experimental designs, with replicated treatment and control areas against which to measure serially correlated environmental effects. Further, probing tests on the replicates can be designed so as to minimize disturbance to overall system performance (for example, fish harder in one area while not fishing in another, so total catch remains nearly constant while two informative disturbances are initiated). I suggest that the design of adaptive control strategies for spatially replicated systems is the single most exciting topic of future research in renewable resource management.

REFERENCES

Aström, K. 1965. Optimal control of Markov processes with incomplete state information. J. Math. Anal. App., Vol. 10, pp. 174-205.

Aström, K. 1969. Optimal control of Markov processes with incomplete state information II. J. Math. Anal. App., Vol. 26, pp. 403-406.

Aström, K. 1976. Introduction to Stochastic Control Theory. Academic Press, NY.

Aström, K. 1983. Theory and Applications of adaptive control--a survey. Automatica, Vol. 19, pp. 471-86.

Athans, M., R. Ku and S.B. Gershwin. 1977. The uncertainty threshold principle: some fundamental limitations on optimal decision making under dynamic uncertainty. IEEE Trans. Aut. Cont., Vol. AC-22, pp. 491-495.

Bar-Shalom, Y. 1981. Stochastic dynamic programming: caution and probing. IEEE Trans. Aut. Cont., Vol. AC-26, pp. 1184-1195.

Bar-Shalom, Y., and E. Tse. 1974. Dual effect, certainty equivalence, and separation in stochastic control. IEEE Trans. Aut. Cont., Vol. AC-19, pp. 494-500.

Bar-Shalom, Y., and E. Tse. 1976. Caution, probing, and the value of information in the control of uncertain systems. Ann. Econ. Soc. Meas., Vol. 5, pp. 323-337.

Bayard, D.S. and M. Eslami. 1985. Implicit dual control for general stochastic processes. Opt. Cont. App. Meth., Vol. 6, pp. 265-279.

Bellman, R. 1961. Adaptive Control Processes: A Guided Tour. Princeton Univ. Press, Princeton, NJ.

Bertsekas, D.P. 1976. Dynamic Programming and Stochastic Control. Academic Press, NY.

Chan, S.S. and M.B. Zarrop. 1985. A suboptimal dual controller for stochastic systems with unknown parameters. Int. J. Control, Vol. 41, pp. 507-524.

Chow, G.C. 1975. Analysis and Control of Dynamic Economic Systems. John Wiley, NY.

Deriso, R. 1986. Risk adverse harvesting strategies. pp. 65-73. IN: M. Mangel (ed.) Resource Management. Springer-Verlag, Berlin. Lecture Notes in Biomathematics 6.

Dersin, P.E., M. Athans, and D.A. Kendrick. 1981. Some properties of the dual adaptive stochastic control algorithm. IEEE Trans. Aut. Cont., Vol. AC-26, pp. 1001-1008.

Fel'dbaum, A.A. 1960,1961. Theory of dual control I-IV. Automatic Remote Control USSR, Vol. 21, pp. 1240-9, 1453-65; Vol. 22, pp. 3-16, 129-43 (in Russian).

Fel'dbaum, A.A. 1965. Optimal Control Systems. Academic Press, NY.

Gessing, R. and O.L.R. Jacobs. 1984. On the equivalence between optimal stochastic control and open-loop feedback control. Int. J. Cont., Vol. 40, pp. 193-200.

Glazebrook, K.D. 1983. Optimal strategies for families of alternative bandit processes. IEEE Trans. Aut. Cont., Vol. AC-28, pp. 858-860.

Hernandez-Lerma, O. 1985. Nonstationary value iteration and adaptive control of discounted semi-Markov processes. J. Math. Anal. App., Vol. 112, pp. 435-445.

Jacobs, O.L.R. and J.W. Patchell. 1972. Caution and probing in stochastic control. Int. J. Cont., Vol. 15, pp. 189-199.

Kolonko, M. and H. Benzing. 1985. The sequential design of Bernoulli experiments including switching costs. Operations Res., Vol. 33, pp. 412-426.

Kumar, P.R. 1983. Simultaneous identification and adaptive control of unknown systems over finite parameter spaces. IEEE Trans. Aut. Cont., Vol. AC-28, pp. 68-76.

Kumar, P.R. and T.I. Seidman. 1981. On the optimal solution of the one-armed bandit adaptive control problem. IEEE Trans. Aut. Cont., Vol. AC-26, pp. 1176-1184.

Ludwig, D. and C.J. Walters. 1982. Optimal havesting with imprecise parameter estimates. Ecological Modelling, Vol. 14, pp. 273-292.

Norman, A.L. 1976. First order dual control. Ann. Econ. Soc. Meas., Vol. 5, pp. 311-321.

Padilla, C.S., J.B. Cruz and R.A. Padilla. 1980. A simple algorithm for SAFER control. Int. J. Cont., Vol. 32, pp. 1111-1118.

Sato, M., K. Abe and H. Takeda. 1985. An asymptotically optimal learning controller for finite Markov chains with unknown transition probabilities. IEEE Trans. Aut. Cont., Vol. AC-30, pp. 1147-1149.

Smith, A.D.M. and C.J. Walters. 1981. Adaptive management of stock-recruitment systems. Can. J. Fish. Aquat. Sci., Vol. 38, No. 6, pp. 690-703.

Streibel, C. 1965. Sufficient statistics in the optimum control of stochastic systems. J. Math. Anal. App., Vol. 12, pp. 576-592.

Streibel, C. 1975. Optimal Control of Discrete-Time Stochastic Systems. Springer-Verlag, Berlin.

Tse, E., Y. Bar-Shalom and L. Meier. 1973. Wide-sense adaptive dual control of stochastic nonlinear systems. IEEE Trans. Aut. Cont., Vol. AC-18, pp. 98-108.

Walters, C.J. 1981. Optimum escapements in the face of alternative recruitment hypotheses. Can. J. Fish. Aquat. Sci., Vol. 38, No. 6, pp. 678-689.

Walters, C.J. 1985. Bias in the estimation of functional relationships from time series data. Can. J. Fish. Aquat. Sci., Vol. 42, No. 1, pp. 147-149.

Walters, C.J. 1986. Adaptive Policy Design in Renewable Resource Management. MacMillan Publ. Co., Inc. NY (in press).

Walters, C.J. and R. Hilborn. 1976. Adaptive control of fishing systems. J. Fish. Res. Board Can., Vol. 33, pp. 145-159.

Walters, C.J. and D. Ludwig. 1986. Adaptive management of harvest rates in the presence of a risk averse utility function. Ms. submitted to Renewable Res. Modelling.

Wenk, C.J. and Y. Bar-Shalom. 1980. A multiple model adaptive dual control algorithm for stochastic systems with unknown parameters. IEEE Trans. Aut. Cont., Vol. AC-25, pp. 703-710.

Wittenmark, B. 1975. Stochastic adaptive control methods: a survey. Int. J. Cont., Vol. 21, pp. 705-730.

Wonham, W.M. 1968. On the separation theorem of stochastic control. SIAM J. Cont., Vol. 6, pp. 312-326.

PARTICIPANT'S COMMENTS

The advantages of using the adaptive control approach to the management of renewable natural resources are lucidly discussed in this paper. I will, therefore, emphasize some of what might be--in my opinion--the limitations of the approach; raise some questions; and point some potential exciting applications. I should point out that I am no expert (by any stretch of the imagination) in adaptive control.

It should be emphasized that analytical solutions to problems in renewable resource management using the adaptive control (AC) approach are (except in the simplest cases) not available. Thus, one can hardly expect to gain some overall qualitative insight into system behavior using this approach. The approach then suffers from the usual drawbacks of simulations, where it is very difficult to "experiment" with all possible scenarios. As such, the AC approach is complementary to the deterministic control theory approach.

Other, more technical, problems are:

1. It is my impression that the white-noise assumption is crucial in searching for sub-optimal solutions using the AC approach. Thus, as opposed to vulnerability analysis (e.g., Grantham and Fisher; this volume) there is no admitting that the model itself may be wrong.

2. A wrong model can be admitted if the AC algorithm experiments with the system (using the balance between probing and caution) with a few alternative models where the parameters in each model need to be estimated. Given the tremendous computational burden at present, I doubt that this could be accomplished.

3. When $J = E\{C\}$, where E denotes expectation and C denotes cost, the following decomposition is essential for application of AC

$$J_{CL} = J_D(k) + J_C(k) + J_P(k)$$

where k denotes the k-th time period and the subscripts CL, D, C and P stand for closed loop, deterministic, caution, and probing, respectively. Given the fact that such a decomposition can be done, if J_C is large compared to J_P, then AC is unnecessary. To what extent J_C dominates in most natural system remains to be seen.

4. For it to be feasible, the AC approach requires that system parameters must be slowly varying. What happens when this is not the case?

5. Could the AC be used when the control affect the parameters? How can this be tested?

6. In the example presented at the workshop, the AC approach resulted in a better management (e.g., higher economic return from a fishery system) over a 50 year time horizon, where managers were to expect an incurred higher loss for the first 10 years. Thus, in applications, the actual time horizon over which it is more profitable to use the AC approach is an important factor. If we are to convince managers to use the approach, would they be willing to use such time horizons as those above? In short, for short time horizons open loop control may actually perform better than closed loop AC (e.g., Bar-Shalom and Tse, 1976).

Some potentially interesting applications of AC could be developed in conjunction with Unifying Foraging Theory (Clark; this volume). Another area of potential application is in general foraging theory and animal behavior. It is well known that animals invest some of their time and energy in play, experimentation and learning. It seems that the AC approach may be invoked to explain and quantify such behaviors.

Yosef Cohen

When I was first introduced to the concepts of adaptive control and its policy prescriptions for probing experiments, my first thought was that fisheries biologists have been trying for years to convince managers to adopt such policies in order to reduce prevailing uncertaintines in estimates of parameters and functional relationships. Almost uniformly they have failed, being unable to respond to the not-unreasonable rejoinder that such a policy will lead to disruption in the fishery, and therefore immediate costs, with rather uncertain future benefits. Armed with these adaptive control formulations, at that time it seemed possible to take explicit account of the manager's qualms and thus present a convincing argument. Further reading has somewhat reduced my naive optimism that adaptive control can solve the world's problems, but reinforced the view that it is a particularly valuable research area. Carl Walter's perspicuous review of the present state-of-the-art in adaptive policy design also left me with similar impressions.

Being only a neophyte in this area, I am a little hesitant to make specific comments, but two related points seemed worth making. First, Carl makes it clear that finding a solution to any half-way realistic problem represents a really formidable computing task, whichever of the five approaches suggested in the paper are used, and that a number of approximations have to be made. Indeed, those who belong to the "realistic modelling" school would probably argue that any computationally feasible adaptive control problem will be based on a resource management model that is at best gross caricature. Given the radical nature of some of the optimal policies, it seems reasonable to query the extent to which reliable quantitative policies can be determined from analysis of an isolated resource management system.

The related point addresses the final research direction suggested in the paper. I strongly agree that experimental manipulation of spatially replicated systems represents probably the most practical means for testing and implementing adaptive control policies. Minimization of disturbance to the overall system both reduces one of the major objections to probing policies and presumably thereby increases the value and frequency of probes. Further, manipulation of actual, rather than model systems reduces the dependence on probably unrealistic models. At this stage, we badly need some demonstrably successful applications of control theory to real resource management.

While there may remain doubt regarding the quantitative adaptive policies determined using the techniques reviewed by Carl Walters, there is no doubt in my mind that the qualitative results obtained are invaluable as indicative prescriptions. From bitter experience, I have found that, however satisfying an optimal policy can be explained after it has been determined, it is very difficult to predict the policy beforehand.

Geoff Kirkwood

COMPUTER-INTENSIVE METHODS FOR
FISHERIES STOCK ASSESSMENT

Donald Ludwig
Institute of Animal Resource Ecology
The University of British Columbia
2204 Main Mall
Vancouver, British Columbia
Canada V6T 1W5

This paper provides an introduction to fishery management problems, with emphasis upon the statistical aspects, and problems resulting from lack of information. Parameter estimation schemes are evaluated by generating series of artifical data, and comparing performance with the case where perfect information is available.

INTRODUCTION

Fisheries provide a great variety of interesting scientific, management and institutional problems. Although control theorists might find it most natural to think of fisheries management as a control problem, other aspects dominate in many situations. As Colin Clark has remarked, no matter how many features you are able to include in your analysis, there is likely to be one which you have neglected, and which invalidates your conclusions.

This essay is primarily concerned with statistical aspects of fisheries stock assessment. In view of the great costs involved in obtaining information about the behavior of fisheries under exploitation, it appears likely that future scientific approaches will have to deal with poor or insufficient information. Although the resulting problems are exceedingly intractable (or perhaps impossible), some optimism is justified: the recent decline in the costs of computation makes it feasible to apply massive computing power to statistical and decision problems.

FISHERIES MANAGEMENT AS A CONTROL PROBLEM

A new era in management began with the systematic application of the methods of control theory, as in Clark (1976). The stated objective is to regulate the harvest in order to maximize a discounted yield, or some other measure of the social utility of the activity. This approach enables general insights from control theory to be brought to bear on biological or economic problems. Even more important, it has enabled workers in diverse fields to communicate by means of a common language and set of concepts.

This series of international meetings has been planned to exploit these possibilities. Nevertheless, I believe that a control-theoretic approach is too limited to be applicable to most fisheries problems, unless it takes into account the conflict between learning about and exploiting a stock.

THE CONFLICT BETWEEN EXPLOITING AND LEARNING

Suppose that the basic problem is to maximize a long-term harvest, and we try to determine a corresponding level of fishing effort. Then we require data which permit a comparison of the effects of large and small efforts and large and small stock sizes upon subsequent harvests. There is no way to know how to regulate the fishery unless we have recent experience of a variety of stock sizes and efforts. On the other hand, economic and social benefits of the fishery often depend upon steady yields or efforts. If we stabilize efforts or yields, we deprive ourselves of the information necessary to regulate efforts or yields in an optimal fashion. There is a conflict between exploiting and learning since information is obtained mainly by performing experiments, i.e., doing things differently than you otherwise would, but economic and social goals usually require steady yields or only incremental changes in policies.

FISHERIES MANAGEMENT AS A STATISTICAL PROBLEM

The control-theoretic approach requires fairly precise knowledge of stock sizes and stock dynamics. Several factors limit that knowledge:

1. Although lengthy sets of data are sometimes available, the older data are of dubious value, for a variety of reasons: there may be environmental alterations such as loss of spawning habitat, or changes in competitive pressures (as have been proposed in the case of the sardine-anchovy interaction), or collapses triggered by environmental fluctuations or human activities.

2. The data available may not be very informative about crucial patterns and processes. Because of the conflict between learning and exploiting, fisheries data generally are lacking in the contrast in important variables which would enable us to make confident predictions of the consequences of actions which are contemplated.

The limited inferences possible from available data may make otherwise plausible objectives quite unattainable. For example, sustained yield may not be attainable or desirable in the longer term, because of the problems resulting from lack of information. Therefore, it is unwise to separate the data collection and analysis from the formulation of management objectives and the solution of the corresponding theoretical control problems.

For similar reasons, scientific and theoretical approaches should not be separated from the political, social and operational context.

DIFFICULTIES IN PARAMETER ESTIMATION

One of the most common and intractable problems in fisheries management is to distinguish between two alternatives. Are we (1) harvesting a small fraction of a large stock, or are we (2) harvesting a large fraction of a relatively small but highly productive stock? In case 1, we expect that the density-dependent effects are controlling the stock size, while in case 2, density-independent effects are controlling the stock size.

How can we distinguish between these two cases? We can expect to observe density-dependent effects only if there are large catches (relative to the stock size) at least part of the time, and if there are large contrasts in catches during the period for which data are available.

Qualitative arguments such as the preceding one, or the earlier argument concerning the conflict between exploiting and learning are useful in forming our intuition, but they are not sufficient to answer quantitative questions. How large is large? How much contrast in catch and effort is required in order to attain acceptable precision in determining optimal efforts? We can provide answers to these questions and related ones about the effectiveness of estimation schemes by extensive computer simulation trials.

THE STRATEGY OF HILBORN, WALTERS AND LUDWIG

Our objective is to match estimation methods and models to data and biological assumptions, and to make our methods accessible to managers and other researchers. Our plan is as follows:

1. Select a variety of possible stock dynamics. Schnute (1985) is a suitable starting point, since it contains a systematic collection of possible models, which are selected by an appropriate choice of parameters.

2. Devise a variety of estimation schemes. Our main present concern is the method outlined below, since it appears to be widely applicable. Another possible approach is given in Ludwig and Walters (1981) and Ludwig and Hilborn (1983). Still other approaches might be based upon time series methods, etc.

3. Compare the performance of the estimation schemes on simulated data. This procedure was followed in Ludwig and Walters (1985), with results which were perhaps surprising: the most realistic model (the one which fits the assumptions under which the data were generated) does not necessarily perform the best. Our present efforts are an attempt to explore this issue more systematically.

4. Carry out the assessment for particular real data sets. Part of this process would be the assignment of support or weight to various hypotheses about the stock and its behavior under various exploitation regimes.

5. Test management strategies (active vs. passive) for their long-term consequences. This phase is addressed more fully in Walters' paper in these proceedings.

6. Make these methods available as a software package. This is a very important part of the enterprise. Nobody should apply theoretical results in a specific problem without experimenting with methods and adapting them to the specific circumstances at hand. It is also much easier to specify a procedure by means of computer code than verbally.

The sequel is mainly concerned with the third and fourth points.

A MINIMAL STATISTICAL MODEL

A management scheme which seeks to regulate human activity should employ data regarding the intensity of these activities, i.e., "effort". The corresponding "catch" measures the effect of these activities upon the target population. It may also be used in determining the value of the harvest. Less aggregated measures might be employed in place of "catch" and "effort", but they will not be considered here.

In order to understand the population dynamics of the exploited population, we must assess the connection between stock size and future recruitment. At a minimum, this will include a measure of the net growth rate at low densities, and an indication of the effect of increased population size upon the growth rate. For our present purposes, a simple model which incorporates these effects is

$$B_{t+1} = S_t \exp(a - b\, S_t + w_t) \tag{1}$$

where S_t is the spawning biomass, and B_{t+1} is the next year's biomass. The parameters a and b are to be estimated, and $\{w_t\}$ are random variables which represent random effects upon recruitment.

The connection with observable quantities is made by

$$C_t = B_t\, (1 - \exp(-q\, e_t)) \quad . \tag{2}$$

Here C_t is the observed catch, q is a catchability parameter to be estimated, and e_t is the effective effort: the observed effort E_t is assumed to be related to e_t by

$$E_t = e_t \exp(v_t) \quad , \tag{3}$$

where v_t represent random discrepancies between the observed effort and the effective effort. We may think of $\{v_t\}$ as "observation errors", while $\{w_t\}$ represent "process errors".

We shall assume that the $\{w_t\}$ are independently distributed normal random variables with mean 0 and variance σ_W^2 .

This completes the description of our formal statistical model. Much more realistic and refined models might be considered. For example, one might have age or size-specific catch data available. In order to make proper use of such data, one must either estimate age or size-specific catchability or make arbitrary assumptions about such parameters. Likewise, age or size-specific recruitment models would require either that a large number of additional parameters be introduced, or that arbitrary assumptions be made.

PARAMETER ESTIMATION

Details of the following scheme will be given elsewhere. In brief, our method consists in approximating the likelihood of the observations by a relatively simple function of the parameters. This is achieved by representing all of the random effects in terms of deviations from a nominal deterministic trajectory. The deviations are modelled by normal random variables, although they will not be normal in non-trivial applications. The errors in the approximation appear only in computing the covariance matrix of the residuals.

The parameters to be estimated are six in all: a, b, q, X_0 (the initial stock size), σ_V^2 and σ_W^2 . Estimates are obtained by maximizing the approximate likelihood as a function of those parameters. The maximum is sought by the "trust region" method (a modified Marquart method), whereby the size of steps is regulated by a parameter, which in turn is adjusted by comparing the expected improvement in the likelihood with the observed improvement.

In the future, we hope to take into account correlations between human actions and past parameter estimates. If such correlations are neglected, they may introduce a bias in future estimates, which tends to provide a false justification for past management actions. By taking these correlations into explicit account, we hope to provide a more reliable method to detect overexploitation of stocks.

SIMULATION TRIALS

In the preliminary set of trials described here, data were generated using the model described above, with a variety of effort sequences and levels of the noise parameters σ_V^2 and σ_W^2. Generally, 100 sets of data were generated for each set of efforts and parameters. Then the estimation scheme outlined above was employed to derive parameter estimates. The results may be used in a variety of ways.

1. One may examine the distribution of the estimates. For example, in Fig. 1, the estimates for σ_W^2, q, a and b are plotted against each other, in pairs. In this case, the effort sequence was lacking in contrast. This is reflected in a high correlation between the estimates of a and b. In Fig. 2, an effort sequence with higher contrast was chosen. The correlations are still present, but not as strong.

2. One may calculate an optimal effort, based upon the parameter estimates. This is a crucial quantity, since efforts will be controlled by management activity. One may determine which parameters have the greatest influence upon the optimal effort, by regressing that effort against the parameters. The best such regression is obtained using q and a. In Fig. 3, the fit in this regression is plotted against the optimal effort, as estimated. The straight line corresponds to a perfect fit.

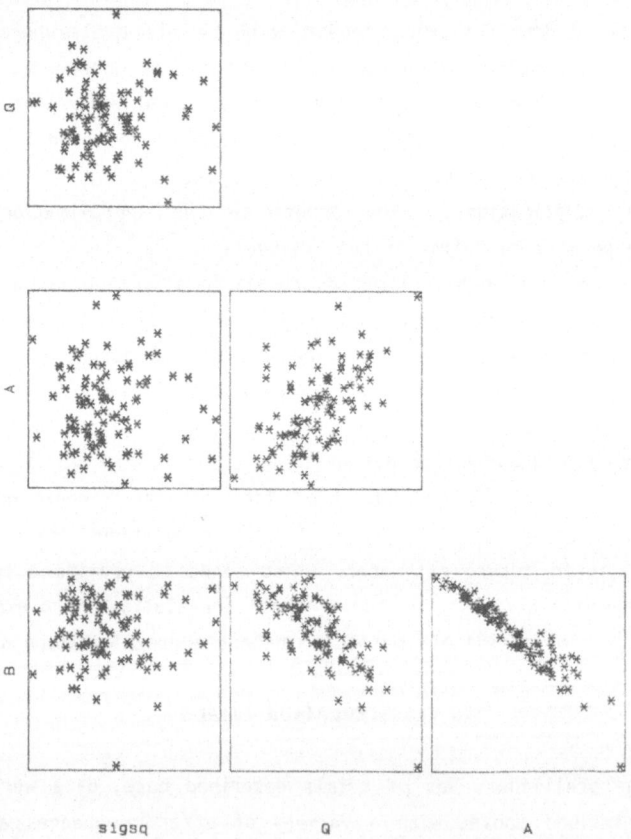

Figure 1.   Pairwise scatter plot of the estimates for the variance σ^2, q, a, and b. This case has relatively poor contrast in efforts, and hence the estimates of a and b are highly correlated.

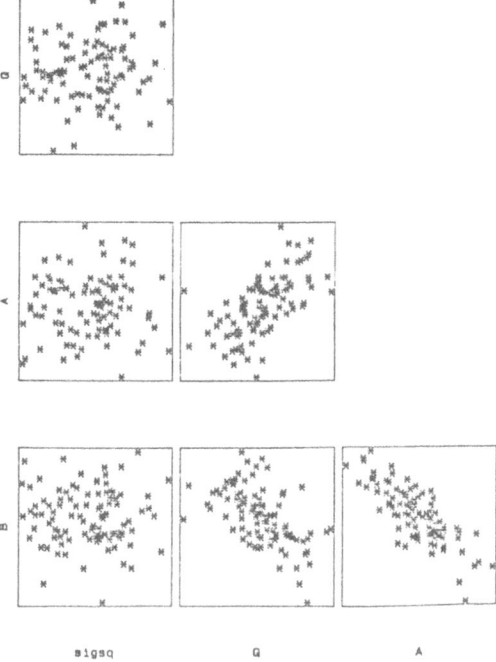

Figure 2.   Pairwise scatter plot as in Figure 1, for a case where more contrast was present in the effort sequence.

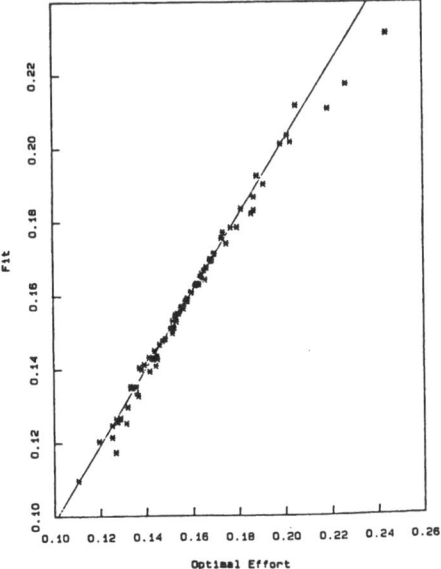

Figure 3.   Scatter plot of the optimal effort, based upon estimates of all parameters, versus the fit of the optimal effort in terms of the parameters q and a. The straight line is a robust regression fit to the points.

3. Since the "true" parameters are known, one may calculate the long term discounted yield obtained by using the optimal effort as estimated in each simulation. One is interested in the distribution of such values. This is easily visualized by means of a boxplot. The "box" in a boxplot encloses the median and the quartiles of the distribution. The dotted lines show the range of the distribution, except that points too far from the median are shown as outliers (with stars). Fig. 4 shows a group of such boxplots, corresponding to differing amounts of contrast in the effort sequence, and differing values of σ_V^2 and σ_W^2 . In this first series of trials, $\sigma_V^2 = \sigma_W^2$, both for generating the data, and when calculating parameter estimates. This restriction is removed later. The first and third of the boxplots correspond to the simulations in Figs. 2 and 1, respectively. The not very striking difference in the earlier figures is magnified in the resulting values. The latter two plots in the sequence show the same phenomenon when the noise level is increased.

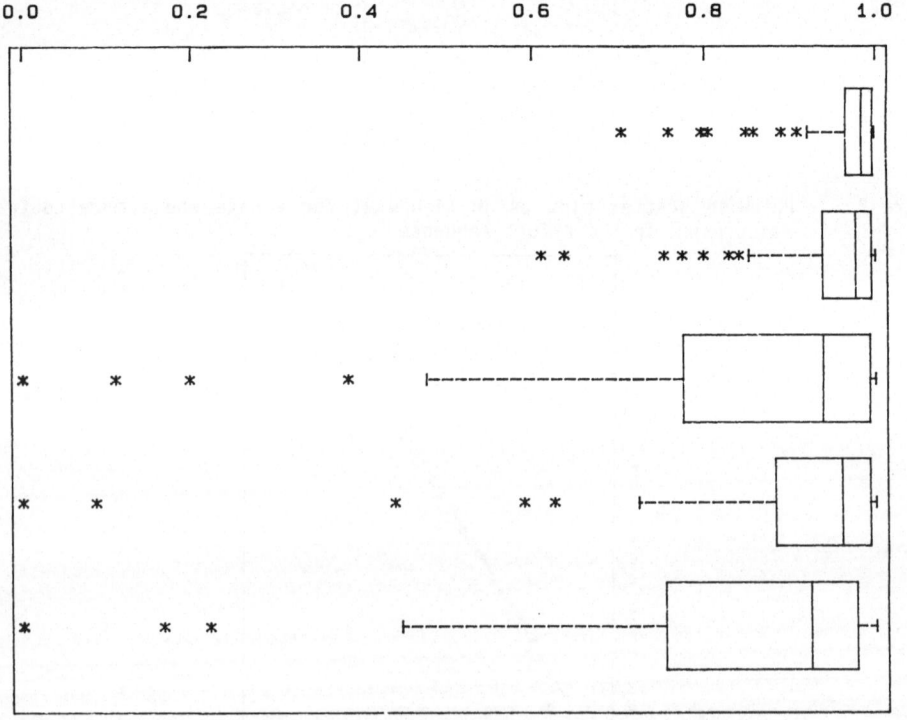

Figure 4.   Boxplots of the value ratio for a variety of effort sequences and noise in the data. The value ratio is obtained by first computing the optimal effort from the parameter estimates. The value of holding the effort at this level is divided by the true optimal effort. The first three boxplots correspond to decreasing levels of contrast in the effort. The last two are a similar sequence, for a higher level of process and observation error.

STOCK ASSESSMENT FROM REAL DATA

The preceding results are not enough to provide a guide to stock estimation. If we have only a single data set available, how much confidence can we assign to our estimate of optimal effort? Likelihood contours have been a fairly robust guide to the precision of our estimates.

In order to apply this procedure, we first add a term to the log-likelihood function, to penalize deviations of the effort from some prescribed value e_0. Then we maximize this penalized likelihood: the resulting parameters will yield an estimate for the optimal effort which differs only slightly from e_0. The unpenalized likelihood at these parameter values will be smaller than the maximum of the unpenalized likelihood. The "likelihood difference" is the difference between these unpenalized likelihoods. Sometimes this likelihood difference is called the support for the effort e_0. Fig. 5 shows this likelihood difference as a function of e_0, for a set of data with contrast intermediate between those in Figs. 1 and 2.

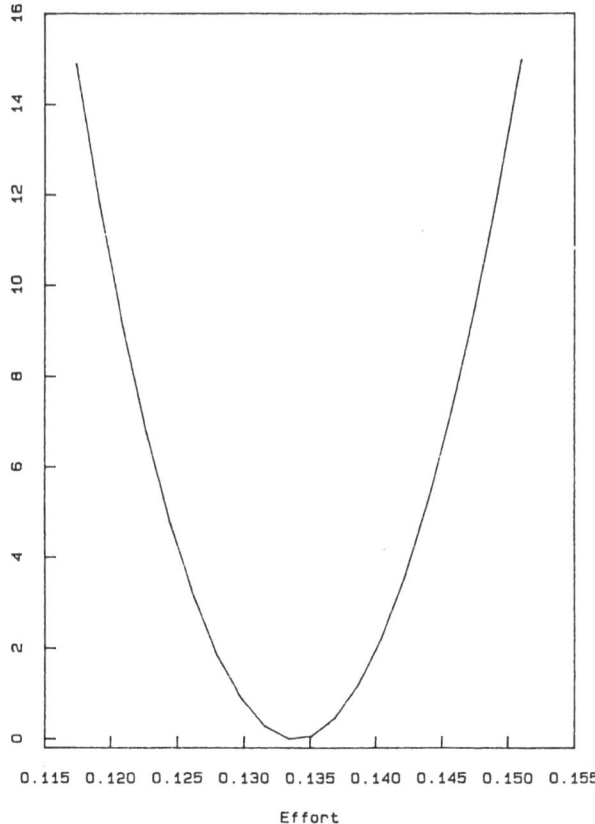

Effort

Figure 5.   The likelihood of various values of the effort, based upon a single set of data.

Now we require some guidance to the expected range of this likelihood difference. This can be obtained from the simulations. In Fig. 6, each boxplot shows the distribution of the difference between the maximum likelihood as computed from the data, and the likelihood of the true parameter values (those which were used in generating the data). The simulations are the same as those shown in Fig. 4. Although the distribution of values in Fig. 4 is very variable, the likelihood differences are gratifyingly consistent. These figures indicate that 75% of the trials have a likelihood difference lying between 0 and 7.5, and nearly all have differences less than 15. Thus, the range shown in Fig. 5 is appropriate.

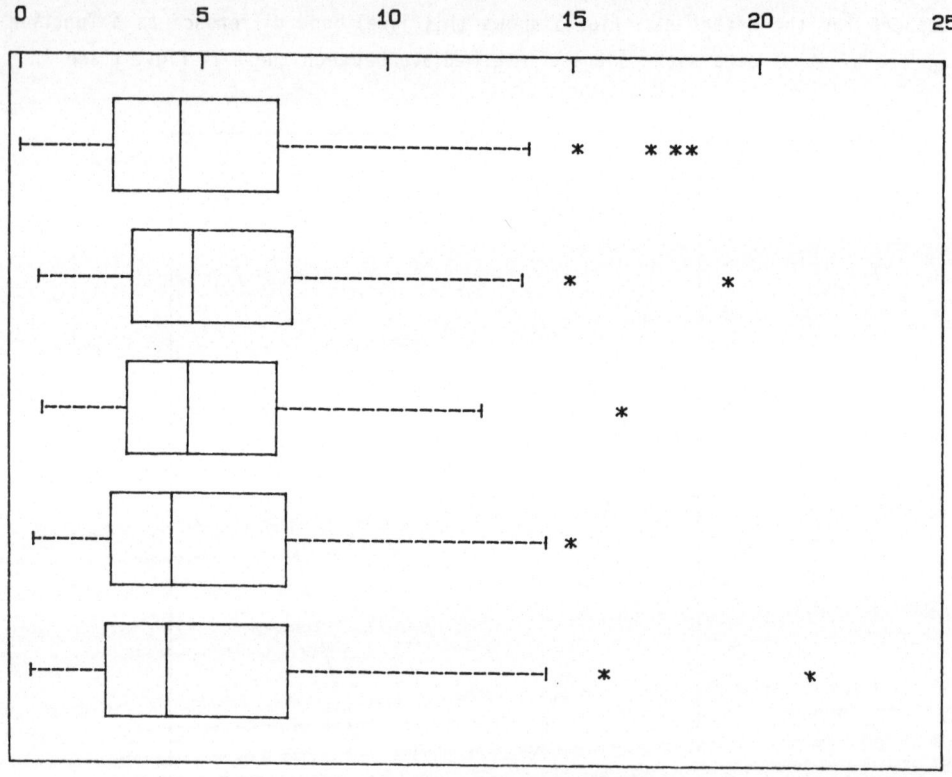

Figure 6. Boxplots of the distribution of the likelihood difference, for the same set of simulation runs as in Figure 4.

A FLY IN THE OINTMENT

Unfortunately, Fig. 6 is not the whole story. In performing the simulations, data were generated with equal values of σ_W^2 and σ_V^2, and this fact was used in calculating parameter estimates. What happens if an incorrect value of the ratio of these variances is chosen? The result is shown in Fig. 7. The boxplots show the same likelihood differences as are plotted in Fig. 6, for a set of simulations where the ratio of σ_V^2 to σ_W^2 was not known to the estimation program. The top row of boxplots corresponds to a case where $\sigma_V^2 = 0$, i.e., no observation errors were present, only "process error" due to fluctuations in recruitment. The middle row corresponds to the case where $\sigma_W^2 = \sigma_V^2$, which was the only case considered earlier. The third row corresponds to the case where $\sigma_W^2 = 0$, i.e., all of the error was due to observation error. The first column corresponds to the assumption that $\sigma_V^2 = 0$ in the estimation, and the second and third columns correspond to the assumptions that $\sigma_W^2 = \sigma_V^2$, and $\sigma_W^2 = 0$, respectively. The last column of Fig. 7 corresponds to a scheme whereby the ratio of σ_W^2 and σ_V^2 was estimated with the other parameters. Although this ratio could not be determined very well from the data, the likelihood difference is much better behaved if this ratio is estimated than if it is not. The only exception to this rule is when this ratio is known from other kinds of data.

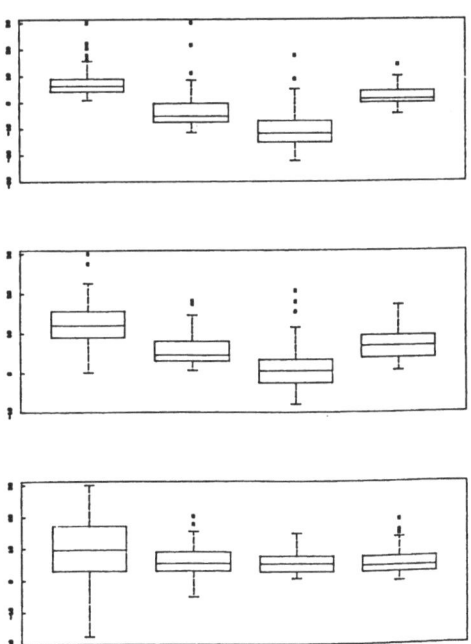

Figure 7.   A series of boxplots of likelihood differences.   Details are given in the main text.

FUTURE OUTLOOK

The results reported upon here are, of course, only preliminary. The technical details are likely to undergo a long evolution. However, the computer-intensive methods seem to offer great promise in understanding the effectiveness of models and estimation procedures in fisheries stock assessment.

REFERENCES

Clark, C.W. 1976. Mathematical Bioeconomics, J. Wiley and Sons, New York.

Ludwig, D. and R. Hilborn. 1983. Adaptive probing strategies for age-structured fish stocks. Can. J. Fish. Aquat. Sci., Vol. 40, pp. 559-569.

Ludwig, D. and C.J. Walters. 1981. Measurement errors and uncertainty in parameter estimates for stock and recruitment. Can. J. Fish. Aquat. Sci., Vol. 38, pp. 711-720.

Ludwig, D. and C.J. Walters. 1985. Are age-structured models appropriate for catch-effort data? Can. J. Fish. Aquat. Sci., Vol. 42, pp. 1066-1072.

Schnute, J. 1985. A general theory for analysis of catch and effort data. Can. J. Fish. Aquat. Sci., Vol. 42, pp. 414-429.

PARTICIPANT'S COMMENTS

The second and third sentences of this paper are "Although control theorists might find it most natural to think of fisheries management as a control problem, other aspects dominate in many situations. As Colin Clark has remarked, no matter how many features you are able to include in your analysis, there is likely to be one which you have neglected, and which invalidates your conclusions." In this discussion, I would like to elaborate on these ideas.

In many ways, natural resources management is a part of operations analysis, which I define here to be the scientific approach to operational problems for which the underlying physical laws are known only vaguely or not at all. Certainly fisheries management fits into this framework. In operational analysis, one is striving for a general understanding of the operational situation in terms of minimal numbers of parameters and assumptions. Operations analysis was brought to life in the US during World War II, mainly by the physicist Philip Morse who lead the Anti-Submarine Warfare Operations Research Group. At the end of the war, Morse and George Kimball published a book entitled "Methods of Operations Research". In this book, they introduce the idea of hemibel thinking. They define hemibel (in the same manner as a decibel) as log to the base 3 (1 = 0 hemibels, 3 = 1 hemibel, 10 = 2 hemibels, etc.) and suggest that operational parameters should be measured in hemibels. They write:

Having obtained the constants of the operation under study in units of hemibels (or to one significant figure), we take our next step by comparing these constants. We first compare the value of the constants with the optimum theoretical value, if this can be computed. If the actual value is within a hemibel (i.e., within a factor of 3) of the theoretical value, then it is extremely unlikely that any improvement in the details of the operation will result in significant improvement...[If] there is a wide gap between the actual and theoretical results... possible means of improvement can usually be obtained by a crude sorting of the operational data to see whether changes in personnel, equipment or tactics produce a significant change in the constants. In many cases a theoretical study of the optimum values of the constants will indicate possibilities of improvement.

These lines are as true today as they were 40 years ago -- perhaps even more so, since today we have the capability to do all sorts of "computer experiments" on operational models that could not be performed 40 years ago.

Ludwig's paper is definitely in the spirit of "hemibel thinking" and I applaud it. It is easy to become seduced by the sexy problems of control, but very often our biggest contribution as modellers and analysts is to introduce simple concepts. (The biggest gains are often made when we can introduce arithmetic to situations where it was not used before.) For systems in which more than one parameter enter into the dynamics and operational models, the study of the estimation of and interaction between these parameters is especially important. The computer provides an ideal tool for such studies and the methods outlined in this paper are very appealing because of the robustness of the approach.

If our objective as renewable resource system modellers is to have impact, rather than to do fancy mathematics, for problems with complicated biological, political and operational interactions, then we must move away from pure "control" approaches into more complex arenas such as the one described in this paper.

Marc Mangel

Ludwig clearly identifies some of the important problems in resource management, and describes an approach to them that is being taken by several prominent and innovative fisheries scientists. This approach involves selection of models on a purely statistical basis from the results of Monte Carlo simulations, rather than on the basis of structural realism (cf. Ludwig and Walters, 1985). In the hope that a contrasting point of view will be a constructive contribution to these proceedings, I point out here several potential problems with this approach.

With regard to his initial comments, I agree that system identification is an important part of the control problem in resource management. Because the learning/exploitation conflict is responsible for the low number of learning strategies in actual use, it seems that more effort should be spent demonstrating its value in practical situations.

The problem that Ludwig associates with long term data sets, that they are of dubious value because of possible episodic changes in the past, is a good argument for mechanistic understanding of resource populations and the environments that affect them. Large scale climatic or oceanographic changes can be detected in ways other than from catch and effort data on the population of direct concern, and potential changes in population dynamics could thereby be identified. It is not clear how Ludwig's approach gets around this problem since these events can occur at any time.

I see two potential problems in Ludwig's approach to resource modeling as outlined in the remainder of his paper: (1) an overly restrictive selection of the data to be used, and (2) heavy dependence on the generality of the simulations to be used (to the extent that performance in the simulations overrides consideration of known mechanisms).

The constraint that fishery managers base decisions on only catch and effort data seems overly restrictive and is responsible for derived behavior of the model selection scheme. The fact that catch and effort data are the only data available for some fisheries is at least in part due to the fact that managers have been told for many years that catch and effort data are adequate for management (e.g., with the logistic model). Allowing only catch and effort data influences the results of model selection because these data are particularly uninformative about the underlying dynamics. For example, catch, even if adjusted for effort, is a poor description of the state of a population because the number in each age class has been averaged out, hence information has been lost. Because there is no a priori reason to constrain ourselves to catch and effort data, and other data exist in most fisheries, I would recommend that the decision of whether to use/collect additional kinds of data be made on an economic basis in terms of their value in allowing more effective harvest?

In Ludwig's scheme, model selection depends on the outcome of many simulations of how the analyst believes the system functions. Adoption of the best model depends critically on the generality of the simulations on which it is based. For management models, if we could be sure that the simulations included all possible occurrences, that we were bound by the data constraints imposed, and that the estimation schemes were the only ones possible, then we could have confidence in using the schemes that proved best in Ludwig's evaluation. However, in resource management problems, we are not certain of these. To paraphrase Ludwig's quote from Colin Clark, no matter how many features you are able to include in your simulations, there is likely to be one which you have neglected, and which invalidates your conclusions.

One particularly bothersome outcome of the proposed model selection strategy is selection of a model that is known not to realistically represent the population. This aspect was emphasized in Ludwig and Walters (1985). Unrealistic, non-mechanistic models are not necessarily bad if one is dealing with a well-defined system whose statistical properties are well known. Indeed, there are widely used statistical estimators that were arbitrarily chosen, yet lead to the best estimate in some sense. However, giving up known structure in an uncertain, poorly understood system is a sacrifice that involves certain risks. These risks would be reduced by better understanding of why the model did not pass the test. In this case, for example, the outcome of selection depends on the assumption that only aggregate data are available.

Rather than a scheme for selection of models that provide a better fit under specific conditions, I would suggest further investigation of models with realistic characteristics such as age and size structure, to determine when the models can be simplified, and yet maintain the essential dynamics (cf. Botsford, 1981). This kind of understanding seems preferable to less critical choice of unrealistic, non-mechanistic models.

References

Botsford, L.W. 1981. More realistic fishery models: cycles, collapse, and optimal policy. In T.L. Vincent and J.M. Skowronski (eds.) Renewable Resources Management, Springer-Verlag, New York.

Ludwig, D. and C.J. Walters. 1985. Are age-structured models appropriate for catch-effort data? Can. J. Fish. Aquat. Sci., Vol. 42, pp. 1066-1072.

Louis W. Botsford

VARIABILITY IN ECOSYSTEM MODELS:
A DETERMINISTIC APPROACH

Michael E. Fisher
Department of Mathematics
University of Western Australia
Nedlands, Western Australia 6009

Models which represent real ecosystems should include a degree of uncertainty in their description due to fluctuations in the environment which may affect the model parameters such as growth rates and carrying capacities. A deterministic approach to assessing the effect of introducing variability into ecosystem models is adopted. An important concept is that of a reachable set, which provides an indication of where in state space the system resides over a given period of time. Methods for calculating or estimating reachable sets for these models are discussed. In particular, one can obtain bounds on the state variables by performing a straightforward nonlinear optimization arising out of an optimal control problem. The methods are illustrated with several examples of discrete-time ecosystem models.

INTRODUCTION

It has long been recognized (Ehrlich and Birch, 1967; May, 1975; Harrison, 1980) that ecosystem models claiming to reasonably represent real ecosystems must include a degree of uncertainty in their description. In real ecosystems, this uncertainty arises due to fluctuations in the environment which may affect the model parameters such as growth rates and carrying capacities. A typical approach often adopted in modelling these uncertainties has been to either consider the model parameters as stochastic variables which fluctuate randomly about their mean values (see, for example, May, 1974; Beddington and May, 1977; Witten, 1978; and Witten and de la Torre, 1984). Alternatively, one could introduce white noise as an additional term in the model equations (Ludwig, 1975; Turelli, 1977; May et al., 1978; Polansky, 1978; and Taljapurkar and Semura, 1979). Both these approaches result in the deterministic model being replaced by a system of stochastic difference or differential equations which may lead, in the continuous-time case, to a complex system of partial differential equations known as the Fokker-Planck equations.

In this paper, deterministic approaches for assessing the effect of introducing variability into ecosystem models are discussed. Deterministic studies in which the model parameters have been functions of time have been undertaken by several authors (for example, Poluektov, 1974; May, 1976; and de Mottoni and Schiaffino, 1981). Harrison (1979, 1980) has used the notion of a persistent set in deterministic studies of systems with uncertainty and those in which precise differential equations are replaced by a set of inequalities. Also, the response of ecosystems to external periodic forcing, has been studied by Silvert and Smith (1981). The approach we adopt is somewhat different in that variability may be introduced into the model in a variety of ways, this variability having no properties other than that of being bounded. For example, instead of assigning fixed numerical values to the model parameters, as a consequence of field observations or laboratory experiments, we can allow the model parameters to belong to an interval of values. In a loose sense, this can be thought of as the deterministic analogue of designating the model parameters as being random variables with pre-assigned means and standard deviations. From a different viewpoint, one can imagine the model as being subjected to external perturbations over which we have little or no control as a result of factors which have not been adequately represented in the model such as climate, migration, etc.

As a consequence of introducing variability into the model, the notion of system trajectories asymptotically approaching stable sets of points such as equilibria or limit cycles no longer provides an adequate description of the system dynamics. Rather, it is usually of greater interest to know whether or not the system is likely to be ultimately confined to some bounded region of state space or, if the system resides in a known region of state space at a given time, where it is likely to be at some future time. If the system without variability (often referred to as the unperturbed system) has, for example, a stable equilibrium point then, provided the variability in the model is not too large, the system trajectories are likely to remain in a neighborhood of this equilibrium point for a reasonable length of time. The set of points in state space which can be visited from a given initial set over a specified period of time is called, in the language of control theory, a reachable set if one thinks of the variable parameters in the model as control variables.

The problem we shall be addressing is that of finding simple but effective methods for estimating reachable sets for a wide range of ecosystem models which can be described by either of the systems of equations

$$\dot{x}(t) = f[x(t),u(t)], \; t > 0 \qquad (1)$$

or

$$x(t+1) = f[x(t),u(t)], \; t = 0,1,\ldots \qquad (2)$$

where $x(t) \in R^n$, with $x(t) > 0$, is the population vector and $u(t) \in R^m$ is a vector representing the variability in the model induced by external perturbations. It is reasonable to expect that the externally induced perturbations and hence the $u_j(t)$ are bounded and so, without loss of generality, we assume that

$$|u_j(t)| < 1, \text{ for } j = 1,2,\ldots,m, \quad t > 0 \qquad (3)$$

As mentioned previously, $u(t)$ can be thought of as representing variability in the model equations themselves or in the model parameters or both due to some externally applied perturbations. It will be assumed that, in the absence of these perturbations, systems (1) and (2) have at least one undisturbed positive equilibrium point **x***, which satisfies

$$f[x^*, 0] = 0 \tag{4}$$

in the case of system (1), and

$$f[x^*, 0] = x^* \tag{5}$$

in the case of system (2). The following example provides a simple illustration of how the **u**(t) can be introduced into an ecosystem model.

<u>Example 1.</u> At time t, let x(t) be the population density (scaled with respect to the carrying capacity, K) of a population with non-overlapping generations and growth rate r. The model of the population is

$$x(t+1) = x(t)\exp\{r[1-x(t)]\}, \quad \text{for } t = 0,1,\ldots \tag{6}$$

which is based on the difference analogue of the logistic differential equation. Variability may be introduced into this model in the following ways:

(a) By including a term to represent the net effect of external perturbations in the time interval [t,t+1]. The model equation then becomes

$$x(t+1) = x(t)\exp\{r[1-x(t)]\} + b_1 u_1(t)x(t), \quad \text{for } t = 0,1,\ldots \tag{7}$$

where $|u_1(t)| \leqslant 1$ and b_1 is a positive constant.

(b) By replacing the carrying capacity K in equation (6) by the term $K(1+b_2 u_2(t))$, where b_2 is a positive constant, so that, after scaling x(t) with respect to K, the model becomes

$$x(t+1) = x(t)\exp\{r[1-x(t)/(1+b_2 u_2(t))]\}, \quad \text{for } t = 0,1,\ldots \; . \tag{8}$$

b_2 then represents the maximum relative fluctuation in K.

(c) By including both the effects of variability in the carrying capacity and external perturbations on the model leads to the equations

$$x(t+1) = x(t)\exp\{r[1-x(t)/(1+b_2 u_2(t))]\} + b_1 u_1(t)x(t), \quad \text{for } t = 0,1,\ldots \; . \tag{9}$$

For a given ecosystem model, the reachable set provides us with a measure of the vulnerability of the system to variability in the model parameters or to externally induced perturbations. The reachable set represents the "worst possible case" in that although the **u**(t) are unknown, we treat them as control variables which are chosen so as to provide the largest possible region in state space over which the system trajectories can roam.

There is another interpretation of the **u**(t) in equations (1) and (2) which is possibly more appropriate to a workshop on renewable resource management. Suppose that the **u**(t) are harvesting parameters and the manager of an ecosystem is able to specify only upper limits on either the catch or intensity of harvesting for each species. (See Vincent, 1981, for an interpretation along these lines.) Within these limits, it is assumed that the manager has no control. Knowledge of the corresponding reachable set for this system then provides an indication of whether or not any of the species is likely to be driven to extinction under this harvesting program.

In section REACHABLE SETS: A SURVEY of this paper, we briefly discuss a variety of existing techniques for computing and/or estimating reachable sets for nonlinear dynamical systems. Estimates of reachable sets in terms of bounds on the state variables can be obtained as solutions of an optimal control problem described in section AN OPTIMAL CONTROL PROBLEM FOR ESTIMATING $R_T(S)$. For discrete-time systems, this optimal control problem is easily solved as a straightforward nonlinear optimization. Examples of several discrete-time ecosystem models are presented and analyzed in section EXAMPLES.

REACHABLE SETS: A SURVEY

Now consider the vectors $u(t)$ in systems (1) and (2) as control variables which satisfy the inequalities (3). Such controls will be referred to as *admissable* controls.

Definitions: (i) Given a set $S \subset R^n$, we say that a point $y \in R^n$ is *reachable from S* if and only if there exists an admissable control law $u(t)$, a time $T > 0$ and a point $x(0) \in S$ such that the solution to (1) or (2) with initial point $x(0)$ satisfies $x(T) = y$. The set of all points reachable from S is said to be the *reachable set from S* and will be denoted by $R(S)$.

(ii) The *reachable set from S in time* $T \geq 0$, denoted by $R_T(S)$, is the set of all points in R^n which are reachable from S in time T.

The following is a survey of some of the methods which have been suggested for calculating or estimating reachable sets for the nonlinear systems (1) and (2).

Continuous-Time Systems

The reachable sets $R_T(S)$ can, in principle, be obtained by solving a minimum time optimal control problem from initial points on the boundary of S and considering all those points which can be reached in the same minimum time T. These points form a closed curve and will correspond to the boundary of $R_T(S)$. Since necessary conditions are employed in minimum time optimal control, only candidates for the boundary of $R_T(S)$ are obtained and these may have to be confirmed by the use of sufficient conditions. This approach is of limited practical use since it can only be applied in one or two dimensions and even in two dimensions a large number of minimum time trajectories may have to be computed in order to obtain an accurate description of the boundary of $R_T(S)$.

For a restricted class of two-dimensional ecosystem models, which includes some prey-predator models, minimum time trajectories from the unperturbed equilibrium, x^*, asymptotically approach and actually trace out the boundary of $R(\{x^*\})$ as a minimum time tends to infinity. This provides a very simple means of generating the boundary of $R(S)$ for these systems when S itself is reachable from x^* (see Fisher and Goh, 1981).

Necessary conditions for system trajectories to lie in the boundary of the reachable set $R(S)$ have been obtained by Grantham and Vincent (1975). These conditions have been successfully applied to generating the boundaries of reachable sets for two-species models by Vincent and Anderson (1979), Vincent (1981), Fisher and Goh (1981) and Grantham (1981a,b). (An analogous result for discrete-time systems has not been developed.) For higher-dimensional systems, this approach is impractical and one is left with the necessity of looking for ways of obtaining approximations to $R(S)$. Grantham (1980, 1981b) has employed Liapunov stability methods to obtain over-estimates of $R(S)$. The problem essentially reduces to a nonlinear optimization whose numerical solution can often be rather difficult to obtain. The method suffers from the same drawbacks as most Liapunov approaches in that the results are highly dependent upon the Liapunov (or Liapunov-type) function chosen and the estimates obtained tend to be rather conservative.

Discrete-Time Systems

As for continuous-time systems, boundaries of the sets $R_T(S)$ for the discrete-time system (2) can be obtained by solving a minimum time optimal control problem from initial points on the boundary of S and joining together all points which can be reached in the same minimum time T. In practice, this approach can only be applied in one or two dimensions and one again needs to compute large numbers of these trajectories in order to be reasonably confident of obtaining an accurate description of $R_T(S)$.

Trajectories that lie in the boundary of the reachable sets $R_T(S)$ at each corresponding time instant T can be generated using a discrete maximum principle provided S consists only of a single point. This principle, which is analogous to the "abnormal" case of the discrete maximum principle for optimal control of systems of the form of (2), is formulated in Grantham and Fisher (1986) and applied to a two-species population model. In essence, the reachable set boundary at time T can be generated from the reachable set boundary at time T-1 by considering only controls $u(t)$ that provide a local maximum of the Hamiltonian for the system. As is the case for continuous-time systems, the method is effectively limited to one- and two-dimensional systems although, in theory, it is applicable to systems of higher dimension.

As previously mentioned, there are no results available for generating boundaries of the reachable sets $R(S)$ for discrete systems. One can, however, use

Liapunov methods, as in Fisher and Grantham (1985), which again lead to conservative over-estimates of R(S).

AN OPTIMAL CONTROL PROBLEM FOR ESTIMATING $R_T(S)$

Due to the lack of precision in most ecosystem models, one is usually not concerned with exactly determining reachable sets but rather in having available methods of easily estimating the region of state space visited by the system trajectories in a particular period of time. For example, in most practical applications, it will probably suffice if accurate bounds on the species population numbers are readily available. These bounds can be obtained, in principle, as solutions of the following optimal control problem.

$$\text{Maximize } \phi(\mathbf{x}(T))$$

subject to either

$$\text{(i) } \dot{\mathbf{x}}(t) = \mathbf{f}[\mathbf{x}(t),\mathbf{u}(t)], \text{ for } t \in [0,T],$$

$$\text{or (ii) } \mathbf{x}(t+1) = \mathbf{f}[\mathbf{x}(t),\mathbf{u}(t)], \text{ for } t = 0,1,\dots,T-1,$$

and the constraints $\mathbf{x}(0) \in S$ and $|u_j(t)| < 1$, for $j = 1,2,\dots,m$, $t = 0,1,\dots,T-1$, where $\phi(\mathbf{x})$ represents any of $\pm x_i$, for $i = 1,2,\dots,n$. This problem may seem rather formidable to solve but in the case of discrete-time systems, its solution is relatively straightforward. For the rest of the paper, we shall only consider ecosystems modelled by the discrete dynamics of equation (2).

The Hamiltonion for the above optimal control problem is

$$H[\mathbf{x}(t),\mathbf{u}(t),\mathbf{p}(t+1)] = \sum_{i=1}^{n} p_i(t+1)f_i[\mathbf{x}(t),\mathbf{u}(t)], \text{ for } t = 0,1,\dots,T-1 .$$

The $\mathbf{p}(t)$ are costate vectors which satisfy the equations

$$p_i(T) = \partial\phi(T)/\partial x_i(T), \text{ for } i = 1,2,\dots,n , \tag{10}$$

$$p_i(t) = \partial H(t)/\partial x_i(t) = \sum_{k=1}^{n} p_k(t+1)\partial f_k[\mathbf{x}(t),\mathbf{u}(t)]/\partial x_i(t),$$

$$\text{for } i = 1,2,\dots,n, \quad t = 0,1,\dots,T-1 . \tag{11}$$

The optimal control $\mathbf{u}^*(t)$ is then chosen to maximize the Hamiltonian whose derivatives with respect to $\mathbf{u}$ are given by

$$\partial H(t)/\partial u_j(t) = \sum_{k=1}^{n} p_k(t+1)\partial f_k[x(t),u(t)]/\partial u_j(t) ,$$

$$\text{for } j = 1,2,\ldots,m, \quad t = 0,1,\ldots,T-1 . \quad (12)$$

The method for solving an optimal control problem of this type is to turn it into a mathematical programming problem. The independent variables in the mathematical programming problem will be the control variables $u_j(t)$, for $j = 1,2,\ldots,m$ and $t = 0,1,\ldots,T-1$, and the initial values of the state vector $x_i(0)$, for $i = 1,2,\ldots,n$. The objective function is $\phi(x(T))$, which is a nonlinear function of the $(mT + n)$ independent variables and so this is a nonlinear programming problem subject to simple bounds on the variables $u_j(t)$ and constraints on the variables $x_i(0)$ (which may also be simple bounds, depending on the nature of the set S). This problem can easily be solved using any of the standard nonlinear programming routines from, for example, the NAG or IMSL libraries. The major steps in the algorithm for solving this problem are:

(i) Choose initial values of the independent variables $u_j(t)$, for $j = 1,2,\ldots,m$ and $t = 0,1,\ldots,T-1$, and $x_i(0)$, for $i = 1,2,\ldots,n$.

(ii) Use a standard nonlinear optimization routine to maximize $\phi(x(T))$ subject to $|u_j(t)| < 1$ and $x(0) \in S$ where

   (a) The state variables $x(t)$, for $t = 1,2,\ldots,T$ are obtained by solving the system equations (2) forwards in time.

   (b) The costate variables $p(t)$, $t = 0,1,\ldots,T$ are obtained from (10) and (11) by solving backwards in time.

   (c) The derivatives of the objective function with respect to the variables $u_j(t)$ are given by (12).

   (d) The derivatives of the objective function with respect to $x_i(0)$ are given by the matrix product.

$$\left[\partial\phi(x(T))/\partial x(T)\right] \prod_{t=1}^{T} J_x(f[x(T-t),u(T-t)]) , \quad (13)$$

where $J_x(f[x(t),u(t)])$ is the matrix of partial derivatives of f with respect to $x(t)$.

This algorithm was implemented using the NAG nonlinear optimization routine E04KBF on a DEC10 computer. Estimates of reachable sets were successfully obtained for a variety of different discrete-time ecosystem models some of which are discussed in the next section. Computer time (CPU) was of the order of a few seconds for most models.

EXAMPLES

Example 2. Consider the ecosystem model of example 1 with variability introduced as in equations (6), (7) and (8). Reachable sets for these three models were computed using the algorithm of the previous section for two values of the growth rate r; that is, r = 1.5, which corresponds to a stable equilibrium point at x* = 1 and r = 2.2, which corresponds to a stable limit cycle at the two points x = 0.497 and x = 1.503.   In all cases, the initial set is defined by the inequality |x(0) - x*| < 0.1 and the constants b_1 and b_2 are given the values 0.1. For one-dimensional models, the algorithm provides an exact description of the reachable sets $R_T(S)$, which are shown in Table 1 for the case T = 5. For larger values of T, the sets $R_T(S)$ differ only marginally from $R_5(S)$ for these models. As can be seen from the results, the introduction of variability through either the carrying capacity or external perturbations has similar effects. The combined effect of the two can lead to the population being almost extinct after only five generations for the case r = 2.2.

Table 1. Reachable Sets $R_5(S)$ for the Models (7), (8) and (9)
with b_1 = b_2 = 0.1.

	(7)	Model (8)	(9)
r = 1.5	[0.797,1.160]	[0.722,1.209]	[0.548,1.285]
r = 2.2	[0.304,1.555]	[0.259,1.660]	[0.0647,1.711]

Example 3. A discrete-time analogue of a krill-baleen whale model introduced by May et al. (1979) is

$$N_1(t+1) = N_1(t)\exp\{r_1[1-N_1(t)/K]-aN_2(t)\} ,$$

$$N_2(t+1) = N_2(t)\exp\{r_2[1-N_2(t)/(bN_1(t))]\} ,$$

(14)

where $N_1(t)$ and $N_2(t)$ represent the krill (prey) and baleen whale (predator) populations, respectively, at time t. Let x_1 and x_2 be the populations scaled with respect to their equilibrium densities and introduce external perturbations through the addition of terms $b_i u_i(t)x_i(t)$, where $|u_i(t)| < 1$, so that the model equations are

$$x_1(t+1) = x_1(t)\exp\{r_1[1-(x_1(t)+dx_2(t))/(1+d)]\} + b_1u_1(t)x_1(t) ,$$

$$(15)$$

$$x_2(t+1) = x_2(t)\exp\{r_2[1-x_2(t)/x_1(t)]\} + b_2u_2(t)x_2(t) ,$$

where b_1 and b_2 are positive constants and $d = abK/r_1$. The parameter values $r_1 = 1$, $r_2 = 0.1$ and $d = 1$, which are plausible values for krill and baleen whales (see May et al., 1979), were chosen. For these values, the undisturbed equilibrium $x = (1 \ 1)^T$ is locally stable and numerical simulation suggests that it is globally stable. Estimates of the reachable sets $R_T(S)$ in terms of bounds on the individual population densities were obtained for S defined by

$$S = \{x \in R^2 : |x_i - x_i^*| < 0.1, i = 1,2\}$$

and with $b_1 = 0.1$ and $b_2 = 0.05$. Some of these sets are illustrated in Figure 1. The growth rates in the model equations (15) were also allowed to vary within ±10% of the values $r_1 = 1$ and $r_2 = 0.1$ but the results differed only marginally from those depicted in the figure.

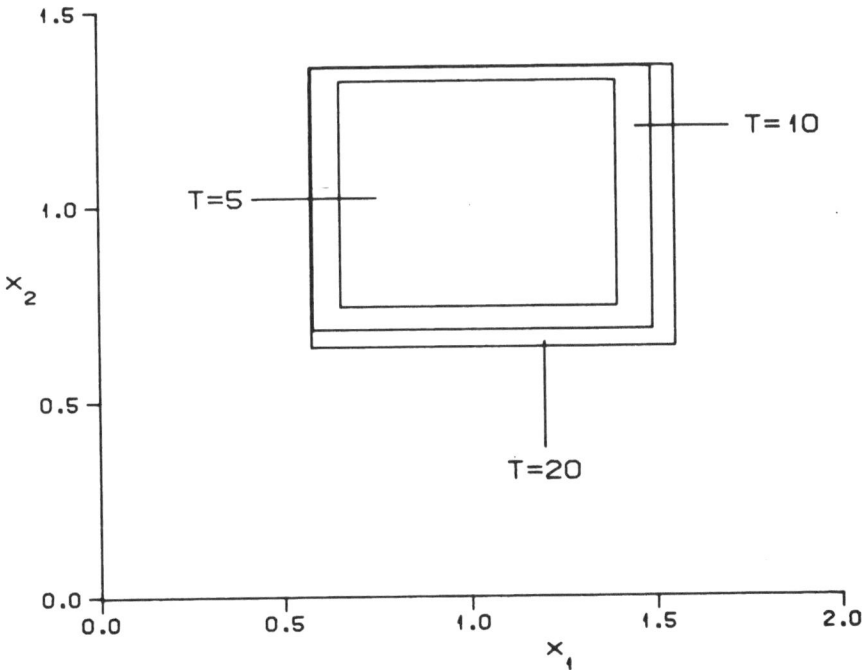

Figure 1.   Bounds on the reachable sets $R_5(S)$, $R_{10}(S)$ and $R_{20}(S)$ for example 3 with $b_1 = 0.1$ and $b_2 = 0.05$.

For values of $T > 20$, the estimates change very little, suggesting that $R_{20}(S)$ is a good approximation to $R(S)$. Figure 2 shows the actual reachable set $R(S)$ generated by using the discrete maximum principle mentioned in REACHABLE SETS: A SURVEY (see Grantham and Fisher, 1986) together with bounds on the individual population densities, as provided by the algorithm, for the same parameter values as before except that $b_2 = 0.01$. Also in Figure 2 is illustrated an estimate of $R(S)$ obtained from a Liapunov method approach using a function of the form

$$V(\mathbf{x}) = \sum_{i=1}^{2} c_i \{[(x_i)^{p_i} - 1]/p_i - s_i \ln(x_i)\} ,$$

(Fisher and Grantham, 1985). The Liapunov approach is computationally more difficult to implement and, as can be seen, it generally provides estimates which are far inferior to those obtained by the variable bound approach depicted in Figure 1.

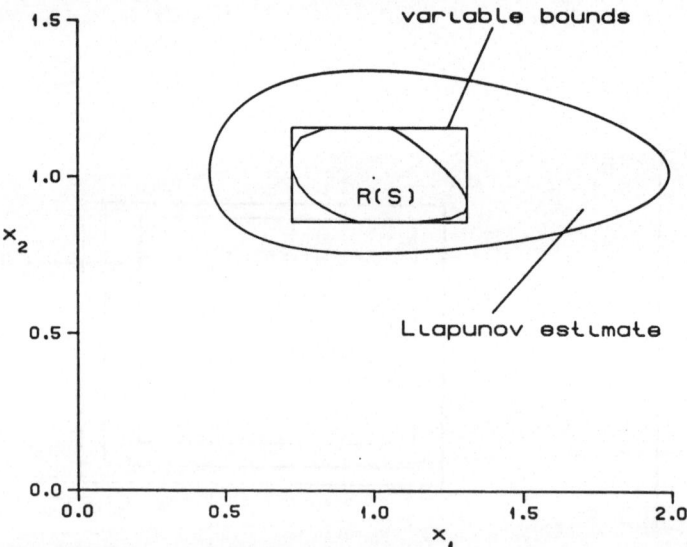

Figure 2.   The reachable set $R(S)$, bounds on the individual populations and a Liapunov estimate for example 3 with $b_1 = 0.1$ and $b_2 = 0.01$.

Example 4.   A prey-predator model with two competing prey species (Comins and Hassell, 1976) is

$$N_1(t+1) = N_1(t)\exp\{r_1 - g_1[N_1(t) + \alpha N_2(t)] - a_1 N_3(t)^{1-m}\}$$

$$N_2(t+1) = N_2(t)\exp\{r_2 - g_2[N_2(t) + \beta N_1(t)] - a_2 N_3(t)^{1-m}\}$$ (16)

$$N_3(t+1) = N_1(t)\{1 - \exp[-a_1 N_3(t)^{1-m}]\} + N_2(t)\{1 - \exp[-a_2 N_3(t)^{1-m}]\} .$$

Consider the special case $r_1 = 1$, $r_2 = 0.5$, $g_1 = 1$, $g_2 = 0.5$, $a_1 = 0.5$, $a_2 = 0.25$, $\alpha = \beta = 0.5$ and $m = 0.5$. (These parameter values are depicted in Figure 5 of Comins and Hassell, 1976.) The corresponding equilibrium populations are given by $N^* = (0.5414, 0.5414, 0.1413)^T$. Let x_1, x_2 and x_3 be the population densities scaled with respect to their equilibrium values. The model equations (16) then become

$$x_1(t+1) = x_1(t)\exp\{1-0.541[x_1(t)+0.5x_2(t)]-0.188[x_3(t)]^{1/2}\} + b_1x_1(t)u_1(t)$$

$$x_2(t+1) = x_2(t)\exp\{0.5-0.271[x_2(t)+0.5x_1(t)]-0.094[x_3(t)]^{1/2}\} + b_2x_2(t)u_2(t) \qquad (17)$$

$$x_3(t+1) = 3.83x_1(t)\{1-\exp\{-0.188[x_3(t)]^{1/2}\}\} + 3.83x_1(t)\{1-\exp\{-0.094[x_3(t)]^{1/2}\}\}$$
$$+ b_3x_3(t)u_3(t)$$

where the terms $b_ix_i(t)u_i(t)$ represent the effect of external perturbations on the model. Now let $b_i = 0.1$, for $i = 1,2,3$, and let the initial set S be such that all populations are within 10% of their equilibrium values. Table 2 shows estimates of the reachable sets $R_5(S)$ and $R_{10}(S)$ in terms of bounds on the scaled population densities x_1, x_2 and x_3.

Table 2. Estimates of the Reachable Sets $R_5(S)$, $R_{10}(S)$ and $R_{20}(S)$ for Example 4.

	x_1	x_2	x_3
T = 5	[0.657,1.352]	[0.614,1.477]	[0.647,1.489]
T = 10	[0.563,1.425]	[0.499,1.605]	[0.597,1.507]
T = 20	[0.515,1.480]	[0.412,1.664]	[0.560,1.508]

REFERENCES

Beddington, J.R. and May, R.M. 1977. Havesting natural populations in a randomly fluctuating environment. Science, Vol. 197, pp. 463-465.

Comins, H.N. and Hassell, M.P. 1976. Predation in multi-prey communities. J. Theor. Biol., Vol. 62, pp. 93-114.

De Mottoni, P. and Schiaffino, A. 1981. Competition systems with periodic coefficients: A geometric approach. J. Math. Biol., Vol. 11, pp. 319-335.

Erlich, P.R. and Birch, L.C. 1967. Balance of nature and population control. <u>Amer. Natur.</u>, Vol. 101, pp. 97-107.

Fisher, M.E. and Goh, B.S. 1981. Nonvulnerability of two species interactions. In Vincent, T.L. and Skowronski, J.M. (eds.), <u>Renewable Resource Management</u>, Springer, Berlin-Heidelberg-New York, pp. 133-150.

Fisher, M.E. and Grantham, W.J. 1985. Estimating the effect of continual disturbances on discrete-time population models. <u>J. Math. Biol.</u>, Vol. 22, pp. 199-207.

Grantham, W.J. 1981a. Estimating controllability boundaries for uncertain systems. In Vincent, T.L. and Skowronski, J.M. (eds.), <u>Renewable Resource Management</u>, Springer, Berlin-Heidelberg-New York, pp. 151-162.

Grantham, W.J. 1981b. Estimating reachable sets. <u>J. Dyn. Syst. Meas. Control</u>, Vol. 103, pp. 420-423.

Grantham, W.J. and Fisher, M.E. 1986. See proceedings for this workshop.

Grantham, W.J. and Vincent, T.L. 1975. A controllability minimum principle. <u>J. Opt. Theory and Appl.</u>, Vol. 17, pp. 93-114.

Harrison, G.W. 1979. Persistent sets via Lyapunov functions. <u>Nonlinear Analysis</u>, Vol. 3, pp. 73-80.

Harrison, G.W. 1980. Persistence of predator-prey systems in an uncertain environment. <u>J. Math. Biol.</u>, Vol. 10, pp. 65-77.

Ludwig, D. 1975. Persistence of dynamical systems under random perturbations. <u>SIAM Review</u>, Vol. 17, pp. 605-640.

May, R.M. 1974. <u>Stability and Complexity in Model Ecosystems</u>, 2nd ed., Princeton University Press, Princeton, New Jersey.

May, R.M. 1976. Models for single populations. In May, R.M. (ed.), <u>Theoretical Ecology</u>, Blackwell Scientific Publications, Oxford, pp. 4-25.

May, R.M., Beddington, J.R., Clark, C.W., Holt, S.J. and Laws, R.M. 1979. Management of multispecies fisheries. <u>Science</u>, Vol. 205, pp. 267-277.

May, R.M., Beddington, J.R., Horwood, J.W. and Shepherd, J.G. 1978. Exploiting natural populations in an uncertain world. <u>Math. Biosc.</u>, Vol. 42, pp. 219-252.

Polansky, P. 1979. Invariant distributions for multi-population models in random environments. <u>Theor. Pop. Biol.</u>, Vol. 16, pp. 25-34.

Poluektov, R.A. (ed.). 1974. <u>Dynamical Theory of Biological Populations</u>. Science Publications, Moscow (in Russian).

Silvert, W.M. and Smith, R.S. 1981. The response of ecosystems to external perturbations. <u>Math Biosc.</u>, Vol. 55, pp. 279-306.

Taljapurkar, S.D. and Semura, J.S. 1979. Stochastic instability and Liapunov stability. <u>J. Math. Biol.</u>, Vol. 8, pp. 133-145.

Turelli, M. 1977. Random environments and stochastic calculus. <u>Theor. Pop. Biol.</u>, Vol. 8, 140-178.

Vincent, T.L. 1981. Vulnerability of a prey-predator model under harvesting. In Vincent, T.L. and Skowronski, J.M. (eds.), <u>Renewable Resource Management</u>, Springer, Berlin-Heidelberg-New York, pp. 112-132.

Vincent, T.L. and Anderson, L.R. 1979. Return time and vulnerability for a food chain model. Theor. Pop. Biol., Vol. 15, pp. 217-231.

Witten, M. 1978. Fitness and survival in logistic models. J. Theor. Biol., Vol. 74, pp. 23-32.

Witten, M. and de la Torre, D. 1984. Biological populations obeying difference equations: The effects of stochastic perturbation. J. Theor. Biol., Vol. 111, pp. 493-507.

PARTICIPANT'S COMMENTS

Management decisions should be based on some model of the systems to be managed. Ecosystem models always contain a great deal of uncertainty and hence special methods must be developed to deal with this uncertainty in order to justify the modeling approach. The concept of reachable sets and its application to ecosystem modeling was introduced by several authors in the previous volume (Springer Verlag, Lecture Notes in Biomathematics, #40, 1981). The most notable disadvantage to a reachable set approach has been associated with its difficulty in providing good estimates of the reachable set in higher dimensional spaces.

In this presentation, the author suggests a particular way of arriving at a very useful approximation. The approximation consists of the bounds on each of the state variables which enclose the actual reachable set. These are arrived at by solving a sequence of optimization problems with the maximum (or minimum) value of one of the states as the performance criterion. As time gets large, the solution to this sequence of problems will result in the proper bound. The author illustrates that, for discrete time systems, the procedure may be reduced to a mathematical programming problem. Thus, a computational procedure for such systems will always be available.

The idea presented here is a good one in that, for higher-dimensional systems, the state bounds obtained are perhaps the most useful way to describe and utilize knowledge of the reachable set. Its advantage over the Liapunov method for approximating the reachable set is that it is never possible to determine how good a Liapunov estimate of the reachable set is without actually determining the reachable set by some other means.

T.L. Vincent

GENERATING REACHABLE SET BOUNDARIES
FOR DISCRETE-TIME SYSTEMS

Walter J. Grantham
Department of Mechanical and Materials Engineering
Washington State University
Pullman, Washington 99164-2920 U.S.A.

and

Michael E. Fisher
Department of Mathematics
University of Western Australia
Nedlands, W.A. 6009 Australia

Computational procedures are presented for determining the reachable set at time k for nonlinear discrete-time control systems, such as a multispecies fishery or other resource management system. In particular, necessary conditions, in the form of a maximum principle, are developed for generating reachable set boundaries at any time k. The maximum principle is analogous to the abnormal case of the discrete-time optimal control maximum principle. However, the results in this paper employ a local maximum principle rather than a global one, and they do not assume convexity of the displacement set. The results are applied to two examples, one of which is a model for a multispecies (krill and Baleen whale) prey-predator fishery.

INTRODUCTION

In the management of a resource system, such as a fishery, many "What if?" questions can be formulated as reachable sets questions. For example, a manager may have no control over the harvesting strategies in a multispecies fishery, beyond the imposition of certain constraints such as bounds on the harvesting efforts. Questions of vulnerability (whether a species might be driven to extinction) in such a system can be addressed by determining the set of population states that are attainable under all possible strategies, that is, by determining what the population levels might be at any given time. Such a "reachable sets" analysis can also be employed when the initial populations themselves are uncertain.

In this paper we present a local maximum principle for the problem of determining reachable set boundaries at time $k \in \Gamma \triangleq \{0,1,2,...\}$ for general nonlinear discrete-time systems of the form

$$x(k+1) - x(k) = f[x(k),u(k)] , \qquad (1)$$

where $x \in E^n$ (Euclidian space of dimension n) is the state, $u \in U \subseteq E^m$ is the control (disturbance, uncertainty, etc.), and U is a specified control constraint set. The function $f(\cdot): E^n \times E^m \rightarrow E^n$ is assumed to be continuously differentiable in x and u, with $|I + \partial f(x,u)/\partial x| \neq 0$, where $|\cdot|$ denotes the determinant and I is the $n \times n$ identity matrix.

The maximum principle presented in this paper is analogous to the "abnormal" case (Vincent and Goh, 1972) of the optimal control maximum principle (Halkin, 1964; Holtsman and Halkin, 1966; Blaquiere and Leitman, 1967; Canon et al., 1970) for discrete-time systems of the form (1). However, our results do not employ any optimal control performance measure. Furthermore, we do not assume convexity of the displacement set $f(x,U) \triangleq \{f(x,u) \mid u \in U\}$. In optimal control theory, an assumption that $f(x,U)$ is convex (Halkin, 1964), "directionally convex" (Holtsman and Halkin, 1966; Canon et al., 1970), or a similar (slightly weaker) assumption (Blaquiere and Leitmann, 1967), is invoked in order to yield a <u>global</u> maximum principle analogous to Pontryagin's maximum principle for continuous-time systems. As we will see, for the problem of generating reachable set boundaries at time $k \in \Gamma$ with $f(x,U)$ not convex, we must consider all controls that provide <u>local</u> maxima of the Hamiltonian. For discrete-time optimal control problems with $f(x,U)$ not convex, the corresponding optimal control local maximum principle is developed in (Canon et al., 1970).

We present results on the geometry of reachable sets, and then develop necessary conditions which govern trajectories that are in the reachable set boundary at each time instant. Two examples are presented to illustrate the process of generating reachable set boundaries. The first example has a nonconvex displacement set. The second example involves a prey-predator multispecies fishery model.

REACHABLE SETS

A control function $u(\cdot): \Gamma \to E^m$ is <u>admissible</u>, denoted by $u(\cdot) \in \Omega$, if $u(k) \in U$ for all $k \in \Gamma$. For $u(\cdot) \in \Omega$, let $x(k) = \xi[k; x_0, u(\cdot)]$, $k \in \Gamma$, denote the solution to (1), satisfying the initial conditions $\xi[0; x_0, u(\cdot)] \triangleq x_0$.

Given an arbitrary set of initial states θ, we define the <u>reachable set from θ at time $k \in \Gamma$</u> by

$$R(k,\theta) \triangleq \{\xi[k; x_0, u(\cdot)] \mid x_0 \in \theta, u(\cdot) \in \Omega\} . \tag{2}$$

We will be concerned with the problem of determining the boundary of $R(k,\theta)$ for any $k \in \Gamma$.

For a set $S \subseteq E^n$ we will use the notations $\bar{S}$, ∂S, Int(S), and Ext(S) to denote the closure, boundary, interior, and exterior of S, respectively. We also introduce the following definition of a tangent vector to a set: a vector $\eta \in E^n$ is said to be <u>tangent to S</u> at $x \in \bar{S}$ if and only if there exists a number $\beta > 0$ and a continuous function $\gamma(\cdot): (0,\beta) \to E^n$ such that

    i) $x + \alpha\gamma(\alpha) \in S$ for all $\alpha \in (0,\beta)$

and

    ii) $\gamma(\alpha) \to \eta$ as $\alpha \to 0$.

A "brute force" approach to generating $\partial R(k,\theta)$ would be to first generate $R(1,\theta)$, by using all possible controls $u \in U$ starting from each state $x_0 \in \theta$. Then $R(2,\theta)$ could be generated by using all possible controls from all states in $R(1,\theta)$. This process could be repeated to generate $R(k,\theta)$, using the relation $R(k,\theta) = R[1, R(k-1,\theta)]$, where $R(0,\theta) = \theta$. The disadvantage of this approach is that it requires much more computational effort than is necessary.

A simpler approach results from noting that, for the purpose of generating $\partial R(k,\theta)$, we only need to consider trajectories for which $x(k-1) \in \partial R(k-1,\theta)$. That is, we can ignore any trajectory having $x(k-1) \in Int[R(k-1,\theta)]$. To verify this conclusion, let $x(\cdot) = \xi[\cdot; x_0, u(\cdot)]$ be any solution to (1), generated by any admissible control $u(\cdot) \in \Omega$. Then we have:

Lemma 1: If $x(k) \in Int[R(k,\theta)]$, then $x(K) \in Int[R(K,\theta)]$ for all $K > k$, $K \in \Gamma$.

Proof: By induction, suppose $x(k+1) \notin Int[R(k+1,\theta)]$. Then $x(k+1) \in \partial R(k+1,\theta)$, from the definition of $R(k+1,\theta)$. Let $\gamma(\cdot)$ generate a vector $\eta(k+1)$ tangent to $Ext[R(k+1,\theta)]$ at $x(k+1)$. That is, $x(k+1) + \alpha\gamma(\alpha) \in Ext[R(k+1,\theta)]$ for all sufficiently small $\alpha > 0$, and $\gamma(\alpha) \to \eta(k+1)$ as $\alpha \to 0$. Define

$$\eta(k) \triangleq [I + \partial f[x(k),u(k)]/\partial x]^{-1} \eta(k+1)$$

and, near $x(k) \in Int[R(k,\theta)]$, let $\bar{x}(k)$ be defined implicitly by

$$\bar{x}(k+1) = \bar{x}(k) + f[\bar{x}(k),u(k)] ,$$

where $\bar{x}(k+1) \triangleq x(k+1) + \alpha\gamma(\alpha)$. Then from Taylor's theorem

$$\bar{x}(k) = x(k) + \alpha\bar{\gamma}(\alpha) ,$$

where $\bar{\gamma}(\alpha) \to \eta(k)$ as $\alpha \to 0$. Then $x(k) \in Int[R(k,\theta)]$ implies $\bar{x}(k) \in Int[R(k,\theta)]$ for all sufficiently small $\alpha > 0$. But $\bar{x}(k+1) \in Ext[R(k+1,\theta)]$, which contradicts the definition of $R(k+1,\theta)$. ■

As an immediate consequence of Lemma 1, we have

Corollary: If $x(0) \in \theta$ and $x(K) \in \partial R(K,\theta)$, then $x(k) \in \partial R(k,\theta)$ for all $k \in \{0,1,...,K\}$.

A "simplified brute force" approach to generating $\partial R(k,\theta)$, $k \in \Gamma$, would be to use all possible controls from each state $x_0 \in \partial\theta = \partial R(0,\theta)$, yielding some set $S \subseteq R(1,\theta)$, with $\partial R(1,\theta) \subseteq \bar{S}$. Then $\partial R(1,\theta)$ could be determined simply by discarding all points $x(1)$ in $\bar{S} \cap Int[R(1,\theta)]$. Having determined $\partial R(1,\theta)$, the process could be repeated by using all possible controls from each state $x(1)$ in $\partial R(1,\theta)$ and then discarding those resulting states $x(2)$ in $Int[R(2,\theta)]$ to determine $\partial R(2,\theta)$, and so on.

This approach still requires more computational effort than is necessary for determining ∂R(k,θ).

A MAXIMUM PRINCIPLE

The "simplified brute force" approach can be further improved by exploring some additional geometric properties of R(k,θ) and ∂R(k,θ). In particular, we will develop a necessary condition, in the form of a first-order local maximum principle, in order for an admissible control function $u^*(\cdot) \in \Omega$ to generate a solution $x^*(\cdot)$: $\Gamma \to E^n$ to (1) having the property that $x^*(k) \in \partial R(k,\theta)$ for all $k \in \{0,1,...,K\}$, $K \in \Gamma$.

At $x \in \partial R(k,\theta)$, $k \in \Gamma$, let $M(x)$ denote the tangent cone to $R(k,\theta)$, i.e., the set of vectors tangent to $R(k,\theta)$ at x. A nonzero vector $\lambda \in E^n$ is said to be an _outward normal_ to $R(k,\theta)$ at $x \in \partial R(k,\theta)$ if and only if $\lambda^T \eta < 0$ for all tangent vectors $\eta \in M(x)$, where $(\ )^T$ denotes transpose. A state $x \in \partial R(k,\theta)$ is a _regular_ point of $\partial R(k,\theta)$ if and only if an outward normal $\lambda \in E^n$ to $R(k,\theta)$ at x exists such that $M(x) = \{\eta \in E^n \mid \lambda^T \eta < 0\}$.

In connection with a solution $x^*(\cdot)$: $\Gamma \to E^n$ to (1) generated by a control $u^*(\cdot) \in \Omega$, let $\eta(\cdot)$: $\Gamma \to E^n$ denote a corresponding solution to the state perturbation equations

$$\eta(k+1) = \{I + \partial f[x^*(k),u^*(k)]/\partial x\}\eta(k) \tag{3}$$

and let $\lambda(\cdot)$: $\Gamma \to E^n$ denote a corresponding solution to the adjoint equations

$$\lambda^T(k+1) = \lambda^T(k)\{I + \partial f[x^*(k),u^*(k)]/\partial x\}^{-1} . \tag{4}$$

Note that for all $k,K \in \Gamma$

$$\lambda^T(K)\eta(K) = \lambda^T(k)\eta(k) . \tag{5}$$

In terms of the adjoint vectors satisfying (4), we have

Theorem 1: Let $u^*(\cdot)$: $\Gamma \to U$ generate a solution $x^*(\cdot)$: $\Gamma \to E^n$ to (1) such that $x^*(k) \in \partial R(k,\theta)$ and $x^*(k+1) \in \partial R(k+1,\theta)$ for some $k \in \Gamma$. If $x^*(k+1) \in \partial R(k+1,\theta)$ is a regular point of $\partial R(k,\theta)$, then there exist nonzero vectors $\lambda(k) \in E^n$ and $\lambda(k+1) \in E^n$, satisfying (4), such that the following conditions hold:

  i) $\lambda(k)$ and $\lambda(k+1)$ are outward normals to $R(k,\theta)$ and $R(k+1,\theta)$, respectively;

  ii) $u^*(k)$ satisfies the first-order necessary conditions for a local maximum of $H[x^*(k),u,\lambda(k+1)]$, subject to the constraints $u \in U$, where

$$H(x,u,\lambda) \triangleq \lambda^T f(x,u) . \tag{6}$$

That is,

$$0 > \frac{\partial H[x^*(k),u^*(k),\lambda(k+1)]}{\partial u} \Delta u \tag{7}$$

for all Δu tangent to U at $u^*(k) \in U$.

Proof: Consider Condition i). Since $x^*(k+1) \in \partial R(k+1,\theta)$ is regular, there exists an outward normal $\lambda(k+1) \in E^n$ to $R(k+1,\theta)$ at $x^*(k+1)$, and the tangent cone to $R(k+1,\theta)$ at $x^*(k+1)$ is given by $M[x^*(k+1)] = \{\eta \in E^n \mid \lambda^T(k+1)\eta < 0\}$. Now, suppose that $\lambda(k)$, defined by (4), is not an outward normal to $R(k,\theta)$ at $x^*(k) \in \partial R(k,\theta)$. Then there exists a vector $\eta(k) \in M[x^*(k)]$, tangent to $R(k,\theta)$ at $x^*(k)$, such that $\lambda^T(k)\eta(k) > 0$. Let $\gamma(\cdot)$ generate the vector $\eta(k)$, that is, $x^*(k) + \alpha\gamma(\alpha) \in R(k,\theta)$ for sufficiently small $\alpha > 0$, and $\gamma(\alpha) \to \eta(k)$ as $\alpha \to 0$. Then the neighboring solution $x(\cdot): \Gamma \to E^n$ to (1), generated by $u^*(\cdot)$ with $x(k) = x^*(k) + \alpha\gamma(\alpha)$, satisfies $x(k+1) = x^*(k+1) + \alpha\eta(k+1) + 0(\alpha)$, where $0(\alpha)/\alpha \to 0$ as $\alpha \to 0$ and $\eta(k+1)$ is given by (3). Since $x(k) \in R(k,\theta)$ implies $x(k+1) \in R(k+1,\theta)$, it follows that $\eta(k+1) \in M[x^*(k+1)]$, from the definition of a tangent vector. Thus $\lambda^T(k+1)\eta(k+1) < 0$. But $\lambda^T(k+1)\eta(k+1) = \lambda^T(k)\eta(k) > 0$, from (5), which is a contradiction.

To establish Condition ii), consider any vector $\Delta u \in E^m$ tangent to U at $u^*(k) \in U$, and let $\delta u(\cdot)$ generate Δu, that is, $u^*(k) + \alpha\delta u(\alpha) \in U$ for all sufficiently small $\alpha > 0$ and $\delta u(\alpha) \to \Delta u$ as $\alpha \to 0$. Then

$$x(k+1) \triangleq x^*(k) + f[x^*(k),u^*(k) + \alpha\delta u(\alpha)]$$

$$= x^*(k+1) + f[x^*(k),u^*(k) + \alpha\delta u(\alpha)] - f[x^*(k),u^*(k)]$$

$$= x^*(k+1) + \alpha\{\partial f[x^*(k),u^*(k)]/\partial u\} \delta u(\alpha) + 0(\alpha) ,$$

where $0(\alpha)/\alpha \to 0$ as $\alpha \to 0$. Since $x(k+1) \in R(k+1,\theta)$ and $x^*(k+1)$ $R(k+1,\theta)$, the function $\gamma(\alpha) \triangleq \{\partial f[x^*(k),u^*(k)]/\partial u\}\delta u(\alpha) + 0(\alpha)/\alpha$ generates a vector tangent to $R(k+1,\theta)$ at $x^*(k+1)$. Thus, since $x^*(k+1)$ is a regular point of $\partial R(k+1,\theta)$ and $\lambda(k+1)$ is an outward normal to $\partial R(k+1,\theta)$, we have,

$$0 > \lambda^T(k+1) \frac{\partial f[x^*(k),u^*(k)]}{\partial u} \Delta u .$$

which is equivalent to (7). ∎

Note that Condition ii) of Theorem 1 is not as straightforward as Pontryagin's maximum principle for optimal control in continuous-time systems. The current control $u^*(k)$ depends on the subsequent adjoint vector $\lambda(k+1)$, which depends, in turn, on the current control $u^*(k)$. In general then, we have a one-step algebraic two-point boundary value problem, and some iteration may be required. In some discrete optimal control problems this difficulty can be eliminated, by propagating the state and adjoint vectors backward in time. For a reachable sets analysis only the forward propagation process can be used, since subsequent reachable sets are unknown. In any case, we note that Condition ii) of Theorem 1 is <u>not</u> a requirement that $u^*(k)$ satisfy the first-order necessary conditions for a local maximum of the function $\bar{H}[x^*(k),u,\lambda(k)] = \lambda^T(k)\{I + \partial f[x^*(k),u]/\partial x\}^{-1}f[x^*(k),u]$.

EXAMPLES

We present two examples to illustrate the use of Theorem 1 for generating reachable set boundaries. The first example involves a nonconvex displacement set and the second example is a krill and Baleen whale fishery model.

Example 1: Nonconvex Displacement Set

Consider the system

$$x_1(k+1) - x_1(k) = \cos u \qquad (8)$$

$$x_2(k+1) - x_2(k) = \sin u , \qquad (9)$$

with the set of initial states $\theta = \{(0,0)\}$ and with the control constraint set $U = \{u \in E^1 \mid |u| < \pi/4\}$. Thus at any time step the state moves a unit distance, in any direction within $\pm 45°$ of the positive x_1-axis direction.

The $H(\cdot)$ function is

$$H(x,u,\lambda) = \lambda_1\cos u + \lambda_2\sin u \qquad (10)$$

and the adjoint equations are

$$\lambda_1(k+1) = \lambda_1(k) \qquad (11)$$

$$\lambda_2(k+1) = \lambda_2(k) . \qquad (12)$$

Condition ii) of Theorem 1 yields the local maximizing control(s)

$$u^*(k) = \begin{cases} \tan^{-1}\left(\dfrac{\lambda_2(k+1)}{\lambda_1(k+1)}\right) & \text{if } \lambda_1(k+1) > 0 \text{ and } |\lambda_2(k+1)/\lambda_1(k+1)| < 1 \\[2ex] \pi/4 & \text{if } \lambda_2(k+1) > |\lambda_1(k+1)| \\[2ex] -\pi/4 & \text{if } \lambda_2(k+1) < -|\lambda_1(k+1)| \\[2ex] \pm\pi/4 & \text{if } |\lambda_2(k+1)/\lambda_1(k+1)| < 1 \text{ and } \lambda_1(k+1) < 0 . \end{cases} \qquad (13)$$

Reachable set boundaries for k = 1,2, and 3 are shown in Figure 1. These boundaries were generated using Theorem 1 in the following fashion. The outward normal $\lambda(0) \in E^2$ at $(x_1,x_2) = (0,0)$ was varied through all possible unit vectors. For each $\lambda(0)$, the state and adjoint vectors were propagated forward to time k via (8)-(9) and (11)-(12), respectively, using a local maximizing control from (13) at each intervening time. This process produced a trajectory with $x^*(k) \in \partial R(k,\theta)$ at each time k. The state was then returned to $x^*(k-1) \in \partial R(k-1,\theta)$ and propagated forward again to time k using any remaining local maximizing controls $u^*(k-1)$ from (13). The backward-forward process was then repeated from k-2 to k, etc., until all local maximizing controls at times from 0 to k-1 had been employed.

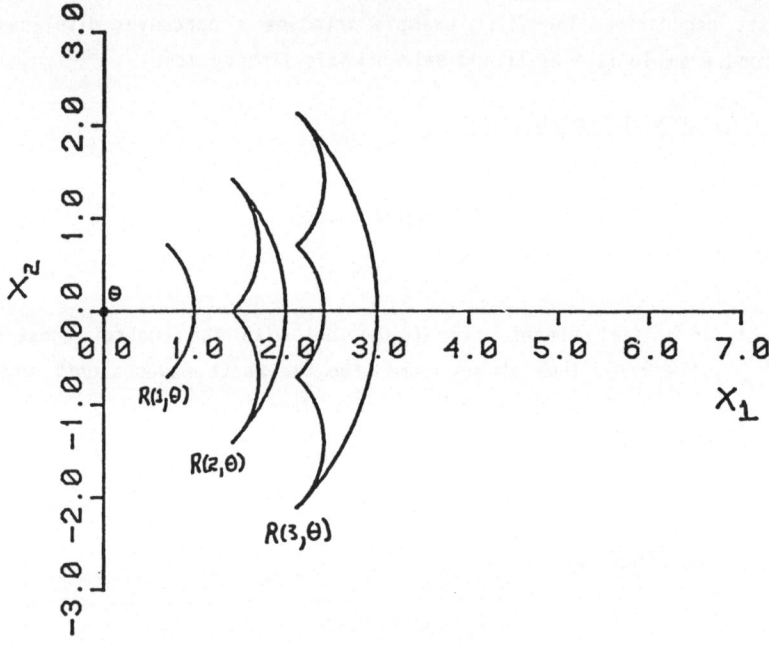

Figure 1. Reachable set boundaries for Example 1.

Note that in this example f(x,U), (e.g., the arc segment $R(1,\theta) = \partial R(1,\theta)$ in Figure 1) is not convex. If we were to employ only global maxima in Condition ii) of Theorem 1, as in the optimal control discrete maximum principle, we would not obtain all of $\partial R(k,\theta)$ for k > 3.

Also note that the endpoints of $R(1,\theta)$ and the cusp and corner points in $\partial R(2,\theta)$ and $\partial R(3,\theta)$ are not regular points; they have many outward normals. But Theorem 1 still applies. In particular, Theorem 1 does not assume that all points in $\partial R(k,\theta)$ are regular points.

Example 2: Multispecies Fishery

Consider a prey-predator fishery management system, with harvesting of both the prey and predator species. Suppose that the managers for each species only have control over the bounds on the harvesting efforts, but not on the actual harvesting effort strategies that the fishermen will employ. Reachable sets analysis can be used to determine the possible population levels at any time, and whether or not one of the species might become endangered.

For illustration purposes consider a krill and Baleen whale fishery (May et al., 1979), modeled by the differential equations

$$\frac{\hat{N}_1}{N_1} = r_1 \left[1 - \frac{N_1}{K} \right] - aN_2 - E_1 \tag{14}$$

$$\frac{\hat{N}_2}{N_2} = r_2 \left[1 - \frac{N_2}{\alpha N_1} \right] - E_2 , \tag{15}$$

where N_1 and N_2 are the krill and whale populations, respectively, K is the krill carrying capacity, r_1 and r_2 are the intrinsic growth rates, α and a are interaction parameters, and E_1 and E_2 are the harvesting efforts. The population levels can be considered either in terms of numbers or biomass, by using appropriate units for the parameters in the model.

For constant harvesting efforts E_1 and E_2, the equilibrium populations are given by

$$\hat{N}_1 = \frac{\frac{K}{r_1} (r_1 - E_1)}{1 + \beta} \tag{16}$$

and

$$\hat{N}_2 = \frac{\frac{\beta}{\alpha} (r_1 - E_1)}{1 + \beta} , \tag{17}$$

where

$$\beta = \frac{K\alpha a}{r_1} \left[1 - \frac{E_2}{r_2} \right] . \tag{18}$$

A nondimensionalized discrete-time representation can be obtained by letting $k = 0,1,\ldots$ denote time, integrating (14) and (15) over one time interval with the right-hand sides held constant, and then adding harvesting terms corresponding to changes from the nominal harvesting efforts. The resulting model is given by

$$x_1(k+1) - x_1(k) = x_1(k) \left\{ \exp\left[\rho_1 \left[1 - \frac{x_1(k) + \beta x_2(k)}{1 + \beta} \right] \right] + u_1 - 1 \right\} \tag{19}$$

$$x_2(k+1) - x_2(k) = x_2(k) \left\{ \exp\left[\rho_2 \left[1 - \frac{x_2(k)}{x_1(k)} \right] \right] + u_2 - 1 \right\}, \tag{20}$$

where $x_1 = N_1/\hat{N}_1$ is the nondimensional krill population, $x_2 = N_2/\hat{N}_2$ is the nondimensional whale population, $\rho_1 = r_1 - E_1$, $\rho_2 = r_2 - E_2$, and u_1 and u_2 are deviations from the nominal krill and whale harvesting efforts, respectively. For $u_1 = u_2 = 0$, the system has a unique equilibrium at $x_1 = x_2 = 1$.

We take the set of initial states as $\theta = \{(1,1)\}$, with $\rho_1 = 1$, $\rho_2 = 0.1$, and $\beta = 1$. The control constraint set is $U = \{(u_1,u_2) \mid |u_i| < w_i\}$ and we will consider two cases: a) $w_1 = 0.1$, $w_2 = 0.01$, and b) $w_1 = 0.1$, $w_2 = 0.05$.

To simplify the notation for specifying controls that yield trajectories in the reachable set boundaries, let $x = x(k)$, $u = u(k)$, $\lambda = \lambda(k)$, $\gamma = \lambda(k+1)$, and $\Delta x = x(k+1) - x(k)$. Then (19) and (20) may be written as

$$\Delta x_1 = g_1(x) + (u_1 - 1)x_1 \triangleq f_1(x,u) \tag{21}$$

$$\Delta x_2 = g_2(x) + (u_2 - 1)x_2 \triangleq f_2(x,u), \tag{22}$$

where

$$g_1(x) = x_1 \exp\left\{ \rho_1 \left[1 - \frac{x_1 + \beta x_2}{1 + \beta} \right] \right\} \tag{23}$$

$$g_2(x) = x_2 \exp\left\{ \rho_2 \left[1 - \frac{x_2(k)}{x_1(k)} \right] \right\}. \tag{24}$$

Now, suppose that we have determined the reachable set boundary $\partial R(k,\theta)$ at time k, with the initial boundary $\partial R(0,\theta)$ being just the point $(1,1)$. We wish to determine the boundary at time $k+1$. We do so by choosing a point $x = x(k)$ in $\partial R(k,\theta)$, with outward normal $\lambda = \lambda(k)$, and then determining a control vector $u = u(k)$ to yield a local maximum for

$$H(x,u,\gamma) = \gamma^T f(x,u)$$

$$= \gamma_1 [g_1(x) - x_1 + u_1 x_1] + \gamma_2 [g_2(x) - x_2 + u_2 x_2], \tag{25}$$

where $\gamma = \lambda(k+1)$ satisfies the adjoint equations

$$\gamma^T = \lambda^T [I + \partial f(x,u)/\partial x]^{-1}, \tag{26}$$

which may be written as

$$\gamma_1 = \frac{\lambda_1 u_2 + \lambda_1 \frac{\partial g_2}{\partial x_2} - \lambda_2 \frac{\partial g_2}{\partial x_1}}{\left[u_1 + \frac{\partial g_1}{\partial x_1}\right]\left[u_2 + \frac{\partial g_2}{\partial x_2}\right] - \frac{\partial g_1}{\partial x_2} \frac{\partial g_2}{\partial x_1}} \tag{27}$$

$$\gamma_2 = \frac{\lambda_2 u_1 - \lambda_1 \frac{\partial g_1}{\partial x_2} + \lambda_2 \frac{\partial g_1}{\partial x_1}}{\left[u_1 + \frac{\partial g_1}{\partial x_1}\right]\left[u_2 + \frac{\partial g_2}{\partial x_2}\right] - \frac{\partial g_1}{\partial x_2} \frac{\partial g_2}{\partial x_1}} \tag{28}$$

The resulting "boundary control" is used to propagate the state and adjoint vectors from time k to time $k+1$, yielding a point $x(k+1) \in \partial R(k+1,\theta)$ and an outward normal $\gamma = \lambda(k+1)$ to $R(k+1,\theta)$. The process can be continued from $k+1$ to $k+2$, etc., and then the whole process can be repeated for the next point in $\partial R(k,\theta)$. However, for the purpose of plotting a sequence of reachable set boundaries it is simpler to generate $\partial R(k+1,\theta)$ completely, by propagating from each state $x(k) \in \partial R(k,\theta)$, and then generate $\partial R(k+2,\theta)$ from $\partial R(k+1,\theta)$, and so on.

The boundary controls for propagating from a point $x = x(k) \in \partial R(k,\theta)$ to a point $x(k+1) \in \partial R(k+1,\theta)$ are given by

$$u_1 = w_1 \text{sgn}(\gamma_1) \tag{29}$$

$$u_2 = w_2 \text{sgn}(\gamma_2) , \tag{30}$$

except for the singular control cases, where $\gamma_1 = 0$ or $\gamma_2 = 0$.

To examine the singular control cases we first note that the controls cannot both be singular, since $\gamma \neq (0,0)$. For this two-dimensional system, the control u_1 (u_2) will be singular at a point $x = x(k) \in \partial R(k,\theta)$ when the outward normal $\gamma = \lambda(k+1)$ at the next boundary point $x(k+1) \in \partial R(k+1,\theta)$ is vertical (horizontal). The singular control for u_1 is arbitrary, i.e.,

$$u_1 \in [-w_1, w_1] \tag{31}$$

and it occurs when $\lambda = \lambda(k)$ is such that

$$0 = \lambda_1 u_2 + \lambda_1 \frac{\partial g_2}{\partial x_2} - \lambda_2 \frac{\partial g_2}{\partial x_1} \tag{32}$$

with $u_2 = \pm w_2$ determined from (30). Similarly, the singular control for u_2 is arbitrary, i.e.,

$$u_2 \in [-w_2, w_2] \tag{33}$$

and it occurs when λ is such that

$$0 = \lambda_2 u_1 - \lambda_1 \frac{\partial g_1}{\partial x_2} + \lambda_2 \frac{\partial g_1}{\partial x_1} \tag{34}$$

with $u_1 = \pm w_1$ determined from (29).

Figure 2 shows reachable set boundaries at various times for Case a), where the control bounds are $w_1 = 0.1$, $w_2 = 0.01$. Each horizontal [vertical] segment of $\partial R(k,\theta)$ was generated by a singular control $u_1(k-1)$ [$u_2(k-1)$], varied over all of its possible values. For this system the displacement set, e.g., the rectangular region $R(1,\theta)$, is convex, so the first-order local maximum condition in Theorem 1 becomes a global maximum condition.

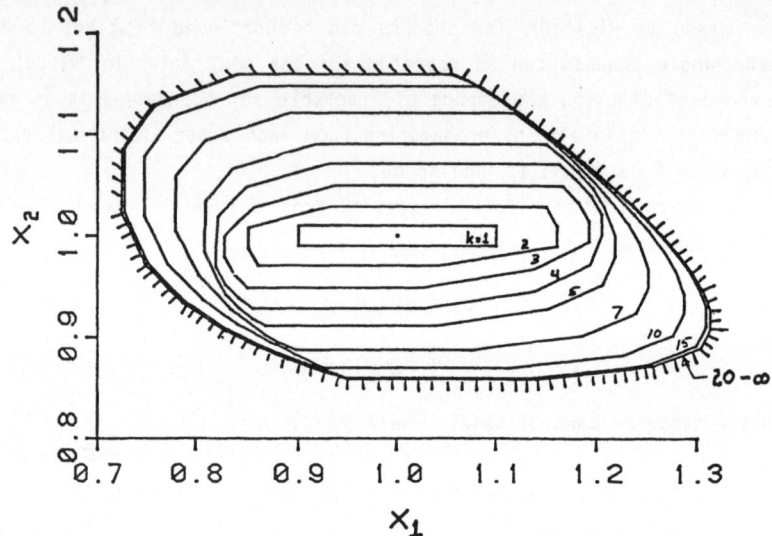

Figure 2. Reachable set boundaries for Example 2a ($w_1 = 0.1$, $w_2 = 0.01$).

Note that for this example system the overall reachable set

$$R(\theta) \triangleq \bigcup_{k \in \Gamma} R(k,\theta)$$

is bounded. In Figure 2 the reachable set boundaries $\partial R(k,\theta)$ asymptotically approach $\partial R(\theta)$ as $k \to \infty$. For continuous-time two-dimensional systems, one way to generate $\partial R(\theta)$ -- discussed in Vincent (1981) for the continuous-time version of this fisheries model -- is to integrate the equations of motion using the analogous boundary control and allow the trajectory to approach $\partial R(\theta)$ asymptotically. This approach does not work for the discrete-time system. The resulting trajectory, shown in Figure 3, consists of discrete points (which evidently repeat themselves). Thus the trajectory does not yield the complete boundary of $R(\theta)$.

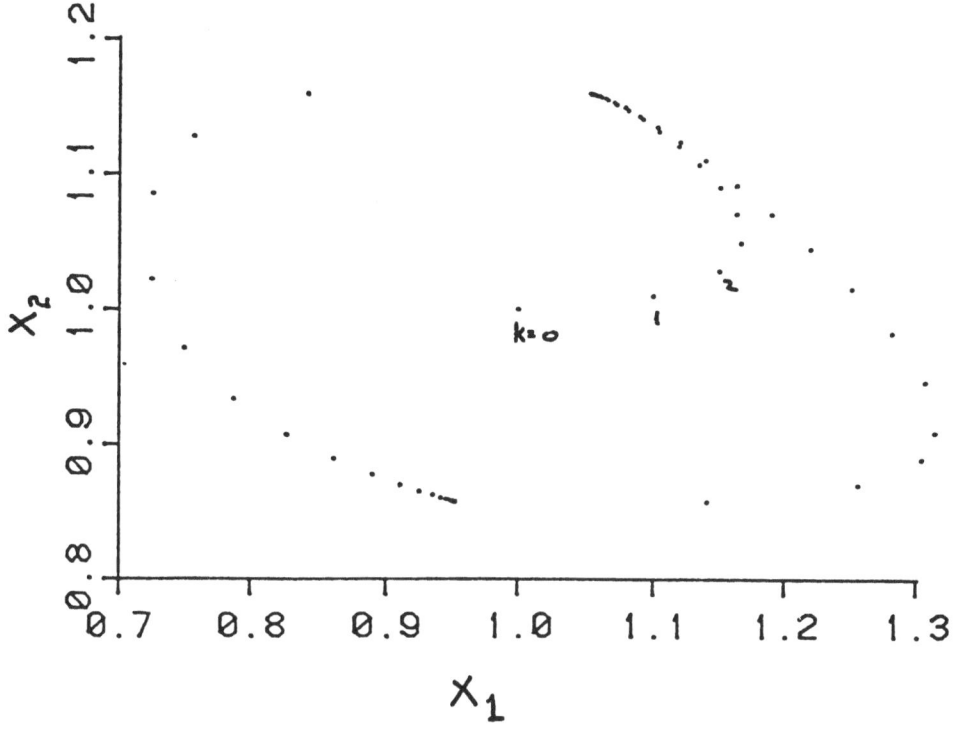

Figure 3. Asymptotic boundary control trajectory for Example 2a.

Figure 4 shows reachable set boundaries for various times for Case b), with control bounds $w_1 = 0.1$ and $w_2 = 0.05$. Again, the overall reachable set $R(\theta)$ is bounded and its (nonconvex) boundary is approached asymptotically by $\partial R(k,\theta)$ as the time k approaches infinity. Figure 5 shows a boundary control trajectory which asymptotically approaches $\partial R(\theta)$. Not only is this trajectory composed of discrete points, so that it does not yield all of $\partial R(\theta)$, but in this case the trajectory approaches a forced equilibrium point in $\partial R(\theta)$ and does not yield any subsequent points in $\partial R(\theta)$ beyond the equilibrium point.

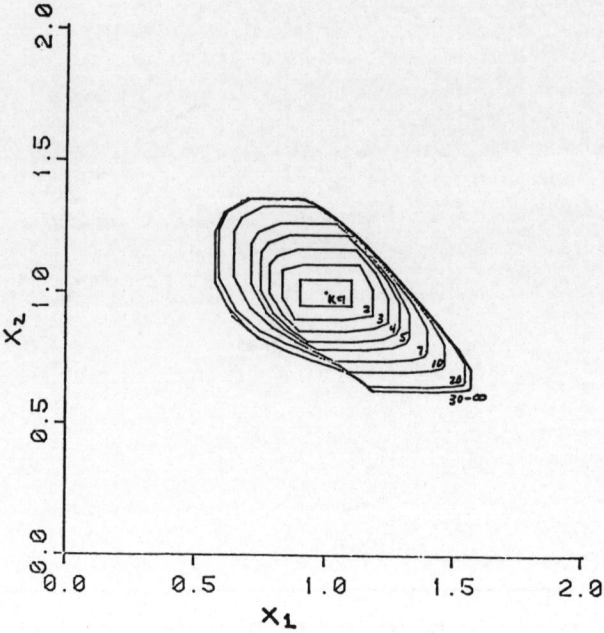

Figure 4. Reachable set boundaries for Example 2b   ($w_1 = 0.1$, $w_2 = 0.05$).

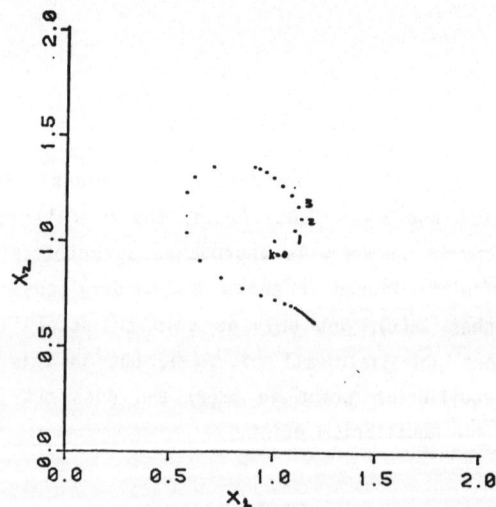

Figure 5. Asymptotic boundary control trajectory for Example 2b.

SUMMARY

For nonlinear discrete-time systems we have presented a first-order local maximum principle for generating reachable set boundaries at any time k. The necessary conditions in Theorem 1 are analogous to the "abnormal" case of the discrete-maximum principle for optimal control. However we have not employed any optimal control performance measure and we have not assumed that the displacement set is convex.

Two examples were presented to illustrate the procedure. The first example dealt with a nonconvex displacement set. For this example the use of a global maximum condition, as in optimal control, would not have generated all of the reachable set boundaries. Instead, one must examine all controls that satisfy a local maximum condition. The second example was a two-species prey-predator fishery model. For this second example, the overall reachable set $R(\theta)$ was bounded. Its boundary was found as the limit of $\partial R(k,\theta)$ as $k \to \infty$, but could not be generated by the analogous continuous-time procedure of simply letting a boundary control trajectory approach $\partial R(\theta)$ asymptotically, since the corresponding discrete-time trajectory consists of discrete points, while $\partial R(\theta)$ is a continuous curve.

REFERENCES

Blaquiere, A. and Leitmann, G. 1967. "Further Geometric Aspects of Optimal Processes: Multistage Dynamic Systems," in Mathematical Theory of Control, A.V. Balakrishnan and L.W. Neustadt, eds., Academic Press, NY, pp. 143-155.

Canon, M.D., Cullum, C.D., and Polak, E. 1970. Theory of Optimal Control and Mathematical Programming, McGraw-Hill, NY, pp. 75-93.

Halkin, H. 1964. "Optimal Control for Systems Described by Difference Equations,", in Advances in Control Systems, Vol. 1, C.T. Leondes, ed., Academic Press, NY, pp. 173-196.

Holtsman, J.M. and Halkin, H. 1966. "Directional Convexity and the Maximum Principle for Discrete Systems," S.I.A.M. J. on Control, Vol. 4, No. 2, pp. 263-275.

May, R.M., Beddington, J.R., Clark, C.W., Holt, S.J., and Laws, R.M. 1979. "Management of Multispecies Fisheries," Science, Vol. 205, pp. 267-277.

Vincent, T.L. 1981. "Vulnerability of a Prey-Predator Model Under Harvesting," in Renewable Resource Management, T.L. Vincent and J.M. Skowronski, eds., Springer-Verlag, NY, pp. 112-132.

Vincent, T.L. and Goh, B.S. 1972. "Terminality, Normality, and Transversality Conditions," J. of Optimization Theory and Applications, Vol. 9, No. 1, January, pp. 32-50.

PARTICIPANT'S COMMENTS

The discrete time approach of vulnerability analysis (discussed as reachable sets in this manuscript) is certainly useful, and should find many applications. This is so since many biological processes such as birth, death, breeding season, and life cycles are essentially event oriented, and therefore may be described as discrete time systems. One immediate application of this approach would be to generate the reachable set for example 2, as presented in the manuscript and determine at which levels of harvest (u_1, u_2) either of the species densities gets too close to zero. I wonder how such an analysis would compare to the continuous case.

It should be emphasized that the general approach of vulnerability analysis, both for continuous and discrete models is especially attractive for the following reasons:

1. It offers a way of dealing with uncertainty with deterministic methods. As such, it simplifies analysis (when compared to the stochastic approach).

2. Unlike most stochastic analyses, the present approach does not make any assumptions about the distribution of the noise; i.e., the analysis admits not only noise, but also modeling errors.

An extension of the present approach would be to derive guaranteed control under uncertainty, similar to the continuous case (Vincent, 1985).

Yosef Cohen

In this paper, the authors have sought to give a further piece of mathematical machinery, an algorithm via a local maximum principle, for the problem of determining reachable set boundaries of nonlinear discrete time systems, that may find application in the management of a resource system.

Possible application has been indicated in the introduction, for example, in the questions of vulnerability of a species under harvesting. The procedure is not entirely straightforward, yet gives information on controls used and is much quicker than generating reachable sets using all possible controls.

How great its worth in application remains to be seen. It, however, is in the spirit of the workshop, seeking to bring to the community's attention a further aspect of control theory that may find significant application.

R.J. Stonier

UNCERTAIN DYNAMICAL SYSTEMS:
AN APPLICATION TO RIVER POLLUTION CONTROL*

C.S. Lee
Department of Mathematics
University of Malaya
Kuala Lumpur 22-11, Malaysia

and

G. Leitmann
Department of Mechanical Engineering
University of California
Berkeley, California 94720, USA

There is an extensive literature dealing with the modelling and
management of water quality systems. However, only a few treatments
include the effects of uncertain disturbances, and those are stochastic
in nature. In the present treatment, we employ a deterministic approach
in dealing with uncertain disturbances. We consider uncertain but
bounded disturbances and allow for two possibilities, the bounds are
known and the bounds are not known.

We introduce controls in the form of effluent rate and in-stream
aeration to manage a simple water quality system whose state consists
of biochemical oxygen demand and dissolved oxygen, and which is subject
to uncertain disturbances. These controls, which are either of the
feedback (memoryless) or of the adaptive (learning) type, require
knowledge of the system's current state; thus, we allow also for
measurement error in the state. The theory is illustrated by numerous
simulations.

INTRODUCTION

As industry expands and population increases, rivers and streams are not
only being used more frequently, but are also being polluted more seriously as both
industrial waste and sewage are often discharged into them. There has been a
growing concern over the increasing pollution level in our rivers and streams. This
concern is reflected not only in the setting up of many government regulatory
agencies and bodies to upgrade and protect the water quality systems in the
environment, but also in a significant number of papers on modelling and management
of water quality systems (Refs. 1-28).

In real situations, a water quality system can be very complex. As has been
pointed out by Hassan and Younis (Ref. 6), there are disturbances to any system at
both the input and output levels. To overcome these drawbacks, Hassan et al.
(Refs. 8,9) have included the effects of input disturbances and output measurement

---

*Research supported by NSF and AFOSR.

errors in their stochastic river pollution models. The objective of this paper is to demonstrate that one can also include the effects of uncertain disturbances at the input and output levels in a deterministic model. To demonstrate this, we will use the Streeter and Phelps model (Ref. 16) for a one reach water quality system having biochemical oxygen demand (B.O.D.) and dissolved oxygen (D.O.) as its state variables. We consider two types of control variables, namely, the effluent discharge rate and the in-stream aeration rate. The uncertain disturbances acting on the system are assumed bounded; they fall into two classes, namely those with known bounds and those with unknown bounds. For the former class of uncertain disturbances, we use a non-adaptive feedback (memoryless) stabilizing control scheme, while for the latter class of uncertainties, we use an adaptive (learning) control scheme to return the system to and then maintain it at some prespecified desired level.

PROBLEM FORMULATION

As has been mentioned earlier, B.O.D. and D.O. are two important indices of water quality in a river. In the process of decomposing organic wastes, micro-organisms consume oxygen dissolved in the water. Thus, the larger the quantity of these wastes, the larger is the population of micro-organisms, and hence the greater is the demand for dissolved oxygen. If the level of D.O. in the stream is below a certain critical value, fish and other aquatic animals as well as many green plants will die.

Let $z(t)$ and $q(t)$ denote the concentration per unit volume of B.O.D. and D.O., respectively, in a reach of a river at time t. Following Singh (Ref. 1), we define a reach as a stretch of a river of some convenient length which receives one major controlled effluent discharge from a sewage or an industrial waste treatment facility. If we assume that the flow rate is constant in the reach and that the water is well mixed, then distance from the top of the reach may be expressed in terms of time, and the parameters may be assumed uniform throughout the reach.

A simple model that describes the down-stream effect of an up-stream pollution load is the well known Streeter and Phelps (Ref. 16) model

$$\dot{z}(t) = -k_1 z(t)$$

$$\dot{q}(t) = -k_3 z(t) - k_2 q(t) + k_2 q^s$$

(1)

where k_1, k_2 and k_3 are positive constants; their magnitudes denote, respectively, the B.O.D. decay rate (day^{-1}), the D.O. reaeration rate (d^{-1}) and the B.O.D. deoxygenation rate (d^{-1}). q^s denotes the saturation concentration of D.O.

Consider a controlled effluent discharge at the top of the reach and let $\dot{m}(t)$ denote its waste discharge rate (concentration of B.O.D. in mg/ℓ per day). Let

$$W(t) = \alpha \, \dot{m}(t)$$

where α is the ratio of the flow rate of effluent to that of the stream. Thus, with this controlled input included, system (1) becomes

$$\dot{z}(t) = -k_1 z(t) + W(t)$$

$$\dot{q}(t) = -k_3 z(t) - k_2 q(t) + k_2 q^s \quad .$$

(2)

Suppose that q^*, the desired steady state value of D.O.[‡], has been specified. Then the corresponding steady state values of B.O.D. and of the effluent discharge rate can be obtained readily:

$$z^* = \frac{k_2 (q^s - q^*)}{k_3}$$

$$W^* = \frac{k_1 k_2 (q^s - q^*)}{k_3} \quad .$$

(3)

That is, if the effluent discharge rate of the waste is set at W^*, and if there are no disturbances acting on system (2), then (z,q) will converge to the steady state (z^*,q^*). In real situations, both the B.O.D. and D.O. rates are subjected to uncertain disturbances along the reach. Let $v_1(t)$ denote the uncertainties that affect the B.O.D. rate of change as, for example, those due to local runoff and scour. Let $v_2(t)$ denote the uncertainties that affect the D.O. rate of change as, for example, those due to photosynthesis and respiration of plants and algae as well as decomposition of mud deposits. Thus, in order to drive the water quality system to and maintain it at the desired steady state in the presence of these uncertain disturbances, we introduce two controls, namely, $u_1(t)$, the additional effluent waste discharge rate and $u_2(t)$, the in-stream aeration rate. System (2) becomes

$$\dot{z}(t) = -k_1 z(t) + W^* + u_1(t) + v_1(t)$$

$$\dot{q}(t) = -k_3 z(t) - k_2 q(t) + k_2 q^s + u_2(t) + v_2(t) \quad .$$

(4)

By the use of the transformation

$$x_1(t) = z(t) - z^*$$

$$x_2(t) = q(t) - q^*$$

[‡]The choice of this steady state value of D.O. depends on the needs of various parties of the community.

(4) is simplified to

$$\dot{x}_1(t) = -k_1 \, x_1(t) + u_1(t) + v_1(t)$$

$$\dot{x}_2(t) = -k_3 \, x_1(t) - k_2 \, x_2(t) + u_2(t) + v_2(t) \ . \tag{5}$$

At this point, we will divide our investigation of the control water quality system into two parts, namely, the one in which bounds of the uncertain disturbances are known and the other one in which bounds of the uncertain disturbances are not known.

NON-ADAPTIVE FEEDBACK (MEMORYLESS) CONTROL

In this section, we state two assumptions and quote a result of Corless and Leitmann (Ref. 29) concerning a class of non-adaptive feedback (memoryless) stabilizing controls which will be utilized for our controlled water quality system. Consider any uncertain system described by

$$\dot{x}(t) = F(t, \, x(t), \, u(t)) \tag{6}$$

$$F \in F \tag{7}$$

where

$$x(t) \in X \subseteq R^n \ ,$$

$$u(t) \in U \subseteq R^m \ ,$$

$$t \in T \subseteq R \ ,$$

and F is some known, non-empty, class of functions which map $T \times X \times U$ into R^n. F satisfies the following assumptions.

Assumption A1. $(F(t,x,\cdot)$ is affine) For each $F \in F$, there exist functions

$f : T \times X \to R^n$ and $B : T \times X \to R^{n \times m}$ such that $\forall \ (t,x,u) \in T \times X \times U$

$$F(t,x,u) = f(t,x) + B(t,x)u \ . \tag{8}$$

Note that if A1 is satisfied, then each $F \in F$ has a unique representation in the form of (8). The unique pair (f,B) is given by

$f(t,x) = F(t,x,0)$

$B(t,x)u = F(t,x,u) - F(t,x,0) \qquad \forall \ u \in U \ .$

Assumption A2. There exist Caratheodory functions[†]

$$f^0 : T \times X \to R^n \quad \text{and} \quad B^0 : T \times X \to R^{n \times m} ,$$

a candidate Lyapunov function $V : T \times X \to R_+$,

a strongly Caratheodory function[‡] $\rho^0 : T \times X \to R_+$,

and a constant $c \in R_+$ such that

(i) V is a Lyapunov function for $\dot{x}(t) = f^0(t,x(t))$.

(ii)   For each unique pair (f,B) of F, there exist Caratheodory functions $e : T \times X \to R^m$ and $E : T \times X \to R^{m \times m}$ such that
$$f = f^0 + B^0 e$$
$$B = B^0 + B^0 E ,$$
and
$$\|e(t,x)\| < \rho^0 (t,x)$$
$$\|E(t,x)\| < c < 1 \quad \forall (t,x) \in T \times X$$

As stated in Ref. 30 and proved in Ref. 31, the following controls assure practical stability[#] for the uncertain system (6)-(7) satisfying Assumptions A1 and A2: Given $\varepsilon > 0$, the control

$$u = p_\varepsilon(t,x) = \begin{cases} -\dfrac{\mu(t,x)}{\|\mu(t,x)\|} \rho(t,x) & \text{if } \|\mu(t,x)\| > \varepsilon \\[3mm] -\dfrac{\mu(t,x)}{\varepsilon} \rho(t,x) & \text{if } \|\mu(t,x)\| < \varepsilon , \end{cases} \qquad (9)$$

where $\mu(t,x) = \rho(t,x) B^{0^T}(t,x) \dfrac{\partial V(t,x)}{\partial x}$

and $\rho(t,x) = \rho^0(t,x) /(1-c) \quad \forall (t,x) \in T \times X$,

practically stabilizes (6)-(7) about zero. For the water quality system (5), we have

---

[†]See Appendix, or just note that if a function is continuous, it is Caratheodory.

[‡]See Appendix, or just note that if a function is continuous, it is strongly Caratheodory.

[#]See Appendix for the definition of practical stability. Loosely speaking, by letting $\varepsilon \to 0$ practical stability can be made to approach asymptotic stability of the zero state.

$$f(t,x) = \begin{bmatrix} -k_1 x_1 + v_1(t) \\ -k_3 x_1 - k_2 x_2 + v_2(t) \end{bmatrix}$$

$$f^0(t,x) = \begin{bmatrix} -k_1 x_1 \\ -k_3 x_1 - k_2 x_2 \end{bmatrix} , \qquad B(t,x) = B^0(t,x) = \begin{bmatrix} 1 & 0 \\ 0 & 1 \end{bmatrix} ,$$

$$e(t,x) = \begin{bmatrix} v_1(t) \\ v_2(t) \end{bmatrix} ,$$

$$\rho(x,t) = \rho^0(x,t \; > \; \sqrt{v_1(t)^2 + v_2(t)^2} \; ,$$

and $V(t,x) = \frac{1}{2} x^T P x$ with

$$P = \begin{bmatrix} \dfrac{2 k_3^2}{k_1(k_1+k_2)} & \dfrac{-k_3}{k_1+k_2} \\ \\ \dfrac{-k_3}{k_1+k_2} & 1 \end{bmatrix} .$$

In the subsequent example, we take $k_1 = 0.32$, $k_2 = 0.20$, $k_3 = 0.32$, and $q^s = 10$ mg/1.[†] Furthermore, we select $q^* = 6.0$ mg/1 so that $z^* = 2.5$ mg/1 and $W^* = 0.8$ mg/1/day. We consider two cases: Control based on current state $x(t)$, and control based on observed (measured) state $\hat{x}(t) = x(t) + w(t)$ where $w(t)$ denotes measurement error. For purposes of simulation, we take $w_1(t) = \frac{1}{4}(\frac{1}{2} - r_1)$ and $w_2(t) = \frac{1}{8}(\frac{1}{2} - r_2)$ as measurement errors in B.O.D. and D.O., respectively, where r_1 and r_2 are random variables that range in $[0,1]$. We consider three disturbances falling into the class of disturbances with known bound $\rho(x,t)$; here we let $\rho(x,t) = 0.56$.

(i) $v_1(t) = 0.5$, $v_2(t) = 0.25$,

(ii) $v_1(t) = 0.5 \sin 2 \pi t$, $v_2(t) = 0.25 \sin 2 \pi t$,

(iii) $v_1(t) = 0.5 - r_3$, $v_2(t) = \frac{1}{2} (0.5 - r_4)$, where r_3 and r_4 are random variables that range in $[0,1]$.

[†] These are the values used by Singh in Refs. 1 and 2.

In each case, we take the same initial conditions, namely $x_1(0) = 2.5$ $(z(0) = 5.0)$ and $x_2(0) = -1.0$ $(q(0) = 5.0)$. Simulations were performed for the following four different system-control combinations.

1) The nominal (undisturbed, uncontrolled) system with steady state effluent discharge rate W^* (that is, (2) with $W(t) = W^*$).

2) The disturbed system with zero control (that is, the uncontrolled system subjected to disturbances (i)-(iii), respectively.)

The third and fourth combinations concern the system subjected to disturbances (i)-(iii), respectively, and with non-adaptive feedback (memoryless) stabilizing control given in (9) with $\epsilon = 0.005$, utilizing either 3) actual state $x(t)$ or 4) measured state $\hat{x}(t)$.

In Figure 1, the B.O.D. and D.O. responses are plotted for the nominal system with effluent discharge rate W^* (that is, system-control combination 1). Figures 2, 3 and 4 contain the B.O.D. and D.O. responses, the control histories and the observed (measured) B.O.D. and D.O. responses for the three disturbed system-control combinations (that is 2, 3 and 4).[†] In each of Figures 2, 3 and 4, the B.O.D. and D.O. responses are plotted in (a) for the disturbed system with zero control (that is, the system-control combination 2), while (b), (c) and (d) display, respectively, the B.O.D. and D.O. responses and the u_1 and u_2 control histories based on actual state (that is, the system-control combination 3). The plots in (e)-(i) display, respectively, the B.O.D. and D.O. responses, the observed B.O.D. and D.O. responses and the u_1 and u_2 control histories, for control based on measured state (that is, the system-control combination 4).

ADAPTIVE CONTROL

In this section, we consider a class of systems in which the disturbance bounds are not known. Some assumptions are stated and a class of adaptive controllers is proposed and utilized. These controllers are based on the theory developed in Refs. 31-34. Consider the class of systems described by

$$\dot{x}(t) = f(t,x(t)) + B(t,x(t)) \, g(t,x(t),u(t)) \qquad (10)$$

where $t \in R$, $x(t) \in R^n$ is the state and $u(t) \in R^m$ is the control; the functions $f : R \times R^n \to R^n$, $B : R \times R^n \to R^{n \times m}$ and $g : R \times R^n \times R^m \to R^m$ are uncertain. These functions are not known but are only assumed to satisfy the following conditions.

[†]Here we present only Figures 1 and 2a-d; the remaining simulation results may be found in Ref. 38.

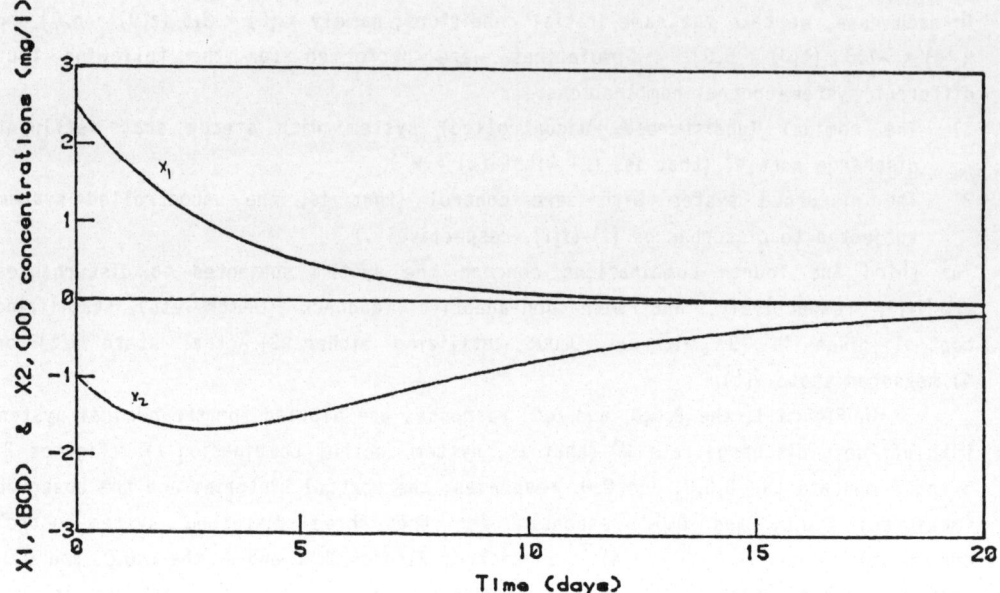

Figure 1. B.O.D. and D.O. responses for the nominal system.

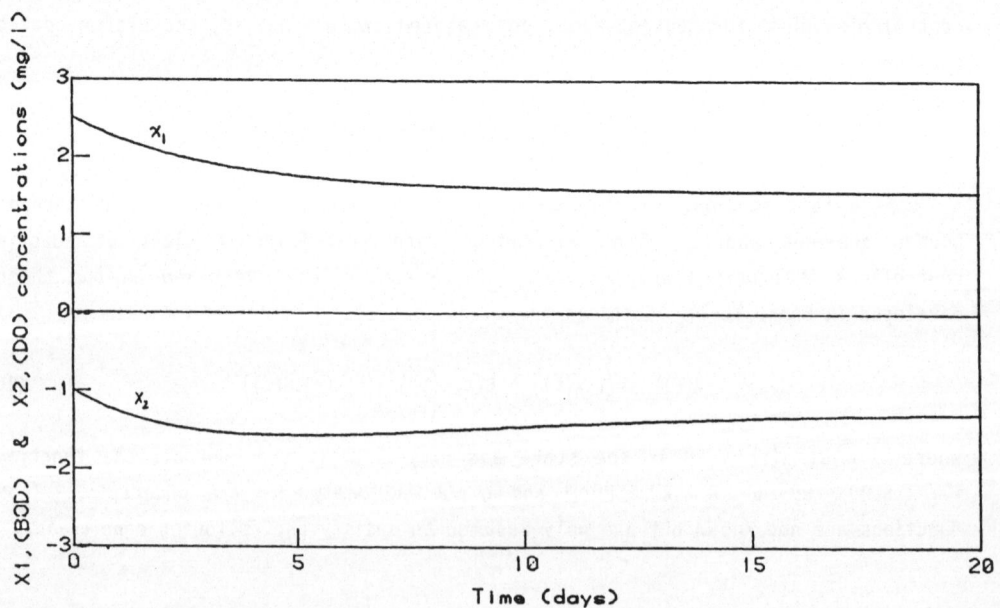

Figure 2(a).   B.O.D. and D.O. responses for the uncontrolled system subject to disturbances (i).

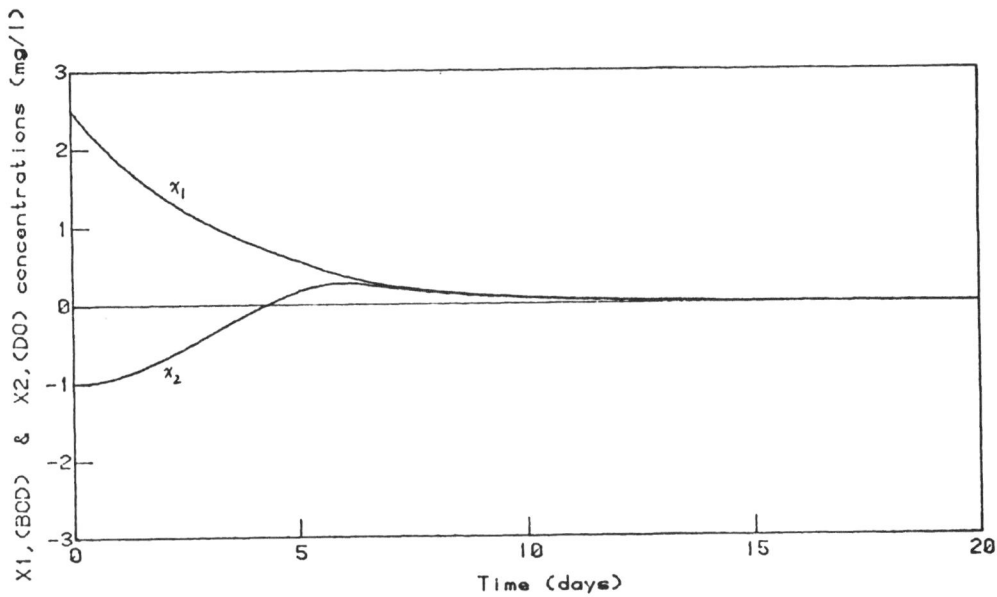

Figure 2(b).   B.O.D. and D.O. responses for the feedback controlled system subject
                to disturbances (i).

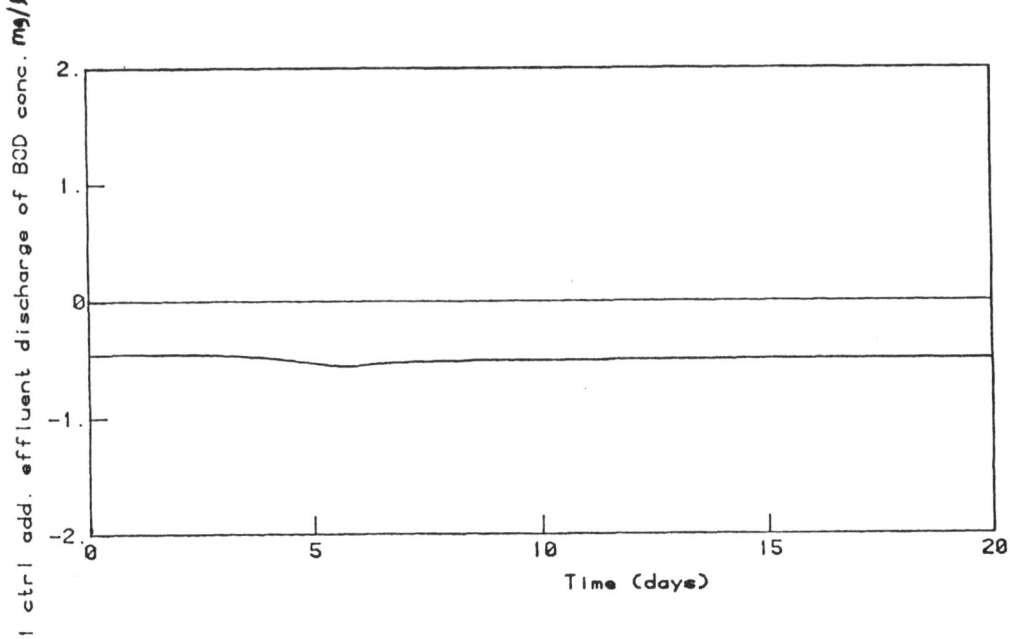

Figure 2(c).   Memoryless effluent control history for the system subject to
                disturbances (i).

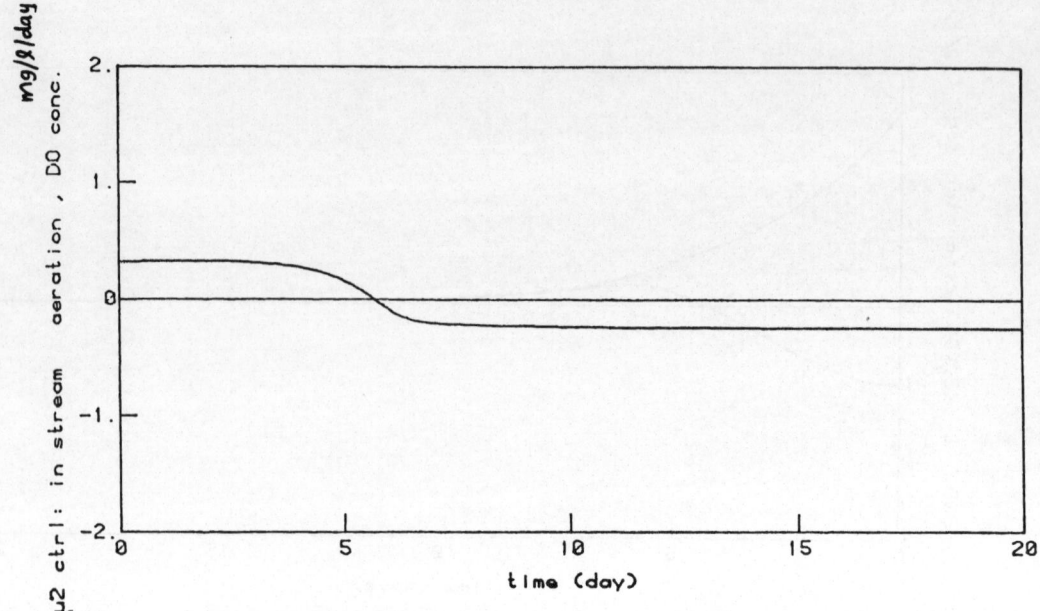

Figure 2(d). Memoryless in-stream aeration control history for the system subject to disturbances (i).

Assumption B1. (i)   f is Caratheodory and $f(t,0) = 0 \ \forall \ t \in R$.

(ii)[†]   There exist a C^1 function $V : R \times R^n \to R_+$ and functions $\gamma_1, \gamma_2, \gamma_3 : R_+ \to R_+$, where γ_1, γ_2 belong to class KR and γ_3 belongs to class K, such that
$\forall (t,x) \in R \times R^n$

$$\gamma_1(\|x\|) \leq V(t,x) \leq \gamma_2(\|x\|) \ ,$$

$$\frac{\partial V(t,x)}{\partial t} + \frac{\partial V(t,x)}{\partial x} f(t,x) \leq -\gamma_3(\|x\|) \ .$$

Assumption B2. (i)   There exist an uncertain function $\rho : R \times R^n \to R_+$ and an unknown constant $\beta_0 > 0$ such that
$\forall (t,x,u) \in R \times R^n \times R^m$

$$u^T g(t,x,u) \geq \beta_0 \|u\| [\|u\| - \rho(t,x)] \ .$$

(ii)[‡]   There exist an unknown constant $\beta \in (0,\infty)^k$ and a known function $\pi : R \times R^n \times (0,\infty)^k \to R_+$ such that
$\forall (t,x) \in R \times R^n$

$$\rho(t,x) = \pi(t,x,\beta) \ .$$

[†]See Appendix for the definitions of class K and class KR.

[‡]That is, ρ depends in a known manner on an unknown constant.

(iii) For each $(t,x) \in R \times R^n$ the function $\pi(t,x,\cdot) : (0,\infty)^k \to R_+$ is C^1, concave, and non-decreasing with respect to each component of its argument β.

Assumption B3.   The function B is Caratheodory and g, π and $\frac{\partial \pi}{\partial \beta}$ are strongly Caratheodory.

Assumption B4.   At each $t \in R$, $\alpha(t,x(t))$ and $x(t)$ are known, where for all $(t,x) \in R \times R^n$

$$\alpha(t,x) = B^T(t,x) \frac{\partial V(t,x)}{\partial x} .$$

For the system (10) satisfying Assumptions B1-B4, it is shown in Ref. 31 that the following controllers assure boundedness and convergence to zero of all system responses (that is, $\lim\limits_{t \to \infty} x(t) = 0$) :

$$u(t) = p\big(t,x(t), \hat{\beta}(t), \varepsilon(t)\big) ,$$

$$p(t,x,\hat{\beta},\varepsilon) = -\pi(t,x,\hat{\beta}) \, s(t,x,\hat{\beta},\varepsilon) ,$$

$$\dot{\hat{\beta}}(t) = L \, \frac{\partial \pi^T\big(t,x(t),\hat{\beta}(t)\big)}{\partial \beta} \, \|\alpha(t,x(t))\| , \tag{11}$$

$$\dot{\varepsilon}(t) = -\ell \, \varepsilon(t) , \quad \ell > 0 ,$$

$$\hat{\beta}(t_0) \in (0,\infty)^k , \quad \varepsilon(t_0) \in (0,\infty) ,$$

where $L \in R^{k \times k}$ is a diagonal matrix with positive elements and $s : R \times R^n \times (0,\infty)^{k+1} \to R^m$ is any strongly Caratheodory function which satisfies

$$s(t,x,\hat{\beta},\varepsilon) \, \|\alpha(t,x)\| = \|s(t,x,\hat{\beta},\varepsilon)\| \, \alpha(t,x)$$

and

$$\|\mu(t,x,\hat{\beta})\| > \varepsilon \Rightarrow s(t,x,\hat{\beta},\varepsilon) = \frac{\alpha(t,x)}{\|\alpha(t,x)\|}$$

where

$$\mu(t,x,\hat{\beta}) = \pi(t,x,\hat{\beta}) \, \alpha(t,x) \quad \forall (t,x,\hat{\beta},\varepsilon) \in R \times R^n \times (0,\infty)^{k+1} .$$

For example, we may select a particular function s to be given by

$$s(t,x,\hat{\beta},\varepsilon) = \begin{cases} \dfrac{\mu(t,x,\hat{\beta})}{\|\mu(t,x,\hat{\beta})\|} & \text{if} \quad \|\mu(t,x,\hat{\beta})\| > \varepsilon , \\[3mm] \dfrac{\mu(t,x,\hat{\beta})}{\varepsilon} & \text{if} \quad \|\mu(t,x,\hat{\beta})\| < \varepsilon . \end{cases} \tag{12}$$

For the water quality system (2.5), we have

$$B(t,x) = \begin{bmatrix} 1 & 0 \\ 0 & 1 \end{bmatrix}, \qquad g(t,x,u) = \begin{bmatrix} u_1 \\ u_2 \end{bmatrix} + \begin{bmatrix} v_1(t) \\ v_2(t) \end{bmatrix}.$$

Thus, if we consider $|v| < \beta$, then

$$u^T g = u^T u + u^T v > u^T u - |u| \, |v| > u^T u - |u| \beta.$$

Hence we may take $\beta_0 = 1$ and $p(t,x) = \beta$; it follows that $\pi(t,x,\hat{\beta}) = \hat{\beta}$. Furthermore, we take $V(t,x) = x^T P \, x$ with

$$P = \begin{bmatrix} \dfrac{1}{2k_1} + \dfrac{k_3^2}{2k_1 k_2(k_1+k_2)} & \dfrac{-k_3}{2k_2(k_1+k_2)} \\[3ex] \dfrac{-k_3}{2k_2(k_1+k_2)} & \dfrac{1}{2k_2} \end{bmatrix},$$

so that

$$\alpha(t,x) = B^T(t,x) \, \frac{\partial V(t,x)}{\partial x} = \begin{bmatrix} \dfrac{x_1}{k_1}\left[1 + \dfrac{k_3^2}{k_2(k_1+k_2)}\right] - \dfrac{k_3 x_2}{k_2(k_1+k_2)} \\[3ex] \dfrac{x_2}{k_2} - \dfrac{k_3 x_1}{k_2(k_1+k_2)} \end{bmatrix}.$$

Numerical Simulation Results

For this simulation, we use the same set of values for the k's and the same functions for the three disturbances as in the previous section. The simulation results for the nominal model with steady state effluent discharge rate W^* and for the disturbed system with zero control obtained previously apply here as well. Therefore, simulations were performed only for the disturbed system with adaptive control given by (11)-(12) with $L = 0.05$, $\ell = 0.10$, $\hat{\beta}(0) = 0.01$ and $\epsilon(0) = 1.0$. Here, Figures 5, 6 and 7 contain the B.O.D. and D.O. responses, the u_1 and u_2 control histories for the disturbed system subject to the disturbances (i), (ii) and (iii), respectively. The B.O.D. and D.O. responses are plotted in (a), while (b) and (c) display, respectively, the adaptive control histories u_1 and u_2. Again, to save space, we present only Figures 5; the remaining ones may be found in Ref. 38.

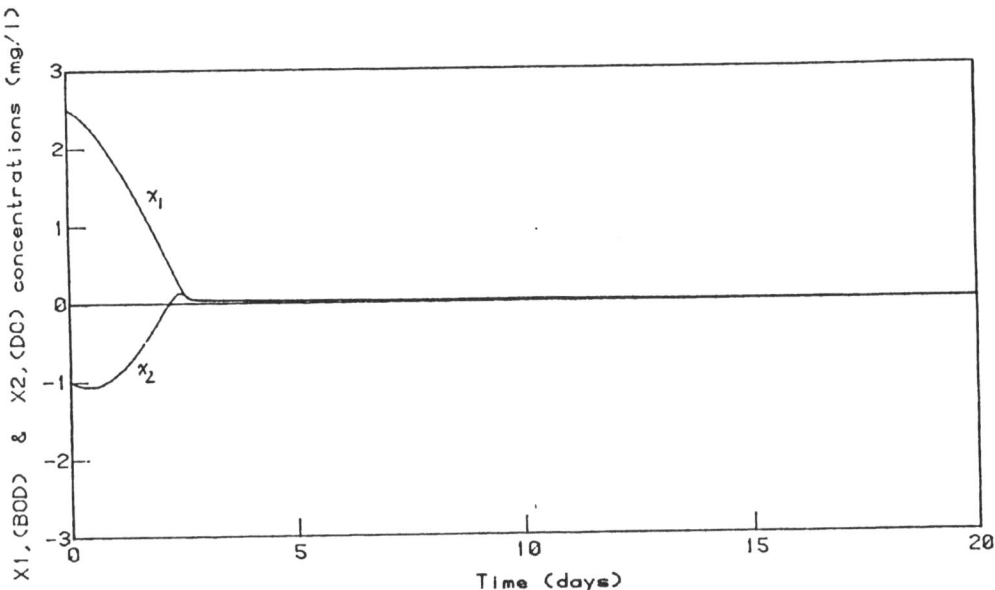

Figure 5(a).   B.O.D. and D.O. responses of the adaptively controlled system subject to disturbances (i).

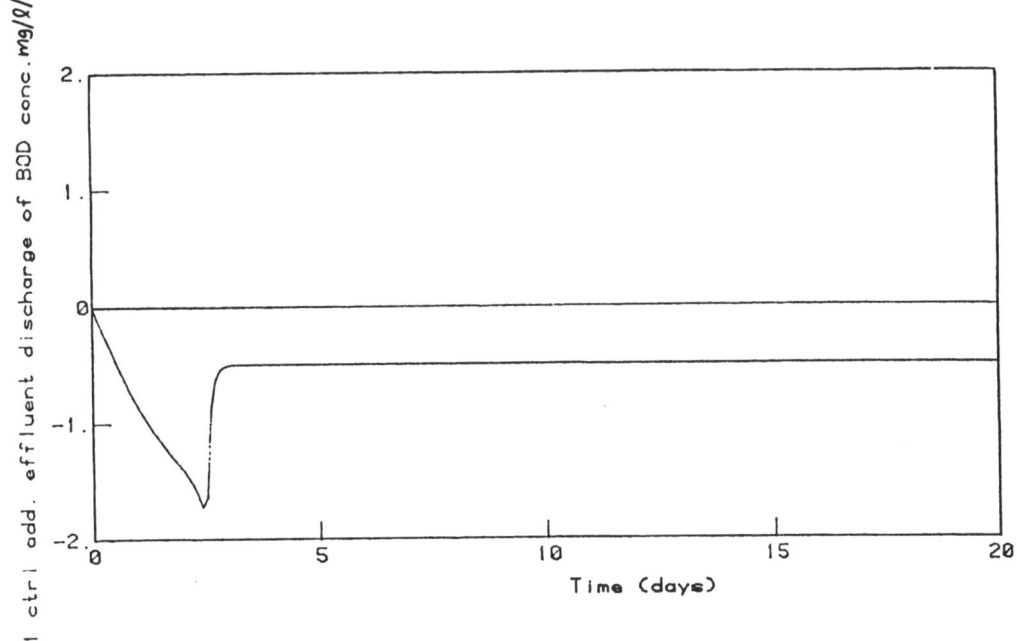

Figure 5(b).   Adaptive effluent control history for the system subject to disturbances (i).

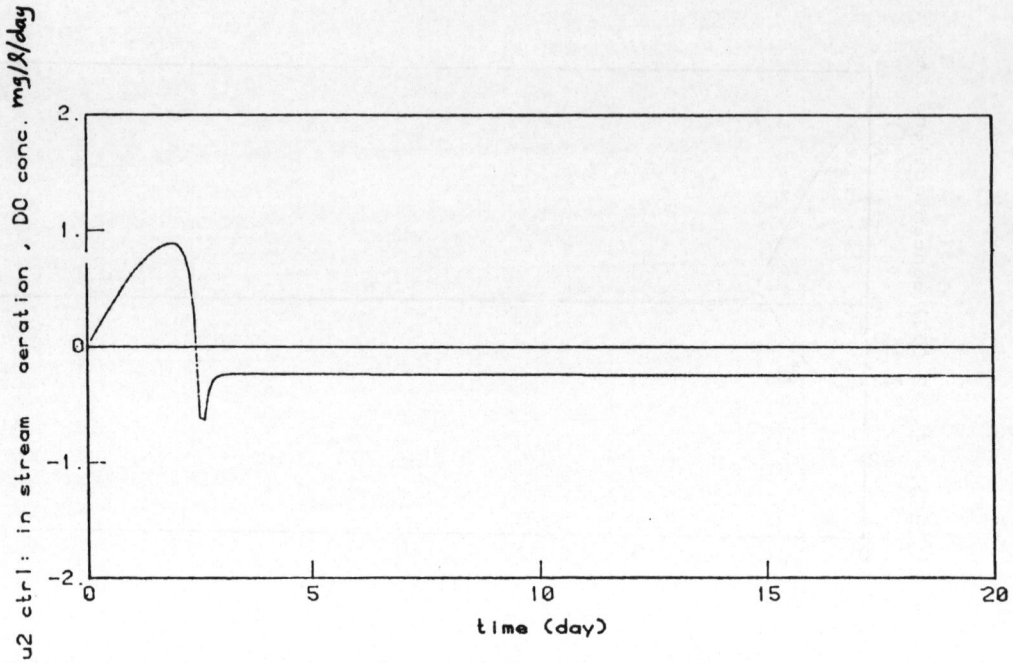

Figure 5(c).   Adaptive in-stream aeration control history for the system subject to disturbances (i).

CONCLUSION

We have endeavored to demonstrate the utilization of a deterministic approach to the control of a water quality system subject to uncertain disturbances at both the input and output levels. Towards that end, effluent as well as in-stream aeration controls based on the knowledge of the system's current state (D.O. and B.O.D.), possibly corrupted by measurement errors, were employed to return the system to and maintain it at a desired water quality level.

We have not considered the economic aspects of implementing these controls; we realize, however, that in-stream aeration is economically less attractive than effluent control. Also, as pointed out by Hassan and Younis (Ref. 6), the measurement of B.O.D. is difficult and time-consuming, taking as long as four days. Thus, a more realistic approach to control would be the sole use of effluent control based on the measured value of D.O. only. Such an approach will be investigated in a subsequent paper.

Furthermore, as mentioned in Ref. 6, the rate of coefficients are not constant but temperature- and hence time-dependent. Such uncertainty in the coefficients can be accommodated readily by the control schemes outlined in this paper. Also, in place of noisy output, state estimation can be employed for the implementation of the control; e.g., see Refs. 35 and 36. Finally, for multi-reach models, it may be efficacious to employ decentralized control, namely, such that the control of each reach is based only on the state of that reach; this case is treated in Ref. 37.

APPENDIX

Definition 1.   (Caratheodory function)   Let T be any non-empty Lebesgue measurable subset of R and X be any non-empty subset of R^n. A function g : T × X → R^p is Caratheodory iff for each t ∈ T, g(t,·) is continuous; for each x ∈ X, g(·,x) is Lebesgue measurable; and, for each compact subset C of T × X, there exists a Lebesgue integrable function $M_C(·)$ such that, for all (t,x) ∈ C, ⏐g(t,x)⏐ < M (t).

Definition 2.   (Strongly Caratheodory function)   A function h : T × X → R^p is strongly Caratheodory iff it is Caratheodory with $M_C(·)$ replaced by a positive constant M_C.

Definition 3.   (Class K functions) A function γ : R_+ → R_+ belongs to class K iff it is continuous and satisfies

$$r_1 < r_2 \Rightarrow \gamma(r_1) < \gamma(r_2) \; \forall r_1 , r_2 \in R_+ ,$$
$$\gamma(0) = 0 , r > 0 \Rightarrow \gamma(r) > 0 .$$

Definition 4.   (Class KR functions) A function γ : R_+ → R_+ belongs to class KR iff it belongs to K and $\lim_{r \to \infty} \gamma(r) = \infty$.

Definition 5.[†]   (Practical stability)   The uncertain dynamical system (6)-(7) is said to be practically stabilizable if given any d > 0 there is a control law p(·) : T × X → U for which, given any Lebesgue measurable uncertainty v(·) whose values lie within a pre-specified bounding set, any initial time t_0 ∈ T and any initial state x_0 ∈ X, the following conditions hold:

(i)   The closed loop system
$$\dot{x}(t) = F(t,x(t), p(t,x(t)))$$

(13)

   F ∈ F

possesses a solution x(·) : $[t_0,t_1)$ → X, x(t_0) = x_0.

[†]For an alternative definition, see Ref. 29.

(ii)   (Uniform boundedness) Given any $r > 0$ and any solution $x(\cdot) : [t_0, t_1] \rightarrow X$, $x(t_0)$, of (13) with $\|x_0\| < r$, there is a constant $d(r) > 0$ such that $\|x(t)\| < d(r)$ for all $t \in (t_0, t_1)$.

(iii)   Every solution of (13) can be continued over $[t_0, \infty)$.

(iv)   (Uniform ultimate boundedness) Given any $\bar{d} > \underline{d}$, and $r > 0$ and any solution $x(\cdot) : [t_0, \infty) \rightarrow X$, $x(t_0)$ such that $\|x(t)\| < \bar{d}$ for all $t > t_0 + t^*(\bar{d}, r)$.

(v)   (Uniform stability) Given any $\bar{d} > \underline{d}$ and any solution $x(\cdot) : (t_0, \infty) \rightarrow X$, $x(t_0) = x_0$, of (13), there is a constant $\delta(\bar{d}) > 0$ such that $\|x_0\| < \delta(\bar{d})$ implies that $\|x(t)\| < \bar{d}$ for all $t > t_0$.

REFERENCES

1. Singh, M.G. 1975. River pollution control. Int. J. Systems Sci., Vol. 6, No. 1, pp. 9-21.

2. Singh, M.G. 1979. Hierarchical methods in river pollution control. In Theoretical Systems Ecology, pp. 419-451, Academic Press Inc.

3. Singh, M.G. 1977. Dynamical Hierarchical Control, North-Holland Publ., Amsterdam.

4. Singh, M.G. 1974. Practical methods for the control and state estimation of large interconnected dynamical systems. Revue Francaise d'Automatique, Informatique et de Recherche Operationnelle J3, pp. 5-45.

5. Singh, M.G. and Hassan, M. 1976. Closed loop heirarchical control for river pollution. Automatica, Vol. 12, pp. 261-264.

6. Hassan, M.F. and Younis, M.I. 1980. Stream quality modelling: A discussion. In: Lainiotis, D. and Tzannes, N.S. (eds), Applications of Information and Control Systems, D. Reidel Publ. Co., Dordrecht, Holland.

7. Hassan, M.F., Salut, G., Singh, M.G. and Titli, A. 1978. A decentralized computational algorithm for the global Kalman filter. IEEE Trans. A.C. Special Issue on Decentralized Control and Large Scale Systems, A.C. 23, No. 2.

8. Hassan, M.F., Hurteau, R., Singh, M.G. and Titli, A. 1978. Stochastic hierarchical control of a large scale river system. Proc. 7th IFAC World Congr., 1978.

9. Hassan, M.F., Singh, M.G. and Titli, A. 1979. Computer aided design of the stochastic controllers for a large scale river system. IFAC Symp. "Computer Aided Design of Control Systems" Zurich, August, 1979.

10. Hassan, M.F., Hurteau, R., Singh, M.G. and Titli, A. 1977. A new three level algorithm for river pollution control. Proc. IFAC Symposium on Systems Approaches to Development, Cairo, Egypt.

11. Tamura, H. 1974. A discrete dynamic model with distributed transport delays and its hierarchical optimization for preserving stream quality. IEEE Trans. Syst. Man., and Cybernetics SMC4 (5), pp. 424-431.

12. Kendrick, D.A., Rao, H.S. and Wells, C.H. 1970. Optimal operation of a system of waste water treatment facilities. Proc. IEEE 9th Symposium on Adaptive Processes, Austin, Texas, pp. XII 3.1-XII 3.5.

13. Young, P.C., Beck, M.B. and Singh, M.G. 1973. The modelling and controls of pollution in a river system. Proc. IFAC Symposium on Water Resource Systems, Haifa.

14. Singh, M.G., Drew, S. and Coales, J.F. 1975. Comparisons of practical hierarchical control methods for interconnected dynamical systems. Automatica Vol. 11, pp. 331-350.

15. Yeh, W.W.G. and Becker, L. 1979. Water resource systems models. In: Leondes, C.T. (ed.), Control and Dynamic Systems, pp. 195-246, Academic Press, New York.

16. Streeter, H.W. and Phelps, E.B. 1925. A study of the pollution and natural purification of the Ohio River. Public Health Bulletin, 146, U.S. Public Health Service, Washington, D.C.

17. Chang, S. and Yeh, W.W.G. 1973. Optimal allocation of artificial aeration along a polluted stream using dynamic programming. Water Resources Bull., Vol. 9, No. 4, pp. 985-997.

18. Beck, B. and Young, P. 1976. Systematic identification of DO-BOD model structure. J. Environ. Eng., Div. ASCE, Vol. 102, No. EE5, pp. 909-927.

19. Beck, M.B. 1973. The application of control and systems theory to problems of river pollution control. Ph.D. Thesis, University of Cambridge, Cambridge, England.

20. Liebman, J.C. and Lynn, W.R. 1966. The optimal allocation of stream dissolved oxygen resources. Water Resources Research, Vol. 2, No. 3.

21. Loucks, D.P., Revelle, C.S. and Lynn, W.R. 1967. Linear programming models for water pollution control. Management Sciences, Vol. 14, No. 4, pp. B166-181.

22. Dysart, D.C. III 1969. The use of dynamic programming in regional water quality planning. Presented at the A.A.P.S.E. Systems Analysis Conference.

23. Tarassov, V.J., Perlis, H.T. and Davidson, B. 1969. Optimization of a class of river aeration problems by the use of multivariable distributed parameter control theory. Water Resources Research.

24. Nairo, M., Takamsatsu, T. and Tamura, H. 1972. Optimum planning of sewage treatment systems for preserving stream quality. Proc. 5th IFAC Congress, Paris.

25. Mahmoud, M.S., Hassan, M.F. and Saleh, S.J. 1985. Decentralized structures for stream water quality control problems. Optimal Control Applications & Methods, Vol. 6, pp. 167-186.

26. Hassan, M.F., Younis, M.I. and Mancy, K.H. 1980. A developed stream water quality model: A case study on the River Nile. Proc. IFAC Systems Approach for Development Conference, Morocco, 413-417.

27. Grenney, W.J. et al. 1978. Characteristics of the solution algorithms for the QUAL II river model. J. Water Pollution Control, Vol. 4, pp. 115-124.

28. Hassan, M.F., Mahmoud, M.S. and Younis, M.I. 1981. A dynamic Leontief modelling approach to management for optimal utilization in water resources system. IEEE Trans. Syst., Man., and Cybernetics, SMC, Vol. 11, pp. 552-558.

29. Corless, M. and Leitmann, G. 1984. Memoryless controllers for uncertain systems. Proc. of the "Meeting on Mathematical Methods for Optimization in Engineering" edited by Gagliardi, F., University of Cassino, Italy.

30. Corless, M. and Leitmann, G. 1981. Continuous state feedback guaranteeing uniform ultimate boundedness for uncertain dynamic systems. IEEE Trans. Automat. Control, AC-26, Vol. 5, pp. 1139-1144.

31. Corless, M. 1984. Control of uncertain systems. Ph.D. Dissertation, University of California, Berkeley.

32. Corless, M. and Leitmann, G. 1983. Adaptive control of systems containing uncertain functions and unknown functions with uncertain bounds. J. of Optimization Theory and Applications, Vol. 41, pp. 155-168.

33. Corless, M. and Leitmann, G. 1984. Adaptive control for uncertain dynamical systems. In: Blaquiere, A. and Leitmann, G. (eds) Mathematical Theory of Dynamical Systems and Microphysics: Control Theory and Mechanics, Academic Press, New York.

34. Corless, M. and Leitmann, G. 1984. Adaptive controllers for a class of uncertain systems. Annales de la Fondation Louis de Broglie, Vol. 9, pp. 65-95.

35. Leitmann, G. 1981. On the efficacy of nonlinear control in uncertain linear systems. J. of Dynamic Systems, Measurement, and Control, Vol. 102, pp. 95-102.

36. Breinl, W. and Leitmann, G. 1983. Zustandrückführung für dynamische systeme mit parameterunsicherheiten, Regelungstechnik, Vol. 31, Heft 3.

37. Leitmann, G., Lee, C.S. and Chen, Y.H. 1986. Decentralized control for a large scale uncertain river system. Proc. IFAC Workshop on Modelling, Decision and Games with Applications to Social Phenomena, Beijing, China.

38. Leitmann, G. and Lee, C.S. 1987. Uncertain dynamical systems: An application to river pollution control. Proc. 1st Bellman Continuum, University of Illinois, Urbana, May 1985.

PARTICIPANT'S COMMENTS

The acknowledgement that uncertainties in process models, parameters and disturbances are significant and persistent in confounding our efforts to model managed natural resources was expressed by many participants in this workshop. Unfortunately these inherent uncertainties in our relative ignorance of their statistical properties are sometimes used as reasons for postponing or even abandoning the application of dynamic systems modeling and control techniques to biological resources. The fact is, nevertheless, that we must and do develop control strategies to manage these dynamic resources in the face of uncertainties, whether those strategies are based on a dynamic mathematical formulation or only a mental model of the resource system and its associated uncertainties. As knowledge of the dynamics of renewable resources has increased, the uncertainty associated with modeling these dynamic systems has been reduced. However, the natural variability within biological populations may preclude being able to determine the processes models or parameters of bioresource dynamics with any more accuracy than some uncertainty bound.

Professors Lee and Leitmann demonstrate now the Streeter-Phelps water quality model can be reformulated into a model which incorporates some of the uncertainties mentioned above. They illustrate how an aeration and effluent control strategy can be employed to bring the Biological Oxygen Demand and the dissolved oxygen content of the river water to within a preassigned deviation from equilibrium values within a finite time. The strategy maintains these variables within those preassigned deviations from equilibrium in spite of only limited knowledge of the uncertainties, namely only the bounds of the uncertainties in the case of memoryless control, and only that bounds exist in the case of adaptive control.

The work of the authors presented here is part of a very active research program in robust control of uncertain dynamical systems. Efforts are proceeding in areas very relevant to the problems which confront renewable resource managers. These include: the inclusion of observers to reconstruct states not directly measurable; the development of control strategies for cases in which complete matching control is not available for all states; and the control of multistage systems. How we might formulate the models and uncertainties and derive the appropriate Lyapunov functions for other problems so that this technique can be fully utilized represents in inviting challenge to researchers in the renewable resource management community.

Ann Lowes Blackwell

This paper, along with the one by Grantham and Fisher, deals with a particularly interesting problem; control of dynamical systems (continuous or discrete) in the face of uncertainties. The paper deals with the important example of control of river pollution. There are two short comments that I would like to make:

1) Costs (or utilities) are not included in the model. Thus, the applicability of this approach remains to be seen.

2) Time delays are not treated.

These two comments are not meant to take away from the viability of this approach, which circumvents many of the difficulties that arise in treating uncertainties statistically (e.g., the control is never guaranteed), but rather to point out necessary extensions. Since I know of no biological processes in which time delays of some magnitude are not involved, the second point is particularly important. If it is possible to add to the bounds of the uncertainty some quantity which is a function of time delays, then the authors' approach remains applicable. If the delays are such that the system becomes chaotic, then control becomes much more difficult.

Yosef Cohen

Reply

The authors agree that time delays, especially for "long" reaches, are of importance. Currently work is being conducted to apply results of deterministic control for uncertain systems with delay (e.g., see Y. Yu, *JOTA*, Vol. 41, No. 3, p. 503) to the problem of this paper.

C.S. Lee and G. Leitmann

REAL TIME MANAGEMENT OF
A RESOURCE CONSUMPTION MODEL

G. Bojadziev
Simon Fraser University
Burmby BC,
Canada V5A 1S6

and

J. Skowronski
University of Queensland
St. Lucia Q. 4067
AUSTRALIA

A real time extinction avoidance controller is introduced for the Lotka-Volterra type model of resource consumption. The conditions are formulated via Liapunov's Direct Method. An example illustrates the results.

INTRODUCTION

Often ecological systems involve components which change behavior abruptly according to an adopted strategy in order to exercise control over the growth of the system. Bojadziev and Skowronski (1985) have discussed a predator (consumer)-prey (resource) model containing a controlling factor which adjusts the number of predators so that a reasonable level of both populations is maintained. A Liapunov design technique, based on a paper by Leitmann and Skowronski (1977), has been introduced to find avoidance conditions of a region for all time so that the admissible action of the populations stabilizes their coexistence. However, in many situations, a requirement of avoidance control during a prescribed time interval may serve as a more realistic approach towards investigating events of this nature.

The objective of this paper is to study the model introduced in Bojadziev and Skowronski (1985) from the point of view of avoidance in stipulated real time. A paper by Leitmann and Skowronski (1983), dealing with avoidance of an anti-target during a specified time interval, is instrumental.

RESOURCE-CONSUMPTION MODEL WITH CONTROL

Consider the model

$$\dot{x} = f(x,u), \qquad \dot{x} = dx/dt, \qquad t > 1 , \tag{1}$$

$$x = (x_1, x_2)^T \in R^2, \qquad x_1, x_2 > 0, \qquad u \in U \subset R, \qquad f = (f_1, f_2)^T,$$

where

$$f_1 = x_1(\alpha_1 - \beta_1 x_2), \qquad f_2 = x_2(-\alpha_2 + \beta_2 x_1) + ux_2^2 . \tag{2}$$

Here, α_i, β_i, $i = 1,2$, are constants, $\alpha_i, \beta_i > 0$, and the control function $u(t)$ ranges in a compact set which will be specified later; x_1 and x_2 measure in some units (number of individuals in population or biomass) the prey (food) and predator (consumer) sizes correspondingly.

The system (1) with (2) is obviously generated by the classical Lotka-Volterra (LV) model

$$\dot{x}_1 = x_1(\alpha_1 - \beta_1 x_2) ,$$

$$\dot{x}_2 = x_2(-\alpha_2 + \beta_2 x_1) ,$$

whose nonzero equilibrium position, $E^0(\alpha_2/\beta_2, \alpha_1/\beta_1)$, is a center. It generates the equilibrium $E(\hat{x}(u))$ of (1), $\hat{x} = (\hat{x}_1, \hat{x}_2)^T$, where

$$\hat{x}_1 = \frac{\alpha_2}{\beta_2} - \frac{\alpha_1}{\beta_1\beta_2} u , \qquad \hat{x}_2 = \frac{\alpha_1}{\beta_1} \tag{3}$$

We require

$$u \in U = \left(-\infty, \frac{\alpha_2\beta_1}{\alpha_1}\right) \tag{4}$$

so that $\hat{x}_1$ given by (3) is positive, hence $E(\hat{x})$ is located in the interior of the first quadrant (the population quadrant).

We denote the system (1) and its equilibrium E correspondingly: for $u > 0$ by $(1)^+$ and E^+, for $u < 0$ by $(1)^-$ and E^-, and for $u = 0$ by $(1)^0$ and E^0. The equilibria E^+, E^-, and E^0 are located on the equilibrium line e with equation $x_2 = \alpha_1/\beta_1$.

The local analysis gives that E^+ is an unstable focus, E^- is an asymptotically stable focus, and E^0 is a center, provided we constrain u in the interval

$$u \in U_f = \left\{2\beta_1\left[-1-(1 + \frac{\alpha_2}{\alpha_1})^{1/2}\right], \ 2\beta_1\left[-1+(1+\frac{\alpha_2}{\alpha_1})^{1/2}\right]\right\} \subset U . \tag{5}$$

If, however, $u \in U\backslash U_f$, E^+ is an unstable node and E^- is an asymptotically stable node.

THE LIAPUNOV FUNCTION

The LV model has a Liapunov function (see, for example, Bojadziev and Skowronski, 1985))

$$V^0(x) = \beta_2(x_1 \frac{\alpha_2}{\beta_2}) = \alpha_2 \ln \frac{\beta_2 x_1}{\alpha_2} + \beta_1(x_2 - \frac{\alpha_1}{\beta_1}) - \alpha_1 \ln \frac{\beta_1 x_2}{\alpha_1} \tag{6}$$

with the following properties:

(A)   The maximum of $V^0(x)$ is attained at the equilibrium E^0.

(B) $V^0(x)$ is monotone increasing about E^0;
$V^0(x) \to \infty$ as $\|x\| \to 0$, $V^0(x) \to \infty$ as $\|x\| \to \infty$.

(C) $\dot{V}^0(x) = \sum_{i=1}^{2} \frac{\partial V^0}{\partial x_i} f_i^0 = 0$, where f_i^0, $i = 1,2$, are given by (2) with $u = 0$.

The equilibrium E^0 is stable. All solutions of the LV model are periodic and in the phase plane (x_1, x_2) they are represented by the family of closed curves $V^0(x) = $ const around the center E^0 (cf. Fig. 1).

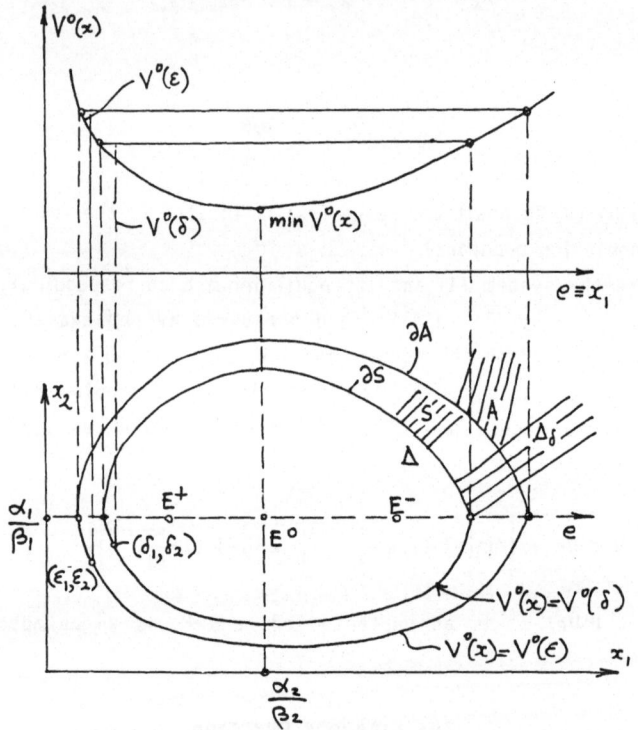

Figure 1. The Liapunov function.

To study the global stability properties of $E(\dot{x})$, taking hint from (6), we introduce the function, Bojadziev and Skowronski (1985),

$$V(x,u) = \beta_2(x_1-\hat{x}_1) - \beta_2\hat{x}_1 \ln \frac{x_1}{\hat{x}_1} + \beta_1(x_2-\hat{x}_2) - \beta_1\hat{x}_2 \ln \frac{x_2}{\hat{x}_2} \tag{7}$$

which may serve as a Liapunov function. For u = 0, (7) reduces to (6).

Properties (A) and (B) remain valid for (7) if $E(\hat{x})$ is considered instead of E^0. To find a similar property to (C) for the function (7), we calculate the derivative of (7) along the solutions of (1)

$$\dot{V}(x,u) = \sum_{i=1}^{2} \frac{\partial V}{\partial x_i} f_i = \beta_1 u(x_2-\hat{x}_2)^2 , \tag{8}$$

where f_i, i = 1,2, are given by (2).

From (8) we see that $E^+(u > 0)$ is an unstable equilibrium and $E^-(u < 0)$ is an asymptotically stable equilibrium. The trajectories of the system $(1)^+$ are flowing away from E^+, while the trajectories of the system $(1)^-$ tend towards E^-. The biological interpretation is that the growth of the predator x_2 is enhanced by increasing the species density (u > 0), and dampened by decreasing if (u < 0).

FINITE-TIME AVOIDANCE

First here, we introduce the avoidance region and a security zone.

<u>Definition 1.</u> The avoidance set A in R^2 is defined by

$$A = \{x \in R^2 \mid V^0(x) > V^0(\epsilon)\} , \tag{9}$$

where $V^0(x)$ is given by (6), $\epsilon = (\epsilon_1,\epsilon_2)^T$, and $\epsilon_1,\epsilon_2 > 0$ (called avoidance parameters) are the "survival distances", small as desired for a particular study.

At the boundary ∂A of A, we have, according to (9), $V^0(x) - V^0(\epsilon) = 0$. The set A encloses all anti-targets to be avoided during a prescribed interval.

<u>Definition 2.</u> The security zone S about A is the set $S = \Delta_\delta \backslash A \neq \phi$, where

$$\Delta_\delta = \{x \in R^2 \mid V^0(x) > V^0(\delta)\} ,$$

$\delta = (\delta_1,\delta_2)^T$, $\delta_1,\delta_2 > 0$ are (security parameters) such that $\delta_i > \epsilon_i$, (i = 1,2 (from property (B) of THE LIAPUNOV FUNCTION we have $V^0(\epsilon) > V^0(\delta)$).

At the boundary of Δ_δ, $\partial\Delta_\delta = \partial S$, we have $V^0(x) = V^0(\delta)$. The set Δ_δ encloses the security zone S and the avoidance region A. The allowable region for the motions of (1) is $\Delta \cup S = R^2 \backslash A$ (Fig. 1).

The design of a strategy that guarantees finite-time avoidance of A is based on the following definition.

<u>Definition 3.</u> The set A is avoidable during a time interval of duration $\tau > 0$ if there is a set $S \neq \phi$ and a control $u \in U$ such that for all $x^0 = x(0) \in \partial S$, the solution $k(x^0, u, t)$ of (1) cannot enter A before τ, i.e.,

$$k(x^0, u, t) \cap A = \phi, \qquad \forall t < \tau . \tag{10}$$

Now we state a theorem that gives sufficient conditions for the avoidance of A.

<u>Theorem.</u> The set A is avoidable during an interval of duration $\tau > 0$ if there are: a strategy of the predator x_2 generating a suitable control $u \in U$, and a C^1-function $\tilde{V}(x)$ defined in an open set containing S such that

(i) $0 < \tilde{V}_{\partial S} < \tilde{V}_{\partial A} < \infty$, $\tag{11}$

(ii) $\dot{\tilde{V}}(x) = \sum_1^2 \frac{\partial \tilde{V}}{\partial x_i} f_i(x, y) < \frac{\tilde{V}_{\partial A} - \tilde{V}_{\partial S}}{\tau}$, $\tag{12}$

where

$$\tilde{V}_{\partial S} = \{\sup \tilde{V}(x) | x(t) \in \partial S\} ,$$

$$\tilde{V}_{\partial S} = \{\inf \tilde{V}(x) | x(t) \in \partial A\} .$$

<u>Proof.</u> Suppose that the set A is not avoidable in finite time τ, i.e., that for some time $t_A < \tau$ the trajectory of (1) starting from $x^0 \in \partial S$ intersects ∂A at the point $x^A = x(t_A)$. Then integrating (12) along the corresponding solution of (2), we get

$$\tilde{V}(x^A) = \tilde{V}(x^0) < \frac{\tilde{V}_{\partial A} - \tilde{V}_{\partial S}}{\tau} t_A$$

or

$$\frac{\tilde{V}(x^A) - \tilde{V}(x^0)}{\tilde{V}_{\partial A} - \tilde{V}_{\partial S}} < t_A / \tau \tag{13}$$

According to assumption (11)

$$\tilde{V}(x^A) - \tilde{V}(x^0) > \tilde{V}_{\partial A} - \tilde{V}_{\partial S}$$

which with (13) gives $\tau < t_A$. This is a contradiction to the assumption $t_A < \tau$. We note that the sufficient conditions (11) and (12) for finite-time avoidance are modifications of those introduced by Leitmann and Skowronski (1983).

THE CONTROL STRATEGY

To design a strategy for finite-time avoidance concerning the system (1), we use the function $V^0(x)$ given by (6) and the Theorem in FINITE-TIME AVOIDANCE. We observe that $V^0(x)$ satisfies condition (11), i.e., $0 < V_{\partial S}^0 < V_{\partial A}^0 < \infty$. In order for $V^0(x)$ to satisfy (12), we require that

$$\sum_{i=1}^{2} \frac{\partial V^0(x)}{\partial x_i} f_i(x,u) = u x_2^2 \left(\beta_1 - \frac{\alpha_1}{x_2}\right) < \frac{V_{\partial A}^0 - V_{\partial S}^0}{\tau} . \tag{14}$$

Note that for $\tau \to \infty$ in (14), we obtain the results in Bojadziev and Skowronski (1985).

The inequality (14) establishes a relationship between the control u and the predator x_2. Both (14) and (4) represent in the plane (x_2, u) two zones, Z_1 and Z_2, with boundaries being the branches z_1 and z_2 of the curve

$$u x_2^2 \left(\beta_1 - \frac{\alpha_1}{x_2}\right) = (V_{\partial A}^0 - V_{\partial S}^0)/\tau$$

and the lines $u = \alpha_2 \beta_1 / \alpha$ and $x_2 = \alpha_1 / \beta_1$ (Fig. 2). Only the boundaries drawn with heavy lines on Fig. 2 belong to the zones.

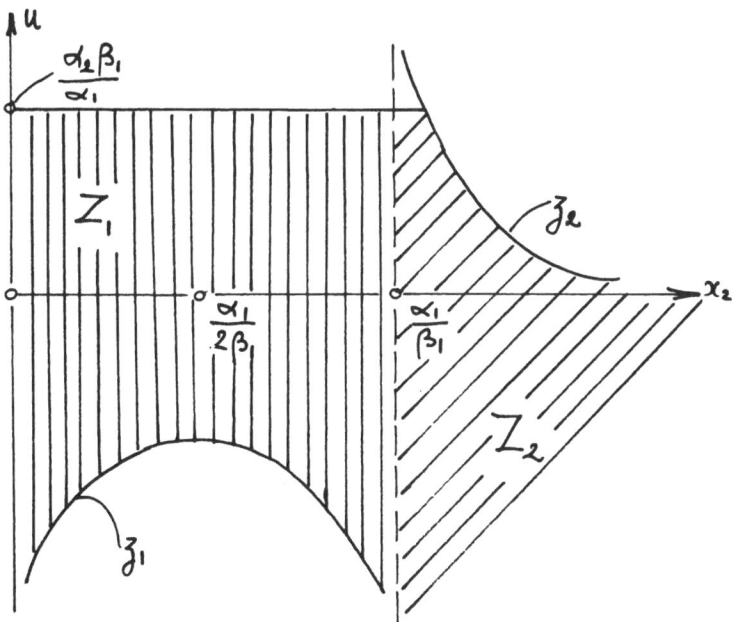

Figure 2. The control function.

Condition (14) requires a strategy of selecting $u \in Z_1$ for $0 < x_2 < \alpha_1/\beta_1$ and $u \in Z_2$ for $x_2 > \alpha_1/\beta_1$. Hence, when a response of (1) enters the security zone S, if a control u is to be exercised at a certain point $x_S \in S$, a control $u \in Z_2$ should be used for x_S above the equilibrium line $x_2 = \alpha_1/\beta_1$ and a control $u \in Z_1$ should be used for x_S below this line (Fig. 3).

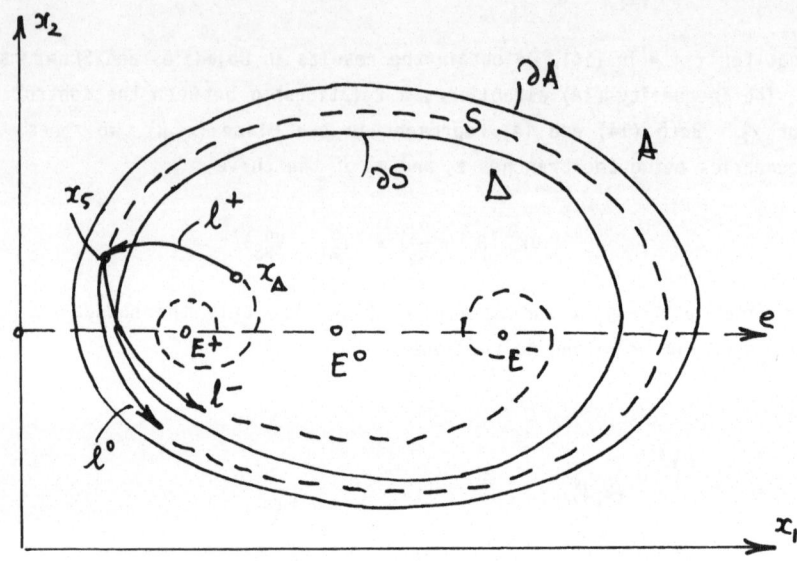

Figure 3. The state-plane.

EXAMPLE

Let us assume that (5) holds and the population behavior is modelled by the system $(1)^+$ with fixed $u(0 < u < 2\beta_1[-1+(1+\alpha_2/\alpha_1)^{1/2}])$ and initial state $x_\Delta \in \Delta$. Consider a response emanating from the point x_Δ (Fig. 3) and moving on the arc $x_\Delta x_S$ of the path ℓ^+ around the equilibrium E^+. The population x_1 decreases as the population x_2 varies. The further move of the response along ℓ^+ may endanger the existence of an acceptable size of the prey population x_1 and consequently may cause an undesirable variation of the number of predators x_2 during a finite time interval. In order to avoid such a situation, or similar situations where either the predator population x_2 is small, or the sizes of both populations are small, the predator can adopt a strategy of self control by restricting the rate of its own growth. In the case shown on Fig. 3, this can be achieved by switching the control value u from positive to negative which changes the path ℓ^+ to ℓ^- at x_S (switching point), and by changing the control zones from Z_1 to Z_2 in order to satisfy condition (14) of avoidance. Hence, the new fixed control value u belongs to the cone $\{2\beta_1[-1+(1+\alpha_2/\alpha_1)^{1/2}] < u < 0, x_2 > \alpha_1/\beta_1\} \subset Z_2$. The new path ℓ^- belongs to the system $(1)^-$ with equilibrium E^-. Also, there is a special case when $u > 0$ changes to $u = 0$ at the switching point x_S. Then the path ℓ^+ changes to ℓ^0 which is part of a periodic orbit of the LV model around E^0.

REFERENCES

Bojadziev, G. and J. Skowronski. 1985. Controlled food consumption. Methods of Operations Research, Vol. 49, pp. 499-506.

Leitmann, G. and J. Skowronski. 1977. Avoidance control. J. Optim. Theory Appl., Vol. 23, pp. 581-591.

Leitmann, G. and J. Skowronski. 1983. A note on avoidance control. Opt. Control Appl. and Methods, Vol. 4, pp. 335-342.

PARTICIPANT'S COMMENTS

At the workshop there appeared to be two professional viewpoints, represented by the "Control Theorists" and the "Resource Scientists." Presumably a main purpose of the workshop was to bring these two groups of specialists together for cross-fertilization of ideas. Ideally, the Control Theorists would present new techniques which would be snapped up by the more practical Resource Scientists, who would, on their part, present real resource management problems that Control Theorists could attack with their theories.

In reality, things were not quite so ideal--each group seemed to be speaking mainly to its own members. The paper of Bojadziev and Skowronski is a case in point. No doubt the results about controlling a two-species predator-prey model so as to avoid undesirable outcomes, as expressed via a Liapunov function, are technically interesting as an advance in control theory. But the authors do not attempt to discuss any actual field applications, either to resource problems or to the understanding of any particular biological phenomena.

In my view, good applied research operates in the exact reverse direction to academic exercises such as this. But unless mathematicians go to the effort of actually talking to biologists, and reading their papers, and trying to understand what their problems are, very little of scientific or economic value is likely to emerge. (A colleague of mine, an algebraist, once asked me if I had any good problems in resource economics which he could solve by nonabelian groups. I'm still not sure whether it was a joke.)

The problem, of course, is that such interdisciplinary research is much more difficult to organize than the traditional kind. It takes many hours, or weeks, to bridge the gap between the disciplines. The mathematician must be prepared to learn and use unfamiliar mathematical techniques. The biologist must be prepared to try to understand whether the mathematical results really solve his problem. Both will be sneered at by their colleagues for their unprofessional behavior.

Workshops such as this can be useful in bringing research workers from different backgrounds tegether. Perhaps a nontraditional workshop format, in which the biologists present surveys of major unsolved problems and the mathematicians discuss possible approaches to resolving them would be appropriate for future workshops.

Colin Clark

Reply

As Colin Clark states, the cross-fertilization between central theorists and resource scientists was the purpose of this workshop - and to such aim both groups had to prepare material which could be cross-fertilized. Indeed, it is exactly what has occurred at the lecture-part of the meeting, and that is why the paper in question (and all the others in the group) present applicable, though not yet applied, control theoretic material. Nobody can deny applicability of the Lotka-Volterra model, even if there is some criticism about it. The cross-fertilization took place later, in round-table discussions, after all the presentations had been made. It seems that this was the proper order of affairs. In the lectures, each group did much better by talking about topics in which they were competent and such that could be used more widely than in a single application.

Jan Skowronski

In this paper, the authors consider a predator (consumer)-prey (resource) model with control which is generated by the predator through its strategy. The model degenerates to the classical Lotka-Volterra prey-predator model in the absence of control. Regions referred to as the avoidance region and security zone are being defined by using the Liapunov function associated with the classical Lotka-Volterra prey-predator model. A Liapunov design technique is used to find conditions that a control must satisfy so that the system will stay away from the avoidance region in a pre-specified finite time.

In the management of a renewable resource, the idea of establishing an avoidance region and a security zone is not only meaningful, but also important. This is owing to the fact that there is always error in the measurement or estimation of the biomass (population) of a renewable resource. Furthermore, with the introduction of a security zone between the avoidance region and the feasible region, coupled with the avoidance control strategy to stabilize the system, the coexistence of both the predator (consumer) and prey (resource) is guaranteed.

C.S. Lee

CONTROL PROBLEMS IN IRRIGATION MANAGEMENT

R.J. Stonier
The Department of Mathematics and Computing
Capricornia Institute of Advanced Education
Rockhampton, Queensland, Australia

In agriculture, loss of nutrient in the vicinity of a growing root due to leaching and evaporation is of a major concern. This problem is addressed by considering the control of the irrigation rate so that the downward movement of the solute peak tracks the root length of a growing plant. A model is developed for which available optimization techniques can be applied to analyze this tracking problem. Other control problems in this area are proposed.

INTRODUCTION

In agriculture, the problem of fertilizer (nitrate) loss when applied to a crop through leaching from the root zone is of major concern (Dayananda et al., 1980). It is reported that less than half the fertilizer applied is usefully recovered in the crop (Dayananda et al., 1980). This has initiated extensive worldwide research with the introduction of the use of mathematical models to describe nitrate and water movement in soils, (cf. Burns and Greenwood, 1983; Rose et al., 1982; Raats, 1975; Levin, 1964; Saxton et al., 1977).

Two types of models are currently used in predicting water and nitrate movement in soil. They are based on dispersion theory and the hydrological layer approach similar to the plate layer theory of chromatography. The latter approach, although less unrealistic, uses parameters that are simple to calculate, and is much easier to use yet gives as good a performance in field tests. We use these models developed by Dyananda et al. (1980) to study the motion of the solute peak in relation to the growth of roots of a given plant.

Very little modelling has been done on root length in relation to growth with respect to time, though it is now considered one of the most important parameters in plant growth (Myers, 1980; Bailey, 1970). We shall assume that the root length (length of the longest root) increases linearly with time. This assumption is realistic for many crops and it is used by farmers to predict the root depth of plants in irrigation management.

The traditional irrigation cycle consists of a brief period of infiltration (irrigation) followed by an extended period of water extraction by the crop and loss by evaporation. In the period of water extraction, the loss of water due to evaporation and transpiration (evapotranspiration) is estimated, and at the next irrigation the irrigation rate is applied to recover the loss to keep a sufficient water content in the soil. This type of policy has resulted due to the modes and cost of irrigation available (e.g., furrow irrigation). To reduce cost, it is desirable to minimize the number of irrigations by increasing the time between them.

To minimize irrigation frequency, the usual goal is to maximize both the quantity of water stored and the proportion of this water used by the crop before the next irrigation. Research has concentrated on storing water in the profile for subsequent use by the crops.

New development in irrigation systems makes it now possible to deliver to the soil small quantities of water as often as desired. As the higher frequency of irrigation is now possible, the infiltration period becomes a more important part of the irrigation cycle (see Rawlins, 1973).

Regardless of the irrigation method, there are two problems of concern:

(i)   Leaching of solute below the root depth by the application of too much water resulting in a high loss of nutrients available for uptake by the roots in the root zone.

(ii)   Loss of nutrients in the vicinity of the root depth due to too little water being applied and loss by evapotranspiration. For as well as nutrient being available over the root length, it is also desirable to have sufficient nutrients in the soil ahead of the growing root.

In this paper, we develop some mathematical models to examine, by use of modern control techniques, the control of movement of water/solute profile to remain in a vicinity of the root depth for the period of time of root growth to the root zone depth D_r.

MATHEMATICAL MODELS OF SOLUTE MOVEMENT AND ROOT GROWTH

Solute Movement

In Dyananda et al. (1980), water movement in soil is studied giving mathematical models to predict the depth of the solute peak with time.

Two cases are discussed, 'wet' and 'dry' cases. In the 'wet' case, the input from irrigation replenishes water in the profile up to the field capacity θ_f, and the 'dry' case occurs when the water content does not reach θ_f.

The equations determining the depth of the solute peak as a function of time are:

Wet Case:

$$\theta_f \frac{dz}{dt} = R_f(I(t) - zE(t)/D_r) \qquad \text{for } z < D_r \qquad (1)$$

$$\theta_f \frac{dz}{dt} = R_f(I(t) - E(t)) \qquad \text{for } z > D_r \qquad (2)$$

Dry Case:

$$\theta \frac{dz}{dt} = R_f I(t)\left[1 - \frac{z}{D_r}\right] \qquad \text{for } z < D_r \qquad (3)$$

and

$$\theta(t) = \theta_f - D_r^{-1} \int_0^t (E(s) - I(s)) \, ds$$

$$\theta_f \frac{dz}{dt} = R_f (I(t) - E(t)) \qquad \text{for } z > D_r . \tag{4}$$

The notation employed here is described below

$z(t)$ is the depth of the peak below the surface at time t.

θ_f is the field capacity of the soil (dependent upon soil properties).

D_r is the depth of the root zone.

R_f is the dimensionless constant which relates the response to a movement of the wetting front in the water profile to the position of the solute peak. In principle, it is dependent upon the water content but may be assumed constant for any particular soil/solute combination (Dayananda et al., 1980). Typically, $R_f < 1$ for the positive sorption, $R_f > 1$ for negative sorption.

$\theta(t)$ is the water content of the soil at time t.

$I(t)$ is the irrigation rate. It is bounded above by the infiltration rate of the soil.

$E(t)$ is the evapotranspiration rate. It may be assumed constant about .8 cm/day under a frequent irrigation cycle with a wet profile in the soil, but in general, it is clearly dependent upon weather conditions, soil, soil moisture content, type of crop, etc. Descriptions such as

$$E(t) = \begin{cases} 0 & , \quad 0 < t < T/2 , \\ 0.8(2t-T)/T & , \quad T/2 < t < T , \end{cases}$$

and

$$E(t) = 0.8t/T \qquad , \quad 0 < t < T ,$$

may be considered, where T is the time period of growth of the plant to reach root depth D_r.

     The processes of water movement, evapotranspiration and root growth are time continuous in reality but not the irrigation process, although this is now 'possible' with the development of high-frequency irrigation systems. In practice, most calculations are done using constant rates for different climatic and soil conditions, and crops. Data are frequently available in discrete form, for example in modelling evapotranspiration. It would seem appropriate to move to a discrete analysis of this problem. However, we shall present the continuous model here as the results obtained yield a good enough approximation to reality. Corresponding discrete models are available for the 'wet' case and 'dry' case above, where the time intervals are typically a day or a month, with spatial and temporal averages being taken over this time, (cf. Raats, 1975; Rose et al., 1982).

Root Growth

In the field, calculation of water uptake by roots is usually done by assuming appropriate constant root growth for a specific plant/crop. So, for the prediction of root growth, we shall use

$$\frac{d\ell}{dt} = \delta \tag{5}$$

where

$\ell(t)$ is the root length,

δ is a constant, typically 2 cm/day.

From data available, other models such as logistic may be fitted for particular crops and used as a prediction of root length. Unfortunately, this data is limited due to the fact that most measurements have been made on root density rather than on root length (Gerwitz and Page, 1972; Taylor and Klepper, 1974; Taylor and Klepper, 1973).

Remark 1

Equations (1) and (3) have been derived on the assumption that evapotranspiration takes place in the root zone $[0, D_r]$ and were used when D_r was small, ~25 cm. When D_r is large, of the order 120-180 cm, it seems more appropriate to model the movement of the solute peak by the equations

$$\theta_f \frac{dz}{dt} = R_f \left[I(t) - \frac{E(t)z}{\ell(t)} \right], \qquad\qquad 0 < z < \ell(t), \tag{6}$$

$$\theta_f \frac{dz}{dt} = R_f \left[I(t) - E(t) \right], \qquad\qquad z(t) \geqslant \ell(t), \tag{7}$$

where $\ell(t)$ is the (variable) depth of the root zone. The equations (6) and (7) then replace (1) and (2).

Typical Field Data

$R_f = 1$

Field capacity $\theta_f = .25$

Growth rate $\delta = 2$ cm/day

Time of growth of crop $T = 60$ days (grain sorghum)

Depth of root zone $D_r = 120$ cm

Constant evapotranspiration rate $E(t) = .8$ cm/day

Infiltration rate of soil $I_m = 1.2$ cm/day

Remark 2

It should be noted that these equations model the movement of the peak of the solute profile. They do not tell us of the spread of the profile which is an important factor when considering water uptake by the roots, (cf. Levin, 1964), for water uptake does not take place solely at the tip of the root but varies over its length.

Remark 3

The root growth model we have taken is clearly an over-simplification of the real thing, yet it is this constant growth rate that is used to make predictions on root growth for irrigation purposes in farm management handbooks.

A lot of quantitative study has been done on root development of numerous plants/crops, (cf. Bloodworth et al., 1958; May et al., 1965; May et al., 1967; Lungley, 1973; Burns, 1980). (Lungley (1973) gives a numerical computer simulation of growth of root systems.)

What is of interest for discussion of the problem in the INTRODUCTION is how irrigation affects root growth. Gardner (1964), Cullen et al. (1972), and Taylor and Klepper (1974) are a few which take up this problem. However, a mathematical model, such as that for the movement of the solute/water peak, which in some way attempts to relate root growth to irrigation rate, has not been seen by the author.

CONTROL PROBLEMS

In this section, we shall present a number of control problems that arise as a result of examining the control of irrigation rate for the movement of the solute peak through the soil. No complete examination has been made of any problem posed. They are presented in the hope that it will interest readers to further application of Control Theory in water irrigation control and in other areas of agriculture.

For presentation, we shall restrict attention to the mathematical modelling of the 'wet' case only. Extension of analysis to handle the 'dry' case does not present major difficulties from the theoretical point of view.

Following Remark 1, we take for presentation:

System equations:
$$\theta_f \frac{dz}{dt} = R_f \left[u(t) - \frac{E(t)}{\ell(t)} z \right], \qquad 0 < z < \ell(t)$$

$$\theta_f \frac{dz}{dt} = R_f [u(t) - E(t)], \qquad z \geq \ell(t)$$

$$\frac{d\ell}{dt} = \delta \qquad \text{for } t \in [0,T].$$

Constraints:
$$z(0) = 0, \quad \ell(0) = \ell_0 \quad \text{and} \quad \ell(T) = D_r.$$

Problem 1 (Capture and Rendezvous)

As a first step towards accomplishing the desired objective described in the INTRODUCTION, we define:

Objective O(1):

Given a positive ε, find the control $u(t) = I(t)$ with $0 < u(t) < u_m$ for which there exists a time t_c such that

$$|z(t) - \ell(t)| < \varepsilon \quad \text{for} \quad t \in [t_c, T]$$

With this objective, we recognize the problem as a capture problem in finite terms with a rendezvous for the remainder of the interval, (cf. Vincent and Skowronski, 1979; Skowronski, to appear).

Current Liapunov theory results for such problems are not applicable unless t_c is specified. Further analysis may be profitable with this objective.

Problem 2 (Water Peak Tracking Root Length)

Another way of examining the problem is to consider it as a classical tracking problem in control. Let us define:

Objective O(2):

Find the control $u(t)$ with $0 < u(t) < u_m$ which minimizes the integral

$$\int_0^T \frac{(z(t) - \ell(t))^2}{2} \, dt$$

In analysis by Pontryagin's Maximum Principle, the Hamiltonian

$$H = \begin{cases} \dfrac{-(z-\ell(t))^2}{2} + \lambda \dfrac{R_f}{\theta_f} \left[u(t) - \dfrac{E(t)z}{\ell(t)} \right], & 0 < (t) < \ell(t) \\[4mm] \dfrac{-(z-\ell(t))^2}{2} + \lambda \dfrac{R_f}{\theta_f} [u(t) - E(t)], & z > \ell(t), \end{cases}$$

with switching function

$$\sigma(t) = \lambda(t) R_f / \theta_f \quad .$$

Since H is linear in control for each set of specified parameters, there exists the possibility of singular control.

For the problem as stated with $z(T)$ free, the transversality condition is $\lambda(T) = 0$. If $z(T)$ was given fixed, say D_r, the solutions would have to be fitted to the boundary conditions numerically.

Problem 3 (Tracking and Cost Control)

Even though the cost of purchasing irrigation equipment is usually a once only cost, the cost of irrigating is becoming a progressively heavy burden on farmers today. To incorporate a minimizing of irrigation cost, we can study the solution to the control problem with

Objective O(3):

Find the control $u(t)$ with $0 < u(t) < u_m$ which minimizes

$$\int_0^T \left[\frac{(z(t) - \ell(t))^2}{2} - C\, u(t) \right] dt \, .$$

A linear cost term $C\, u(t)$ has been incorporated into the performance integral.

This problem, like Problem 2, also can yield switching from maximum control u_m to zero and the possibility of singular control.

Remark 4

Shortage of time has not permitted the numerical analysis of Problems 2 and 3, either in the solution to this continuous formulation or its equivalent discrete formulation for a given set of parameters. It is research to be further completed. What is of interest here is the determination of the frequency of irrigation, deciding when to irrigate and when not, if the problem is bang-bang with no singular control and to compare them with field data on predictions of irrigation frequency for given crops.

Remark 5

In the problems posed above, the selection of the class of control functions has not been discussed. It would seem appropriate to restrict the analysis to control functions that are piecewise continuous functions, for in practice the irrigation rate is usually set to a constant value over the irrigation period.

Problem 4

In many crops, the yield from a plant is not highly correlated with the weight of a plant. This is not so with a crop such as sugar cane, and in forest plantations yielding usable timber.

Many mathematical models exist to describe the weight of a plant as a function of time, (cf. Thronley, 1976; Hayhoe, 1981). However, to the author's knowledge, the mathematical modelling of how irrigation rate affects weight has received little attention due to complex plant physiology.

How to extend the modelling to be able to examine the maximization of yield of an entire crop at harvest under input irrigation is a further area of open research.

CONCLUSION

As in the case of fishery management, the use of mathematical models as a quantitative approach to plant and crop physiology has only in recent decades gained wide acceptance (Thronley, 1976). However, in this area of agriculture, there has been little application of mathematical machinery such as Control Theory to obvious control problems that arise, for example irrigation control and plant yield as discussed here.

It is hoped that this paper may stimulate further research in the area of modelling in the biological/agricultural community and in the application of mathematics by mathematicians/engineers.

A comforting thought is knowing that in agriculture, we have an can readily obtain data to measure model parameters!

REFERENCES

Bailey, K.P. 1970. The configuration of the root system in relation to nutrient uptake. Adv. Agron., Vol. 22, pp. 159-201

Bloodworth, M.E., Burleson, C.A. and Cowley, W.R. 1958. Root distribution of some irrigated crops using undisrupted soil cores. Agronomy Journal, Vol. 50, pp. 317-320.

Burns, I.G. 1980. Influence of the spatial distribution of nitrate on the uptake of N by plants: A review and a model for rooting depth. Journal of Soil Science, Vol. 31, pp. 155-173.

Burns, I.G. and Greenwood, D.J. 1983. Principles and practice of modelling nitrate and water movement in agricultural soils. In Leaching and Diffusion in Rocks and Their Weathering Products, Prof. S.S. Augustithis (ed.), Theophrastus Pub. S.A., Athens, pp. 543-562.

Cullen, P.W., Turner, A.K. and Wilson, J.A. 1972. The effect of irrigation depth on root growth of some pasture species. Plant and Soil, Vol. 37, pp. 345-352.

Dayananda, P.W.A., Winteringham, F.P.W., Rose, C.W. and Parlange, J.Y. 1980. Leaching of a sorbed solute: A model for peak concentration displacement. Irrigation Science, Vol. 1, pp. 169-175.

Gardner, W.R. 1964. Relation of root distribution to water uptake and availability. Agronomy Journal, Vol. 56, pp. 41-45.

Gerwitz, A. and Page, E.R. 1972. An empirical mathematical model to describe root systems. J. Appl. Ecology, Vol. 11, pp. 773-782.

Hayhoe, H. 1981. Analysis of a diffusion model for plant root growth and an application to plant soil water uptake. Soil Science, Vol. 131, No. 6, pp. 334-343.

Levin, I. 1964. Movement of added nitrates through soil columns and undisturbed soil profiles. Trans. 8th Int. Cong. of Soil Science, Bucharest, Romania, pp. 1011-1022.

Lungley, D.R. 1973. The growth of root systems - A numerical computer simulation model. Plant and Soil, Vol. 38, pp. 145-159.

May, L.H., Chapman, F.H. and Aspinall, D. 1965. Quantitative studies of root development. 1. The influence of nutrient concentration. _Aust. Journal of Biological Sciences_, Vol. 18, pp. 23-35.

May, L.H., Randles, F.H., Aspinall, D. and Paleg, L.G. 1967. Quantitative studies of root development. II. Growth in the early stages of development. _Aust. Journal of Biological Sciences_, Vol. 20, pp. 273-283.

Myers, J.K. 1980. The root system of a grain sorghum crop. _Field Crops Research_, Vol. 3, pp. 53-64.

Raats, P.A.C. 1975. Distribution of salts in the root zone. _Journal of Hydrology_, Vol. 27, pp. 237-248.

Rawlins, S.L. 1973. Principles of managing high frequency irrigation. _Soil Science Society of America Proc._, Vol. 37, pp. 626-629.

Rose, C.W., Hogarth, W.L. and Dayananda, P.W.A. 1982. Movement of peak solute concentration position by leaching in a non-sorbing soil. _Aust. J. Soil Res._, Vol. 20, pp. 23-36.

Rose, C.W., Chichester, F.W., Williams, J.R. and Ritchie, J.T. 1982. A contribution to simplified models of field solute transport. _J. Environ. Qual._, Vol. 11, No. 1, pp. 146-150.

Saxton, K.E., Schuman, G.E., and Burkwell, R.E. 1977. Modelling nitrate movement and dissipation in fertilized soils. _Soil Science of America Journal_, Vol. 41, pp. 625-771.

Skowronski, J.M. To appear. Control for collision with rendezvous or capture. _JOTA_.

Taylor, H.M. and Klepper, B. 1974. Water relations of cotton. 1. Root growth and water use as related to top growth and soil water content. _Agronomy Journal_, Vol. 66, pp. 584-588.

Taylor, H.M. and Klepper, B. 1973. Rooting density and water extraction patterns for corn. _Agronomy Journal_, Vol. 65, pp. 965-968.

Thronley, J.H.M. 1976. _Mathematical Models in Plant Physiology_, Academic Press, New York.

Vincent, T.L. and Skowronski, J.M. 1979. Controllability with capture," _JOTA_, Vol. 29, No. 1, pp. 77-86.

PARTICIPANT'S COMMENTS

The author addresses the problem of "optimizing" irrigation policy in order to (i) minimize the loss of nutrients available for uptake by the roots of a plant due to leaching of solute below the root depth; and (ii) minimize nutrient loss due to loss by evapotranspiration caused by too little water in the vicinity of the growing root.

The first step in an attack on this important resource allocation problem is clearly in valid mathematical model for the growth mechanism of the plant under consideration. The author chooses Dyananda's model from a variety of available models, and makes some reasonable assumptions concerning root growth with time.

The first problem posed by the author is that of controlling irrigation rate to achieve matching of root length and depth of solute peak during some time interval after beginning of irrigation. The second problem posed and solved, at least in principle, is that of water peak tracking root length. The former is conceived of as a so-called "capture and rendezvous" problem, whereas the latter is proposed as a linear-quadratic optimal control problem; two versions of that problem are given. The final problem posed in detail is the same as the tracking one but with cost of irrigation included.

All of the problems posed, but not solved, are of interest. Since the validity of the mathematical model is of prime importance in arriving at a meaningful solution, more attention might be paid to seeking for and employing the "best" available model. Also, since optimizing a single criterion might be too restrictive, some thought should be given to multicriteria optimization in the context of irrigation management.

G. Lietman

Russell Stonier's paper introduces optimal control into irrigation management. To appreciate the significance of his paper, it is worthwhile to take stock of the problem in its complexity.

The question is how to vary the supply of water in order to modify the growth of a plant species so that the yield or profit are maximized.

A closer scrutiny reveals some features setting this problem apart from the simpler applications of control theory.

1. The control variable (supply of water), and the state variable relevant to the objective (i.e., to the commercially valuable part of biomass), are not linked directly. Irrigation influences crops through a mechanism with complicated dynamics. Understanding of these dynamics involves study of soil hydrology and biological growth.

2. A general and reliable theory describing the movement of water in soil is yet to be developed. 'Because of complexity of transport processes in soils, all models describing water and nitrate movement are basically empirical in nature' (Burns and Greenwood, 1983).

3. The not-so-well-known dynamics of moisture movement in soils is further complicated by the action of growing roots actively seeking moisture.

4. Agricultural yield and profit in general depend on the past history of the plant growth. This relationship is specific for a given species: compare the growth of sugar care, rice and peanuts, for example.

If we accept the validity of these points, it becomes obvious that to use optimal control in irrigation management, the problem must be considerably simplified.

Russell Stonier has cut through the Gordian knot by assuming that plants grow best when their roots, while extending at a constant rate, are in the region with maximal concentration of nutrient. These assumptions dispose of all the complications listed above except the second; and this last difficulty is removed by relying on the work of Rose et al. (1982).

Justification of these assumptions calls for an extensive analysis of the results produced by the models and for their comparison with reality. Unfortuantely, the Hamiltonian formulation used leads to two point boundary value problems whose solution in general requires numerical techniques. The analysis of numerical solutions is notoriously cumbersome, and their comparison with experimental evidence is tedious.

In the following, I shall demonstrate how a technique proposed by Clark and DePree (1979) could overcome some of these difficulties.

Consider a generalized problem which contains Stonier's problems 2, 3, 4 as special cases:

$$\text{Minimize} \int_0^T \{(1/2)\,(z-\ell(t))^2 + C(u)\}\,dt$$

$$\text{subject to } \dot{z} = Au - F(t,z),$$

where A is a constant, $C(u)$ is a cost of irrigation, and $F(t,z)$ is a function describing the retardation of the solute peak. We assume $A > 0$, $C(u) > 0$, $F(t,z) > 0$ for the values u, t, z which are of interest.

The corresponding Euler-Lagrange equation is

$$z-\ell + (1/A)C'F' - (1/A^2)C''(\ddot{z} + F'\dot{z} + \dot{F}) = 0 \ ,$$

where C', F', C" denote the corresponding derivatives with respect to z, and $\dot{F} = \partial F/\partial t$. The solution of this equation can be sometimes easily obtained as we shall demonstrate.

Let us consider two special cases:

(i)   Constant marginal cost of irrigation: $C'' = 0$, $C' = c_1 = $ constant.

The Euler-Lagrange equation reduces to

$$z - \ell + (1/A)\,c_1 F' = 0$$

which defines the singular path. The corresponding optimal control is

$$u^* = (1/A)\,(\delta + F(t,z) - (1/A)c_1 \frac{dF'}{dt}).$$

We observe that as $(1/A)c_1 F' \to 0$ the solution approaches $z = \ell$, $u = (1/A)\,(\delta + F(t,L))$.

The optimal irrigation regime thus consists of an initial period during which the maximum rate of irrigation is applied while z increases from zero to $\ell(t) - (1/A) - (1/A)c_1 F'$, followed by irrigation with the rate u^*. This rate could be expected to increase with time until u_m is reached. Thus, contrary to expectation, the optimal irrigation regime does not have a switching character in this case.

(ii)   Variable marginal cost of irrigation: $C'' \neq 0$.

To facilitate analysis, we shall also assume that the solute peak is located deep and hence $F' = 0$. We shall relax this assumption later.

The Euler-Lagrange equation can be written

$$\ddot{z} - (A^2/C'')z = -(A^2/C'')\,\ell(t) + \dot{F} \ .$$

We can distinguish two cases:

a)   Convex cost function (capacity cost present): $C'' > 0$

The Euler-Lagrange equation has the solution

$$z = k_1 \exp(\lambda t) + k_2 \exp(-\lambda t) + \gamma(t)$$

where $\lambda = \sqrt{A^2/C''}$ is a characteristic frequency; k_1, k_2 are arbitrary constants, and $\gamma(t)$ is a particular integral of the Euler-Lagrange equation. For $\dot{F} = 0$, the particular integral is simply $\ell(t)$.

b)   Concave cost function (economies of scale): $C'' < 0$

$$z = k_1 \sin(\lambda t + k_2) + \gamma(t) \ .$$

In the simple case when $F = bD_r$, the irrigation is described by

$$z = k_1 \sin(\lambda t + k_2) + \ell_0 + \delta t$$

$$u^* = (1/A) \left(k_1 \lambda \cos(\lambda t + k_2) + \delta + bD_r \right)$$

For appropriate values of parameters, u^* can oscillate between u_m and zero and hence the economies of scale can entail a 'quasi-switching' optimum irrigation regime.

This agrees with the finding of Clark (1976) who found that pulse fishing can be attributed to economies of scale. Similarly, pulse thinning of forest plantations is an optimal strategy when economies of scale are present (c.f., Lesse and Anderson, 1987).

The existence of a connection between cost of irrigation and frequency of the irrigation cycle has been pointed out by Rawlins (1973).

When the solute peak is not deep enough, $F' \neq 0$ and the Euler-Lagrange equation becomes a general non-linear. This can considerably modify the character of the solution, of course. However, a quadratic cost function with a linear function F merely introduce a damping effect.

References

Burns, I.G. and Greenwood, D.J. 1983. Principles and practice of modelling nitrate and water movement in agricultural soils. 543-562, in: Leaching and Diffusion in Rocks and Their Weathering Products. Augustithis, S.S. (ed.), Theophrastus Publications, Athens.

Clark, C.W. 1976. Mathematical Bioeconomics, Wiley, London.

Clark, C.W. and DePree, T.D. 1979. A simple linear model for the optimal exploitation of renewable resources. Appl. Mathematics and Optimization, Vol. 5, pp. 181-196.

Lesse, P.F. and Anderson, A.E. 1987. In preparation.

Rawlins, S.L. 1973. Principles of managing high frequency irrigation, Soil Science of America Proceedings, Vol. 37, pp. 626-629.

Rose, C.W., Hogarth, W.L. and Dayananda, P.W.A. 1982. Movement of peak solute concentration position by leaching in a non-sorbing soil, Aust. J. Soil. Res., Vol. 20, pp. 23-36.

P.F. Lesse

A DIFFERENTIAL GAME BETWEEN TWO PLAYERS HARVESTING FROM A DIVIDED FISHERY

Brodie Nicol
Department of Mathematics
University of Queensland
St. Lucia, Brisbane
Queensland 4067 AUSTRALIA

A model of a divided fishery is developed where we assume a logistic growth function and diffusion between the two regions. The optimal steady state stock levels for a player are calculated given the other player's control variable and its value. This is applied to a herring fishery.

INTRODUCTION

We consider a nonlinear dynamic game of two players harvesting the same species of fish. The fishery is divided by an international boundary and each sub-fishery is harvested by a sole-owner. This paper is motivated by the need for a sufficiently well-developed theory to support the decision-making by the management of transboundary resources. We extend the work done by Kaitala and Hämäläinen (1982) by modifying their model and considering subfisheries of unequal size. The Nash equilibrium solutions are determined for different policy variables. Then a two-stage game is analyzed in which the players determine firstly the reaction curves to each pair of policy variables and then a non-cooperative game is played to determine the policy variables. A-posteriori values for the sizes of the subfisheries may be determined to meet certain criterion.

A MODEL OF A DIVIDED FISHERY

In each subfishery, we assume the growth function is given by the well-known Schafer model (Clark, 1976; Munro, 1979). The size of the fishery is determined by the carrying capacities. Further, we consider some diffusion between the two regions. This term is assumed proportional to the difference between the two population densities. The density, defined as the ratio of biomass x_i to the carrying capacity k_i for the subfishery i, is thought to give some measure of the ease of finding food. The dynamics of the harvested growth are given by the equations

$$\dot{x}_i = F_i(x_i) - \sigma \left[\frac{x_i}{k_i} - \frac{x_j}{k_j} \right] - h_i(t) \tag{1}$$

$$= f_i(x_i,x_j,t) \qquad \begin{array}{l} i,j = 1,2 \\ i \neq j \end{array}$$

where $x_i(t) > 0$ is the biomass of subfishery i, $F_i(x_i)$ is the growth rate, σ is the diffusion coefficient, k_i is the carrying capacity, and $h_i(t)$ is the harvesting rate.

The logistic growth equation gives

$$F_i(x_i) = rx_i \left(1 - \frac{x_i}{k_i}\right) \tag{2}$$

where r is the intrinsic rate of growth. The harvest rate $h_i(t)$ is assumed to be linearly dependent on the fishing effort E_i. Thus, see Clark (1976), we have

$$h_i(t) = q_i E_i \psi(x_i) \tag{3}$$

where $\psi(x_i)$ is some non-linear function of x_i, and q_i is the catchability coefficient of subfishery i. The harvesting cost is a linear function of the fishing effort E_i, hence

$$c_i(E_i) = c_i E_i . \tag{4}$$

The cost can be expressed as a function of the harvest rate by equation (3). Thus

$$c_i(h_i) = \frac{c_i}{q_i \psi(x_i)} h_i . \tag{5}$$

The net revenue from the subfishery i is

$$\pi_i(x_i, E_i) = pq_i E_i \psi(x_i) - c_i(E_i) , \tag{6}$$

or in terms of h_i,

$$\pi_i(h_i) = ph_i - c_i(h_i) , \tag{7}$$

where p is the price of the fish.

OPTIMAL STOCK LEVELS

We determine, using the Pontragrin maximization principle (Leitmann, 1981), the optimal steady-state stock levels for a player given the other player's control variable and its value. There are three possible nominations for the control variable: namely, harvest rate, stock level, and effort level. Each player is assumed to have perfect knowledge of the population dynamics as given by equation (1). The optimal steady-state reaction curve is defined by the optimal steady-state stock levels for each value of the control variable of the other player. The performance criterion, for player 1, is given by

$$J_1(h_1(x_1)) = \int_0^\infty e^{-\delta_1 t} \pi(h_1) \, dt \tag{8}$$

where δ_1 is the discount rate.

Firstly, consider the reaction of player 1 to a constant harvest rate by player 2. The optimization reduces to an optimal control problem which is to find

$$\max\ J_1(h_1(x_1)) \tag{9}$$

$$h_1 \in (0, h_1^{max})$$

$$\text{subject to } \dot{x}_i = f_i(x_i, x_j, t), \quad i = 1, 2, \quad i \neq j \ .$$

We assume that h_1^{max} is such that the control constraints are not active in the steady-state solution. The Hamiltonian is

$$H(x_1, x_2, h_1, \lambda_1, \lambda_2, t) \quad = e^{-\delta_1 t}\ (p - c_1(x_1))h_1 + \lambda_1 g_1 + \lambda_2 f_2$$

$$= e^{-\delta_1 t}\ \bar{H}(x_1, x_2, h_1, \mu_1 \mu_2) \tag{10}$$

where $\mu_i = e^{\delta_1 t}\ h_i$, λ_i are the adjoint variables. Thus,

$$\frac{d\lambda_i}{dt} = -\frac{\partial H}{\partial x_i} \tag{11}$$

and hence

$$\frac{d\mu_i}{dt} = \delta_1\ \mu_i + e^{\delta_1 t}\ \frac{d\lambda_i}{dt}$$

$$\frac{d\mu_i}{dt} = \delta_1 \mu_1 - \frac{\partial \bar{H}}{\partial x_i} \tag{12}$$

The singular solution is obtained as the trajectory along which the switching function is zero:

$$\frac{\partial H}{\partial h_1} \equiv 0\ . \tag{13}$$

This gives

$$p = c_1(x_i) \equiv \mu_1(t)\ . \tag{14}$$

Differentiating, we get

$$\frac{d\mu_1}{dt} = -c_1'(x_1)\ \frac{dx_1}{dt}\ . \tag{15}$$

However, steady-state solutions imply $dx_1/dt = 0$, thus

$$\frac{d\mu_1}{dt} \equiv 0\ . \tag{16}$$

From equation (12), we have

$$\delta_1 \mu_1 = \frac{\partial H}{\partial x_1}$$

$$= -c_1'(x_1)h_1 + \mu_1(F_1'(x_1) - \frac{\sigma}{k_1}) + \mu_2 \frac{\sigma}{k_1} . \tag{17}$$

Since $d\mu_1/dt \equiv 0$, differentiation equation (17) gives

$$\frac{\partial \mu_2}{dt} \equiv 0 . \tag{18}$$

Hence,

$$\delta_1 \mu_2 = \mu_1 \frac{\sigma}{k_1} + \mu_2(F_2'(x_2) - \frac{\sigma}{k_2}) . \tag{19}$$

Thus, from equations (14), (17), and (19), we get

$$F_1'(x_1) - \frac{\sigma}{k_1} - \frac{c_1'(x_1)h_1}{p - c_1(x_1)} = \delta_1 - \frac{\mu_2}{\mu_1} \frac{\sigma}{k_1}$$

$$\Leftrightarrow F_1'(x_1) - \frac{\sigma}{k_1} - \frac{c_1'(x_1)\left[F_1(x_1) - \sigma\left(\frac{x_1}{k_i} - \frac{x_2}{k_2}\right)\right]}{p - c_1(x_1)}$$

$$= \delta_1 + \frac{\sigma^2}{k_1 k_2\left[F_2'(x_2) - \frac{\sigma}{k_2} - \delta_1\right]} \tag{20}$$

using equation (1) in the steady state. Also,

$$F_2(x_2) - \sigma\left(\frac{x_2}{k_2} - \frac{x_1}{k_2}\right) = h_2 . \tag{21}$$

Equations (20) and (21) define a candidate (x_1^*, x_2^*) for the optimal steady-state singular solution. Thus the reaction curve can be written as $x_1^*(h_2)$. The optimal harvest rate h_1^* and fishing effort E_1^* are obtained from the equations

$$h_1^* = F_1(x_1^*) - \sigma\left(\frac{x_1^*}{k_1} - \frac{x_2^*}{k_2}\right) \tag{22}$$

and

$$E_1^* = \frac{h_1^*}{q_1 \psi(x_1^*)} . \tag{23}$$

Now, consider a constant fishing effort E_2 by player 2. The problem becomes

$$\max \; J_1(x_1,E_1) \qquad (24)$$

$$E_1 \; \epsilon \; (0,E_1^{max})$$

subject to $\dot{x}_i = f_i(x_i,c_j,t), \qquad i = 1,2 \qquad i \neq j \;.$

As before, we assume that the control constraints are not active. After a similar analysis as for the constant harvest case, we obtain

$$F_1'(x_1) = \frac{\sigma}{k_1} - \frac{c_1'(x_1)\left[F_1(x_1)-\sigma\left(\frac{x_1}{k_1} - \frac{x_2}{k_2}\right)\right]}{p-c_1(x_1)}$$

$$= \delta_1 + \frac{\sigma^2}{k_1 k_2 \left[F_2'(x_2) - \dfrac{\sigma}{k_2} - q_2 E_2 - \delta_1\right]} \qquad (25)$$

and

$$\left[F_2(x_2) - \sigma\left(\frac{x_2}{k_2} - \frac{x_1}{k_1}\right)\right]/q_2\psi(x_2) = E_2 \;. \qquad (26)$$

The steady-state singular values (x_1^*,x_2^*) satisfy equations (25) and (26). In this case, the reaction curve may be written as $x_1^*(E_2)$. The corresponding values of h_1^* and E_1^* are given by equations (22) and (23).

Now, let us take the alternative control variable from those mentioned, namely the biomass level. The problem becomes

$$\max \; J_1(x_1,h_1) \qquad (27)$$

$$x_1 \; \epsilon \; (0,k_1)$$

subject to $\dot{x}_i = f_1(x_1,x_2,t)$

where x_2 is some constant. The performance index may be written as

$$J_1(x_1,h_1) = \int_0^\infty e^{-\delta_1 t} \; (p-c_1(x_1))(g(x_1,x_2) - \dot{x}_1) \; dt$$

$$= \int_0^\infty \emptyset(x_1,x_2,t) \; dt \qquad (28)$$

where $f_1(x_1,x_2,t) = g(x_1,x_2) = h_1(t)$. We apply the classical Euler equation $\frac{\partial \emptyset}{\partial x_1} = \frac{d}{dt}\left[\frac{\partial \emptyset}{\partial \dot{x}_1}\right]$. Thus

$$\frac{\partial \emptyset}{\partial x_1} = e^{-\delta_1 t} \left[-c_1'(x_1)(g - \overset{\circ}{x}_1) + (p - c_1(x_1))g^1 \right]$$

and

$$\frac{\partial \emptyset}{\partial \dot{x}_1} = -e^{-\delta_1 t}(p - c_1(x_1)) .$$

Hence

$$\frac{d}{dt}\left[\frac{\partial \emptyset}{\partial \dot{x}_1}\right] = e^{-\delta_1 t}(p - c_1(x_1)) + e^{-\delta_1 t} c_1'(x_1) \dot{x}_1 .$$

Equating gives

$$-c_1'(x_1)g + (p - c_1(x_1))g' = \delta_1(p - c_1(x_1))$$

$$\leftrightarrow F_1'(x_1) - \frac{\sigma}{k_1} - \frac{c_1'(x_1)\left[F_1(x_1) - \sigma\left(\frac{x_1}{k_1} - \frac{x_2}{k_2}\right)\right]}{p - c_1(x_1)} = \delta_1 . \tag{29}$$

Here, the prime denotes differentiation with respect to x_1. The steady-state singular value x_1^* satisfies equation (29) with x_2 = constant. In this case, we get a reaction curve $x_1^*(x_2)$. Again h_1^* and E_1^* are given by equations (22) and (23).

The three above problems may be solved similarly for player 2 when player 1 announces his policy variable and its value. They can be posed in a symmetric form in which the variables k_i, δ_i, c_i and q_i are the same for both subfisheries or an antisymmetric form. Hämäläinen and Kaitala (1982) considered the problem for $k_1 = k_2$, $\delta_i = 0$, $q_1 = q_2$ and $c_1 \neq c_2$. Here, we are particularly interestedin $k_1 \neq k_2$.

There are no reasons why both players should use the same control variable. For each pair of control variables, we get a nonlinear infinite horizon differential game where the Nash equilibrium is given by the intersection of the two reaction curves. If it is assumed that each player does not have knowledge of the others catchability coefficient, then the effort level cannot be used as a control variable. Thus, there are two control variables, biomass level x_i and harvest rate h_i.

ILLUSTRATIVE APPLICATION TO A HERRING FISHERY

Following Hämäläinen and Kaitala (1982), we now consider a particular example. The parameter values were estimated from a herring fishery in the Gulf of Finland. They are as follows: $r = .46$, $p = .92$, $q_1 = q_2 = 9 \times 10^{-5}$, $\sigma = .6 \times 10^5$, $c_1 = c_2 = 4.0$, $k = 4 \times 10^5$. The value of the diffusion coefficient σ requires some comment. Since the diffusion term used here is not the same as that used by

Hämäläinen and Kaitala (1982), the numerical value of the diffusion term is different. Here $\sigma = \hat{\sigma}/2$, where $\hat{\sigma}$ is the value used by the above authors. Thus, in the case of equal sized subfisheries, the two terms agree numerically. Also, the cost coefficients used here are illustrative only and are of the same order of magnitude as the values from the herring fishery. This is because the purpose of this study is to examine the effect of antisymmetry in the subfishery sizes only. The discount rate used is $\delta_1 = \delta_2 = .04$. We let $\psi(x_i) = x_i$ since we have parameter values for this case.

Consider the Nash equilibrium with (x_1, x_2) being the policy variables. For $x_2 = $ constant, we have equation (29) which becomes, after substitution for the growth function and cost function,

$$\delta = r\left[1 - \frac{2x_1}{k_1}\right] - \frac{\sigma}{k_1} + \frac{c}{qx_1^2} \frac{\left[rx_2\left(1 - \frac{x_1}{k_1}\right) - \sigma\left(\frac{x_1}{k_1} - \frac{x_2}{k_2}\right)\right]}{\left(p - \frac{c}{qx_1}\right)} . \tag{30}$$

Rearranging, multiplying by k_2, and writing $\dfrac{k_1}{k_2} = R$, we get

$$r\left[k_2 - \frac{2x_1}{R}\right] - \frac{\sigma}{R} + \frac{cr}{q} \frac{\left[k_2 - \frac{x_2}{R}\right]}{\left(px_1 - \frac{c}{q}\right)} - \frac{c\sigma\left(\frac{x_1}{R} - x_2\right)}{qx_1\left(px_1 - \frac{c}{q}\right)} = \delta k_2 . \tag{31}$$

Now $k_2 = \dfrac{k}{R+1}$, where $k = k_1 + k_2$ is the total carrying capacity, hence

$$r\left[\frac{k}{R+1} - \frac{2x_1}{R}\right] - \frac{\sigma}{R} + \frac{cr}{q} \frac{1}{\left(px_1 - \frac{c}{q}\right)} \frac{k}{R+1}$$

$$- \frac{cr}{q} \frac{\frac{x}{R}}{\left(px - \frac{c}{q}\right)} - \frac{c\sigma}{qx} \frac{\frac{x_1}{R} - x_2}{\left(px - \frac{c}{q}\right)} = \frac{\delta k}{R+1} . \tag{32}$$

Similarly, for $x_1 = $ constant, the reaction curve is

$$F_2'(x_2) - \frac{\sigma}{k_2} - \frac{c_2'(x_2)\left[F_2(x_2) - \left(\frac{x_2}{k_2} - \frac{x_1}{k_1}\right)\right]}{p - c_2(x_2)} = \delta . \tag{33}$$

After manipulation, this becomes

$$r\left(\frac{k}{R+1} - 2x_2\right) - \sigma + \frac{cr}{q}\frac{1}{\left(px_2 - \frac{c}{q}\right)}\frac{k}{R+1} - \frac{cr}{q}\frac{x_2}{\left(px_2 - \frac{c}{q}\right)}$$

$$-\frac{c\sigma}{qx_2}\frac{\left(x_2 - \frac{x_1}{R}\right)}{\left(px_2 - \frac{c}{q}\right)} = \delta\frac{k}{R+1} \, . \tag{34}$$

Thus, equations (33) and (34) give two equations in (x_1, x_2) which define two curves in this plane. The intersection is the Nash equilibrium and is dependent on R. These reaction curves are plotted in Figures 1-3 for three values of R (R = .5, 1 and 2).

Next, consider the Nash equilibrium with (h_1, h_2) being the policy variables. For h_2 = constant, the reaction curve is given by equation (20). This becomes

$$r\left(\frac{k}{R+1} - \frac{2x_1}{R}\right) - \frac{\sigma}{R} + \frac{cr}{q}\frac{1}{\left(px_1 - \frac{c}{q}\right)}\frac{k}{R+1} - \frac{cr}{q}\frac{\frac{x_1}{R}}{\left(px_1 - \frac{c}{q}\right)} - \frac{c\sigma}{qx_1}\frac{\left(\frac{x_1}{R} - x_2\right)}{\left(px_1 - \frac{c}{q}\right)}$$

$$= \frac{\delta k}{R+1} + \frac{\sigma^2}{R\left[r\left(\frac{k}{R+1} - 2x_2\right) - \sigma - \frac{\delta k}{R+1}\right]} \, . \tag{35}$$

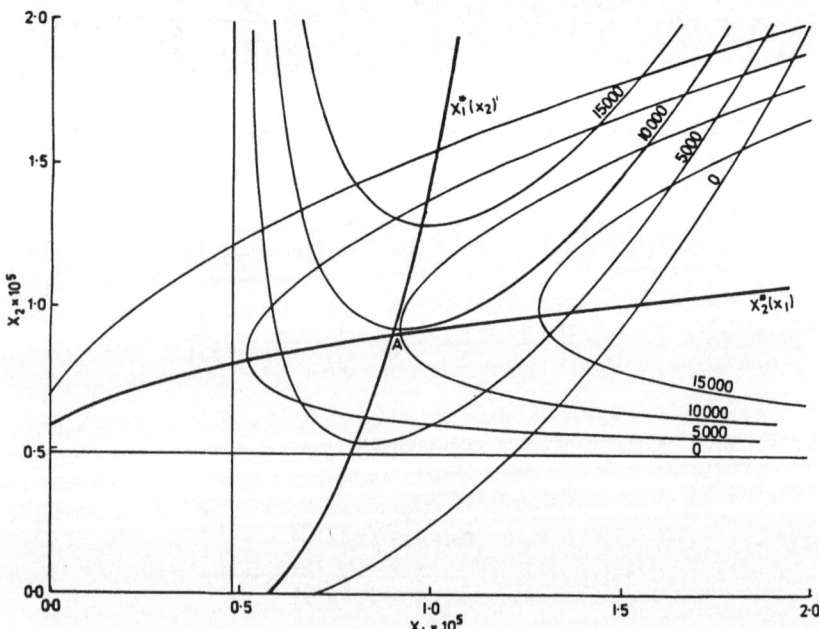

Figure 1.   The Nash equilibrium solution is given by point A. The ratio of the carrying capacities, denoted by R, is one. Curves of constant new revenue, for the system in equilibrium, are also shown.

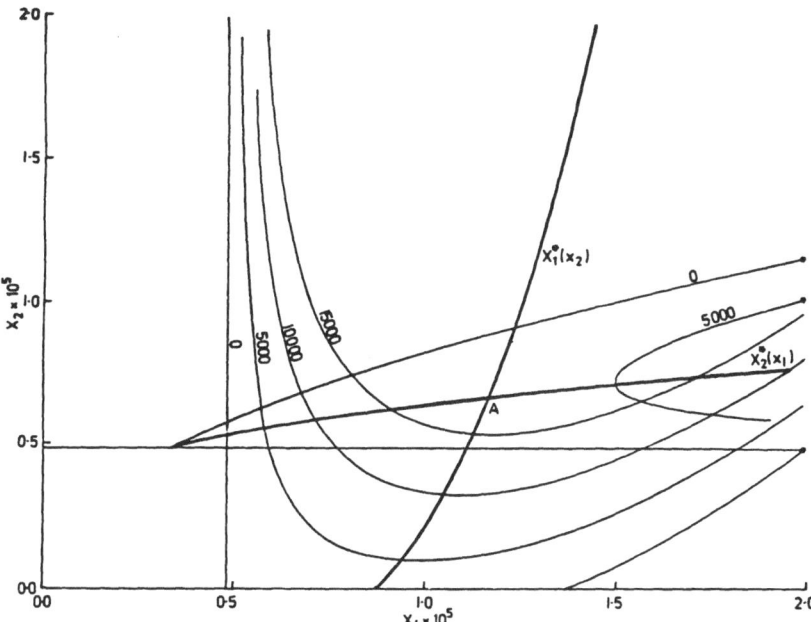

Figure 2.   The same figure as Figure 1 but the ratio of the carrying capacities equals one-half.

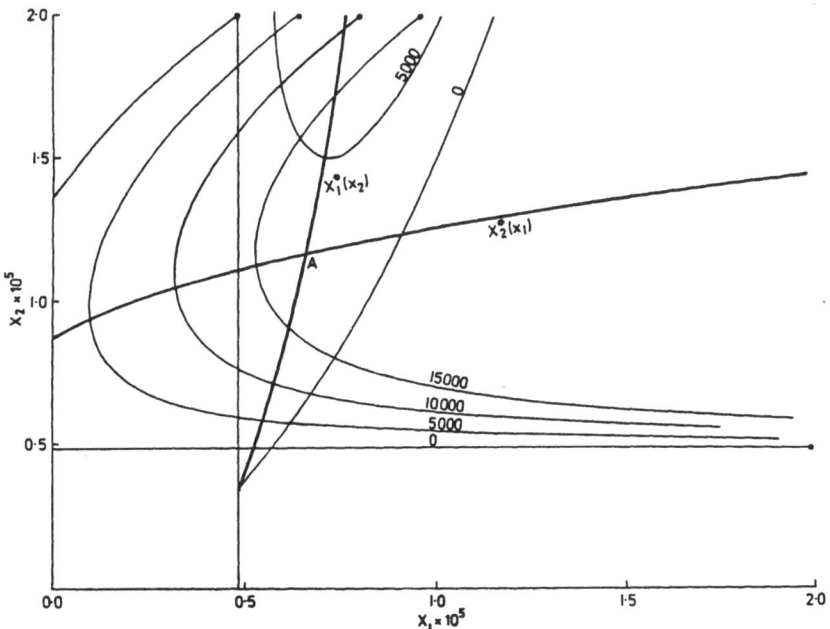

Figure 3. Here, the ratio of the carrying capacities equals one-half.

For h_1 = constant, the reaction curve is

$$F_1'(x_2) - \frac{\sigma}{k_2} - \frac{c'(x_2)\left[F_2(x_2) - \sigma\left(\frac{x_2}{k_2} - \frac{x_1}{k_1}\right)\right]}{p - c(x_2)} = \delta + \frac{\sigma^2}{k_1 k_2\left[F_1'(x_1) - \frac{\sigma}{k_1} - \delta\right]} . \tag{36}$$

After similar algebra as before, we get

$$r\left(\frac{k}{R+1} - 2x_2\right) - \sigma + \frac{cr}{q} \frac{1}{\left(px_2 - \frac{c}{q}\right)} \frac{k}{R+1} - \frac{cr}{q} \frac{x_2}{\left(px_2 - \frac{c}{q}\right)}$$

$$- \frac{c\sigma}{qx_2} \frac{\left(x_2 - \frac{x_1}{R}\right)}{\left(px_2 - \frac{c}{q}\right)} = \frac{\delta k}{R+1} + \frac{\sigma^2}{\left[r\left(\frac{kR}{R+1} - 2x_1\right) - \sigma - \frac{\delta kR}{R+1}\right]} . \tag{37}$$

Equations (35) and (37) define two curves each dependent on R which are the optimal steady-state reaction curves. The intersection defines the Nash equilibrium for these policy variables.

When the policy variables are for player 1, harvest rate, and for player 2, biomass, then the Nash equilibrium is defined by the intersection of the curves given by the equations (32) and (37). The Nash equilibrium for the fourth choice of policy variables, namely (x_1, h_2), is defined by the intersection of the curves given by equations (34) and (35).

Consider the curves of constant revenue. From equation (8), after substitution of equation (5) and (7), we get

$$J_1 = \left(p - \frac{c}{qx_1}\right) \frac{h_1}{\delta} \tag{38}$$

at equilibrium points. Since δ = constant, write $\hat{J} = J_1\delta$, and substitution for h_1 gives

$$\hat{J} = \left(p - \frac{c}{qx_1}\right) \left[F(x_1) - \sigma\left(\frac{x_1}{k_1} - \frac{x_2}{k_2}\right)\right] . \tag{39}$$

Now, writing $\frac{k_1}{k_2} = R$, and substituting for the growth term, we get

$$\hat{J}_1 = \left(p = \frac{c}{qx_1}\right) \left[rx_1 \, 1 - \frac{x_1(R+1)}{kR} - \sigma\left(\frac{x_1}{R} - x_2\right) \frac{R+1}{k}\right] . \tag{40}$$

This gives a curve in the (x_1, x_2) plane along which the net revenue is constant for player 1. Similarly, for player 2,

$$\hat{J}_2 = (p = \frac{c}{qx_2}) \left[rx_2 \left(1 - \frac{x_2(R+1)}{k} \right) - \sigma\left(\frac{x_1}{R}\right) \frac{R+1}{k} \right]. \tag{41}$$

These curves are shown in Figures 1-3. By substituting the appropriate values of (x_1, x_2) at any Nash equilibrium point, the value of the game to both players is found. Thus, a matrix game between policy variables can be played for any given R.

DISCUSSION

The decision making thus follows a two-stage process whereby the reaction curves are determined and then a matrix game is played between policy variables. The value of R used would be determined from estimates of the sizes of the subfisheries. The value for this ratio clearly depends on the location of the international boundary. It may be desirable to determine the ratio R in such a way that the Nash equilibrium is in some desired subset of the (x_1, x_2) phase plane.

The information structure is important. If there is no knowledge of the opponent's catchability coefficient q_i, then effort cannot be used as a policy variable. Also, if there is no knowledge of the other player's performance criterion, then it is not possible to negotiate cooperatively (pareto solutions).

The model used here is more illustrative than realistic. The Schafer model, although extensively used, has been widely criticized and the reliability of the parameter estimates is debateable.

REFERENCES

Clark, C.W. 1976. Mathematical Bioeconomics: The Optimal Management of Renewable Resources, Wiley-Interscience, New York.

Hämäläinen, R.P. and V. Kaitala. 1982. Proc. 21st IEEE Conf. on Control and Decision Making.

Leitmann, G. 1981. The Calculus of Variation and Optimal Control: An Introduction, Plenum Press, New York.

Munro, G.R. 1979. The optimal management of transboundary renewable resources. Can. J. Economics, Vol. 12, pp. 355-376.

PARTICIPANT'S COMMENTS

Idealized models are only useful if they lead to insights that are independent of particular functional forms or illustrative numerical results. The two-player transboundary fisheries problem crops up in a number of economically two-player transboundary fisheries problem crops up in a number of economically important and politically contentious situations, and any insights gained from analyses of such models can be used to negotiate solutions to real problems. It is in this context that the contribution of Brodie should be judged.

Extending the Kaitala-Hämäläinen two-player formulation to subfisheries of unequal size is worthwhile, especially in elucidating the role that the relative size parameter R (see equation (31)) plays in determining each player's strategy. Unfortunately, as discussed below, the model proposed by Brodie appears to be fatally flawed. Furthermore, even if the model were acceptable, Brodie's numerical study is limited to finding the Nash equilibria only for the cases R = 1 and R = 1/2 (note that Figure 3 is the rotated mirror image of Figure 2) so that little insight is obtained into how the solution depends on R.

In the Kaitala-Hämäläinen formulation, the diffusion (migration) term in the fisheries equations (cf. equation (1)) is given by $\sigma(x_i = x_j)$ which implies that the diffusion coefficient σ has the dimensions of time^{-1}. Brodie's extension divides the biomass density variables x_i by the carrying capacity coefficients k_i to obtain the dimensionless proportion x_i/k_i. Thus, in Brodie's equation (1) the diffusion coefficient σ has the dimensions of biomass × time^{-1}. Because σ is assumed to be constant, in Brodie's model the total diffusion rate at time t is independent of biomass densities $x_i(t)$. Intuitively this makes little physical sense. In addition, the Kaitala-Hämäläinen formulation itself is limited by the fact that if we regard $x(t) = x_1(t) = x_2(t)$ as the biomass variable of a fishery, where x_1 and x_2 represent the biomass on either side of the boundary line, then by adding the two equations for i = 1,2 in (1) we obtain

$$\frac{dx}{dt} = F_1(x_1) + F_2(x_2) - h_1 - h_2 .$$

This equation is only expressable in the form

$$\frac{dx}{dt} = F(x) - h_1 - h_2$$

if the functions satisfy $F(x_1 + x_2) = F_1(x_1) + F_2(x_2)$; which excludes all the usual nonlinear population models.

A second question that needs clarification is the nature of the matrix game that would be played between policy variables after the Nash equilibria are found for a priori designated policies. If the Nash equilibria correspond to situations in which all constraints are inactive; then it is not clear to me why the values would be different between policy variables. The three policy variables are related by (3), which together with the equilibrium relationships completely (though not necessarily uniquely) determines one variable in terms of the others.

Wayne Getz

Despite the worldwide acceptance of 200 mile exclusive economic zones (EEZ's), many fish stocks move between national fishing zones. Nearly all fisheries in the North Sea, the tuna fisheries of the Atlantic and Eastern Pacific, and most fisheries in South-East Asia share species which move between economic zones. The number of international agencies which either regulate or study such fisheries is quite large, an explicit recognition that 200 mile EEZ's have not eliminated the concern for interaction between national fisheries.

The traditional theory of open access fishing suggests that the stock will be driven down to a level where there is no profit to be made. In fisheries where stocks move between countries, there is a need for understanding the interaction of fishing policies adopted by the two countries since these situations conform neither to the traditional theory of open access fisheries, nor to the theory of a sole owner.

This paper is, therefore, important in addressing this problem and providing both general insight into the structure of the problem, and a quantitative mechanism for analysis of fisheries of this type. The most important point made in this paper has, perhaps, been understated, the Nash equilibrium points yield stocks and profitabilities well below the potential optimum: the optimum biological stock sizes would be at 2 (roughly double the Nash equilibrium values), and the stock sizes which maximize profits would be even greater.

The current trends in international fisheries management make dynamic game theory an important tool, since many fisheries that had been open access can now be considered two-player games, and the methods used in this paper could be applied to many fisheries problems.

Ray Hilborn

PART III

MANAGEMENT/REAL PROBLEMS

GROUNDWATER-BASED AGRICULTURE IN THE ARID AMERICAN WEST: MODELING THE TRANSITION TO A STEADY-STATE RENEWABLE RESOURCE ECONOMY

Robert McKelvey
Department of Mathematical Sciences
University of Montana
Missoula, Montana USA

In the years since World War II, a flourishing large-scale irrigation agriculture has developed in the arid American West, based on the pumping of groundwater from vast underground pools. Water withdrawals, far in excess of natural recharge, had seemed to make inevitable a cyclic "bust" following the present agricultural "boom", much as happened in the earlier gold-silver-copper "rush" in the Mountain West. However, remarkable technological advances, in irrigation techniques and in the development of genetically-adapted dryland crops, now offer at least the hope of a "soft" transition to a steady-state renewable-resource based economy. Whether this occurs, or whether on the other hand the marvelous technological gains are dissipated away and only briefly delay agricultural collapse in this harsh and erratic climate, may depend on whether farmers and agricultural planners continue to ignore basic economic institutional forces: the destructive effects of uncoordinated common property exploitation of water, compounded by the high degree of irreversibility in capital investment in pumping and distribution systems and in regional infrastructure. These issues are explored with the aid of simple mathematical models.

INTRODUCTION

The concept of a "steady-state" society has fascinated utopian social theorists, environmental conservationists, and practical resource managers alike; e.g., K.E. Boulding (1966). A difficulty, always, has been in managing the transition - it has never been clear how one might get from here to there. This article explores the problem of transition to a sustainable steady-state in a contemporary regional agricultural economy based on a renewable natural resource - water.

The setting is the arid American West, where a flourishing irrigation agricultural industry has developed, based on large-scale pumping of exhaustible groundwater sources. Up to now the groundwater use has to be characterized as "mining", with annual use greatly exceeding annual natural recharge. We examine here the notion of transition from this mining mode to a steady-state mode, and explore in particular how this transition might be affected by two interacting characteristics of groundwater use. First, groundwater is a common property resource, being pumped competitively by independent irrigators from a common pool. Second, much capital investment in groundwater pumping and distribution systems, as well as in regional infrastructure, is irreversible. It cannot be reallocated to alternative uses.

We shall try to understand the implications of these interacting aspects of western irrigation agricultural practice. To this end, our principle tool will be the construction of simple, analytically tractible mathematical models.

In the years since World War II, a large-scale irrigation agriculture has developed in the arid American West, based on the pumping of groundwater from vast underground pools. One of these pools, the Ogalalla Aquifer, underlies parts of eight states in the High Plains, stretching from Oklahoma, West Texas, and New Mexico on the south, through the plains of eastern Colorado and western Kansas and Nebraska, northward into Wyoming and South Dakota.

Once called the Great American Desert, and believed to be unfit for human settlement, this region became in the 1860's and 70's the setting for the legendary cattle drives of the Cowboy West. Subsequently, after the arrival of barbed wire fencing and the self-governing windmill, the High Plains were plowed to wheat, initiating an era of dryland agriculture that at times could bring prosperity and even wealth (when prices were good and the rains came) but which periodically brought ruin and widespread out-migration. In the 1930's, when both markets and the weather failed, the High Plains became the Dust Bowl, and experienced the worst that the Great Depression had to offer.

But then, in the early 1950's, introduction of deep-well turbine pumps and improved well-drilling techniques made possible pumping from the underground water pool on the scale necessary to sustain a high-productivity irrigation agriculture. There followed, in the late '50s, '60s and '70s a period of prosperity and stability, as the High Plains became one of the richest and most productive agricultural areas of North Americal.

Unfortunately, this intensive agricultural development was built, literally, on sand. The waters, contained in the sand and gravel layers of the Ogalalla, recharge only slowly, and irrigation withdrawals soon came to overwhelm the natural influx.

As the water table has been drawn down, in some places by hundreds of feet, well depths have had to be extended and pumping costs have risen, sometimes to prohibitive levels. In places wells have gone dry, and irrigation agriculture has had to be abandoned altogether. People are beginning to perceive, as a consequence of the "mining" of groundwater, the possibility of a boom-or-bust cycle in irrigation agriculture, not unlike that which attended earlier hard-rock mining in the Mountain West, and which has left as its legacy only placer-gouged stream beds and a scattering of abandoned ghost towns.

Whether something like this is destined to happen on the Ogalalla remains to be seen. An important difference is that water from the Ogalalla is, at a certain level of utilization, a renewable resource. Currently, remarkable technological changes in the efficiency of water delivery systems and in newly developed genetic strains of arid land crops have made possible a shift in the Plains agriculture to a less intensive dependence on water (see High Plains Associates, 1982; and Suppola, et al. 1982).

Thus, an orderly transition to a steady-state, low water-use agricultural economy now seems at least possible. Whether these hopes are realized and the "technological fix" indeed will lead to a soft landing for High Plains irrigation agriculture also remains to be seen. Very likely the manner and timing of the transition will be crucial. But these are the issues that we wish to explore.

A pattern of rapid buildup and precipitate decline appears to be characteristic of regional development that is based on the exploitation of an exhaustible natural resource—the so-called boom-and-bust economy. Resource economists have undertaken to study this phenomenon from a broad theoretical perspective, hoping to arrive at general principles for public policy concerning the conservation and appropriate temporal pattern of utilization of a valuable and increasingly scarce natural resource (Cummings and Schulze, 1978).

Much of the theoretical analysis has been based on the examination of aggregated regional economic growth models, which explicitly treat the role of the natural resource input. But mathematical analysis of these models has proved to be difficult. This is because such models, to be meaningful, must incorporate a characteristic feature of regional resource development, one that vastly complicates the dynamic behavior of the models and hence complicates the task of their mathematical analysis. This feature is capital "immalleability": the circumstances that much of the investment, both in primary extractive processes and in infrastructure, is irreversible. That is, the capital cannot later be recovered for other uses (see, for example, Howe, 1984).

The importance of capital immalleability to long-term public investment planning was strongly emphasized by Arrow and Kurtz in their well-known book (1970). Explicit consideration of capital constraints in groundwater modeling was first proposed by Cummings and Burt, also in 1970. The tie-in with the regional development literature has been made by Charles Howe (1984).

The most successful mathematical analysis to date of the effects of irreversible investment in an _optimally_ managed renewable resource industry may be that of Clark, Clarke, and Munro (1979), using sophisticated methods from optimal control theory. (See also Kamien and Schwartz, 1977). Solution trajectories of these models exhibit a characteristic inertia, maintaining an initially high rate of resource extraction until resource stock levels have fallen to well below optimal long-term steady-state levels. The resource then recovers from this depleted state only very slowly.

This sort of system behavior is explained as a "sunk-capital" effect: once the initial investment commitment has been made for the purchase of durable capital, then subsequent decisions concerning operating levels within these capital capacity limits will depend on operating cost considerations alone. The capital equipment will not be allowed to sit idle, so long as gross revenue exceeds the operating costs alone. Only when capital equipment has aged, and continued operation will require its replacement, will investment costs once again enter into operational decisions.

This sunk capital effect is an essential part of the boom-town phenomenon, and has also been cited, e.g., by Clark and Lamberson (1982), as a root cause of the disasterous overexploitation of the great whales by international whaling fleets in mid 20th century. In groundwater exploitation, investment in well-drilling, pumps, water distribution systems, roads and rail-lines, and community "infrastructure," all represent investment in immalleable capital. While this may be required to achieve a productive agricultural system, it at the same time imparts an inertia to the economy which later will inhibit rapid adjustment of the pattern of groundwater exploitation.

Furthermore, in the case of groundwater utilization, the sunk capital effect can be complicated by a significant common property effect, the consequence of competitive pumping of groundwater from a jointly utilized groundwater pool. Indeed, resource economists have often cited competitive utilization of groundwater as a prototypical example of the "common property externality"; see Dasgupta, 1982.

The concept is that, since everyone can draw upon the common resource pool, therefore no one can sequester any part of it for his own future use. So, no one has an incentive to be thrifty about its current use. The predicted result is a more rapid draw-down of the pool than is socially desirable or economically optimal. In the process, the "rents" (royalties) that the resource is capable of providing will be entirely dissipated through an inefficient development pattern.

While there is recent evidence (e.g., Gisser and Sanchez, 1980; Beattie, 1981) that common property effects sometimes may be relatively unimportant in groundwater development, there is also evidence that under certain circumstances, both in early stages of development (e.g., Worthington et al., 1985) and when the water is near to physical exhaustion (Nieswiadomy, 1983), these common property effects can be very substantial. (For detailed examination of these issues, see McKelvey, 1987.)

Our theme in the present article is that common property effects, interacting with capital immalleability effects, can also be of great significance in a period like the present in the Ogalalla, when technological advance holds out the prospect of a ultimate, sustainable steady-state irrigation-based agriculture.

In modeling the withdrawal of water from a single underground pool, one may, at first approximation, abstract from the spatial structure of the underground aquifer and simply record the total volume x of water in the pool at time t. Then x(t) changes according to

$$dx/dt = r - \beta y . \tag{1}$$

Here, r is the rate of natural recharge, y is the rate of pumping from all wells, and β is the fraction of pumped water which is consumed, i.e., which does not find its way back into the underground pool by infiltration from the irrigated lands.

The rate y at which water can be pumped is limited by the level of previous investment in drilled wells, pumps, and distribution systems, all of which we shall lump together as "pumping capital" K. Most simply, we may assume that

$$0 < y < \Omega \cdot K \tag{2}$$

The proportionality factor Ω may be considered to be approximately constant when one is pumping from a deep homogeneous aquifer, that is, one in which each well casing is immersed in a considerable depth of water. But as the water table is drawn down, so that the underground pool becomes shallow, the yield from a given well begins to drop off. This is because the cone of depression in the water-table surface around the well, which forms as a dynamic response to the pumping, begins to intersect the bedrock underlying the aquifer. Then Ω drops off, approximately linearly with the depth of the aquifer. Equivalently, assuming spatial homogeneity, $\Omega(x)$ drops off linearly with decreasing volume x in the pool. Thus, we have $\Omega(x) = \min[mx, \Omega_0]$ for certain geophysical parameter values m and Ω_0.

It follows that, as the water table is drawn down, it becomes necessary, in order to maintain a given volume of flow, first to drill wells deeper and install heavier pumps, and then, nearer exhaustion, to drill additional wells. All of this we lump into increased pumping capacity K. Current pumping capacity is the resultant of investment I in new capital, balanced against the aging of the old:

$$dK/dt = I - \gamma K \tag{3}$$

Here we assume a simple "evaporative" model of capital depreciation, with losses proportional to existing capital. This is a convenient and conventional device; in principle it would be better to keep track of the vintage of capital in the system, perhaps assuming a fixed lifetime for each item in the inventory.

Immalleability is subsumed in an assumption of irreversible investment: $I < 0$. A less drastic assumption would be to allow some disinvestment, but at a cost penalty.

Stock variables x and K describe the instantaneous state of the aquifer development. The rates y of pumping and I of investment are the controls. These are adjusted over time in a way which reflects the management regime that governs the resource development.

For example, "socially optimal" management might undertake to maximize the discounted sum of the flow of "rents" from groundwater exploitation. That is, it would maximize, with respect to y(t) and I(t), the integral

$$\int_0^\infty e^{-\delta t} [U(y) - W(x)\dot{y} - cI] \, dt \quad . \tag{4}$$

Here we shall assume that the utility $U(y)$ of applying water to irrigation is an increasing but concave function of y, recognizing a saturation level beyond which additional water confers no additional benefits in crop production on a fixed land base. The cost $W(x)$, of pumping a unit of water to the surface, increases with depth to water table, hence increases with decreasing x. Per-unit investment costs are assumed to be constant. Finally, δ is the instantaneous discount rate for comparing net returns which are distributed over time.

From the maximum principle of control theory, optimal management will require choosing a rate $y(t)$ of pumping which balances the marginal benefit $p(y) = U'(y)$ of pumped water against total marginal costs of pumping:

$$p(t) = W(x) + \lambda \quad . \tag{5}$$

Here λ is the user cost (or shadow value) of the marginal unit of water in the ground. It includes the increased future costs of pumping that result from the lowering of the water table due to the current pumping. The marginal rule (5) applies unless the capacity constraint (2) binds; in that case, the pumping rate is extremal.

The optimal investment decision is made by comparing the unit cost c of investment with the marginal value μ of expected future returns from the investment. Here

$$\mu(t) = \int_t^\infty e^{-(\tau - t)(\delta + \gamma)} [p(y) - \lambda - W(x)]^+ \, \Omega(x) \, d\tau \, ,$$

where the superscript "+" indicates "positive part". The exponential factor $e^{-(\tau - t)\gamma}$ reflects the depreciating productivity over time τ of investment made initially at time t. The investment rule is dichotomous: when $\mu < c$, then the investment rate is zero; when $\mu > c$, a pulse investment occurs. Positive finite investment can occur only at a steady state, where $\mu(t) \equiv c$.

When, as is the norm, investment and pumping decisions are made, not by a central manager, but by individual irrigators who are competitively withdrawing water from a common underground pool, then $y(t)$ and $I(t)$ will be set at levels which are suboptimal, resulting in total dissipation of the royalty value of the groundwater. In effect, each individual (and the agricultural community as a whole) will set $\lambda(t) \equiv 0$, ignoring the future value of water in the ground. (They will, on the other hand, fully value future returns from capital investment, hence will continue to take into account the rates of discount and depreciation.)

Figure 1(a) and 1(b) are phase plane patterns of the evolution of groundwater volume x(t) and pumping-capital K(t), first under optimal central management and second under a common property competitive regime when investment is fully irreversible. Figure 2 shows the compensating effect of positive depreciation. Note that in the common property case, both initial investment and terminal water-table depression are excessive, as compared with the optimum. For a more complete account of the analysis, and further interpretations, see McKelvey (1987).

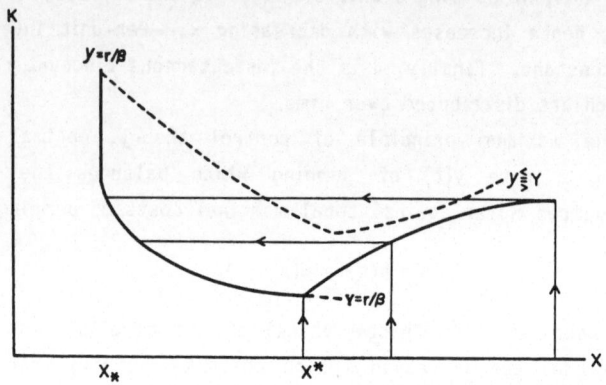

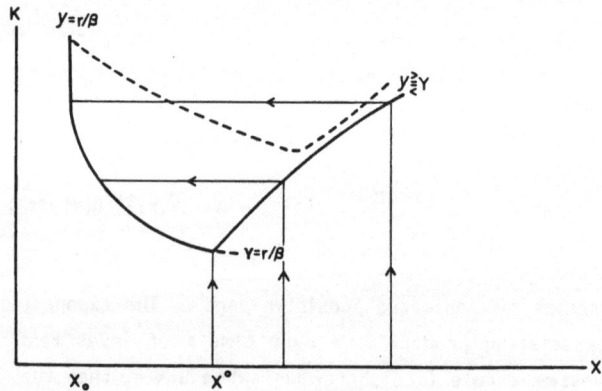

Figure 1.   Groundwater development with immalleable capital.   No depreciation: $\gamma = 0$.   Phase-plane trajectories, with X = groundwater stock level and K = capital investment, contrasting exploitation trajectories (a) under common property use, and (b) with socially optimal management. Pumping rate is $y = \min[y,Y]$ where $Y = \Omega(x)K$, the capacity-constrained rate, and y is determined by the marginal rule $P(y) = W(x)$ (for common property exploitation) or $P(y) = W(x) + \lambda$ (at social optimum). (Yield Trends in the Ogalalla (bu/acre, using best technology), Source: Supalla et al., 1982.)   An initial investment pulse is followed by drawdown of the water table, at first rapidly and then, as pumping costs rise, more slowly until finally withdrawals just balance recharge.   But the equilibrium watertable level is depressed because of sunk capital.

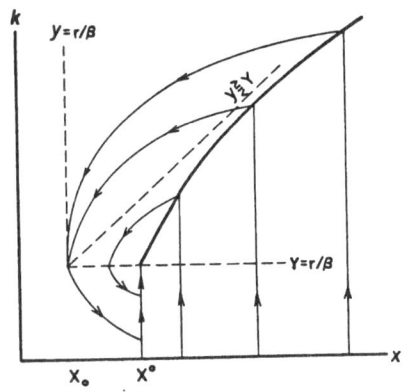 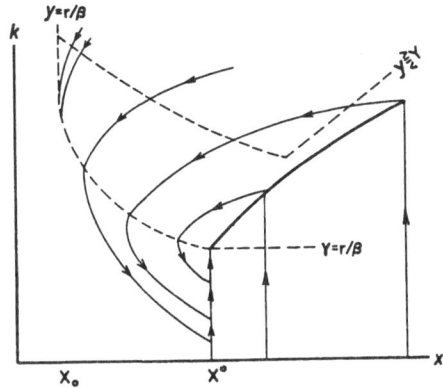

Figure 2.   Common property groundwater exploitation with capital which is immalleable but subject to positive depreciation: $\gamma > 0$. Trajectories shown are, respectively, (a) for deep and (b) for shallow aquifer. Unlike the previous case ($\gamma = 0$), there is, in the long term, a very slow recovery of watertable level, and eventual capital replacement, to achieve a steady state which balances capital costs.

TECHNOLOGICAL CHANGE

Recent studies, [The Ogalalla High Plains Study, 1982; Supalla et al., 1982] have documented remarkable changes, almost revolutionary in character, that are beginning to affect the basic terms of agricultural activity in the High Plains. By a combination of improved irrigation efficiency, i.e., getting more of the water to the crop roots, and genetic improvements in the plants themselves, per-acre yield trends have been going up at a phenomenal rate (see Table 1.) This trend affects both irrigated and dryland crops. In particular, it is estimated that by the year 2020, dryland may be producing nearly as much crop yield per acre as irrigated land does today.

Table 1

	1950	1980	2020 (projected)
Irrigated Corn	75	137	203
Dryland Sorghum	35	76	112

As a result, while by 2020 one acre in three of currently irrigated agricultural land is expected to revert to dryland as a result of physical or economic exhaustion of groundwater, still over-all irrigated acreage is projected to increase by 25%! This will occur because irrigation investment is coming to be profitable in new, generally more northerly, High Plains regions at the same time as wells are going dry in the more arid south.

To water policy analysts, these predictions provide hope that the agricultural boom of the '60s and '70s need not, after all, give way to a bust in the '90s. On the other hand, one is entitled to wonder whether the stunning new technological breakthroughs represent anything more than a breathing spell--one whose benefits might be dissipated as quickly and as inevitably as those of the earlier era. Let us return to our model to examine this issue a bit more closely.

In terms of the model, the effect of technological change is, crudely, to alter the demand curve for groundwater in agricultural use, somewhat like in Figure 3. There, $p_-(y)$ is marginal return to water use prior to technological advance, and $p_+(y)$ after. At the same time, the saturation level of profitable water use will be lowered. In most cases, sustainable yield levels of water use, matched to natural recharge, will gain significant value: that is, $p_+(r/\beta) \gg p_-(r/\beta)$.

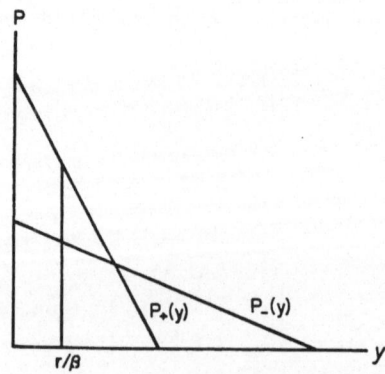

Figure 3.   Technological change, as represented by a changing marginal utility curve. $p_-(y)$ is before change; $p_+(y)$ is after, with y the rate of discharge, and p(y) its marginal value in irrigation agriculture.

Let us suppose that the schedule $p_+(y)$ will replace $p_-(y)$ instantaneously at a certain time T. If this "technological fix" is put in place without advance warning, or if (as when there is common property exploitation of the aquifer) individuals do not feel justified in modifying useage in anticipation of the change, then the transition time T can be treated as an initial point of a trajectory of Figure 1. Most likely $x(T) > x_+^0$, so that that conceivably there will be an investment pulse at time T. (On the other hand, existing equipment may be able to deliver the smaller pumped flows now called for by the new marginal rule.) Certainly, in the long run, water tables will be lowered to a new, depressed equilibrium level, consistent with $p_+(r/\beta) > p_-(r/\beta)$.

With some foreknowledge of impending technological change (as indeed exists at the present time), optimal management would seem to call for a transitional period of conservation of groundwater, in anticipation of the expected appreciation in its value. To explore this, let us examine a short-run approximation to the model of withdrawing water from a single underground pool, with the prices p_+ and p_- now taken to be constant, and capital capacity to be fixed.

Formally, one undertakes to solve

$$\max J = \int_0^\infty e^{-\delta t} [p(t) - W(x)] \, y(t) \, dt$$

where

$$p(t) = \begin{cases} p_- & \text{for} \quad t < T \\ \\ p_+ & \text{for} \quad t > T \end{cases}$$

and

$$0 < y < y_{max} \quad .$$

As is well known [Clark 1976, Section 3.3], such a linear model does predict a period of conservation (with $y \equiv 0$) prior to time T, so that groundwater level initially will rise, before being drawn down (with $y = y_{max}$) to its long-term steady state level.

Under a common property regime, no such initial rise of water table will be expected (see Figure 4). Furthermore, extrapolating back to conditions of immalleable investment, one would expect initial capital capacity y_{max} to be larger than optimal under the common property regime, so that common property drawdown would be more rapid than optimal.

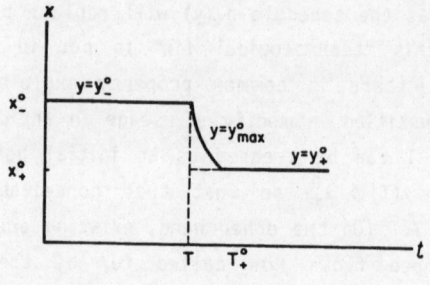

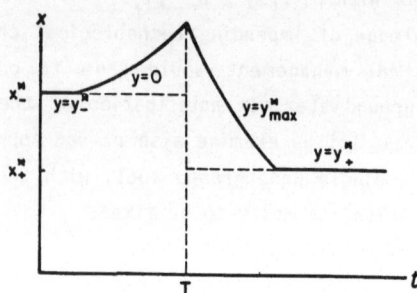

Figure 4.   The effect of sudden technological advance in a linearized groundwater
model, contrasting the situation of (a) common property utilization with
(b) socially optimal management.   Here, groundwater stock level is
plotted against time, and y is the rate of discharge.

Furthermore, there is more opportunity for (and need for) conservation than
just in the initial transition period to the high technology regime. This is because
technological adaptation to the dry climate cannot be expected to be absolute. A
return to a semblance of dryland agriculture on the High Plains means living once
again with the drought cycle pattern that created the Dust Bowl.

In the linear model, this state-of-affairs can be represented by a succession
of extensive periods of constant value p_+ for water, broken by shorter periods
(drought periods) in which water value would rise to $p_{++} > p_+$. Optimal management
would require conservation, and a rising water table, in periods prior to drought,
something like in Figure 5.

However, with common property exploitation, the model predicts that these
efforts at conservation would not occur. Then there would be less water available
to get through the hard times, and more of a scramble to use what there was. In
reality, of course, periods of drought will not be predictable in advance, so that an
added conservation cushion optimally would be maintained at all times, in
anticipation of random fluctuation in recharge. (Indeed, a proper model ought to be
stochastic.)

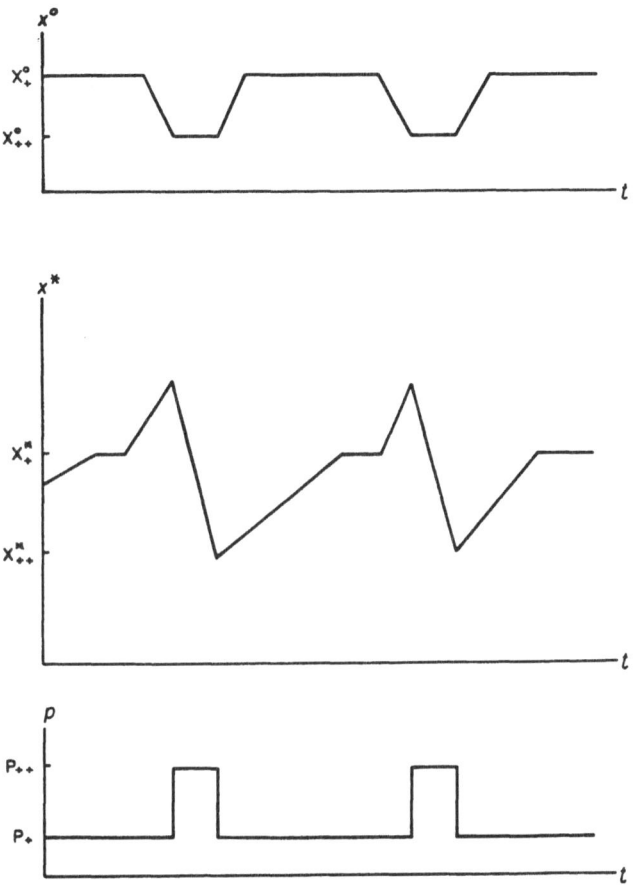

Figure 5.   Water storage across drought cycles in the linearized model. (a) shows
common property exploitation, and (b) shows the social optimum.

Finally, let us emphasize that we have had to abstract from capital immalleability in order to make the model tractible. Returning to the full model of withdrawing water from a single underground pool, an extrapolation to non-autonomous, even stochastic, $p(y,t)$ would suggest that under common property use irrigators would have (or acquire) excessive levels of pumping capacity at the onset of technological change, and would maintain excessive levels, in expectation of drought periods. The result should be, as always with common property exploitation, an excessive lowering of water tables and increased incidence of dry wells.

In brief, the long-run state-of-affairs might be no better than that of today; all the gains from the technological miracle having been dissipated away.

Thus, we are led back to our original queries, still unanswered: How feasible is a steady-state irrigation-based agriculture on the High Plains? How can one best manage the transition to it? To these, we must add a new query, suggested by our modeling exercise: How can a steady-state High Plains agriculture best cope with the natural variations inherent in a harsh and unstable physical environment? As usual, more data and further analysis will be needed in order to say. Perhaps simple modeling can illuminate the way.

REFERENCES

Arrow, Kenneth J. and Mordecai Kurtz. 1970. Public Investment, The Rate of Return, and Optimal Fiscal Policy, Resources for the Future, Johns Hopkins Press, Baltimore.

Beattie, Bruce R. 1981. Irrigated agriculture and the Great Plains: Problems and policy alternatives. W. Journal of Agri. Econ., Vol. 6, pp. 289-299.

Boulding, K.E. 1966. The economics of the coming spaceship Earth, in Environmental Quality in a Growing Economy, H. Jarrett (ed.), Johns Hopkins Press, Baltimore.

Clark, Colin. 1976. Mathematical Bioeconomics: The Optimal Management of Renewable Resources, John Wiley, New York.

Clark, Colin W., Frank Clarke, and Gordon Munro. 1979. The optimal exploitation of renewable resource stocks: Problems of irreversible investment. Econometrica, Vol. 47, pp. 25-49.

Clark, Colin W., and R.H. Lamberson. 1982. An economic history and analysis of pelagic whaling. Marine Policy, Vol. 6, pp. 102-120.

Cummings, R.G. and Oscar Burt. 1969. The economics of production from exhaustible resources. American Economic Review, Vol. 59, pp. 985-990.

Cummings, R.G. and W.D. Schulze, 1978. Optimal investment strategies for boomtowns. American Economic Review, Vol. 68, pp. 374-385.

Dasgupta, Parthan. 1982. The Control of Resources, Harvard University Press, Cambridge, Massachusetts.

Gisser, Micha and D.A. Sanchez. 1980. Competition vs. optimal control in groundwater pumping. Water Resources Research, Vol. 16, pp. 638-642.

High Plains Associates. 1982. Six-State High Plains Ogalalla Aquifer Regional Resources Study. Camp Dresser and McKee, Inc. 3445 Executive Center Drive, Suite 200, Austin, Texas.

Howe, Charles W. 1984. On the Theory of Optimal Regional Development Based on an Exhaustible Resource, University of Colorado Economics preprint.

Kamien, M.L. and N.L. Schwartz. 1977. Optimal accumulation and durable goods production. Zeitschr. fur. nationalökonomie, Vol. 37, pp. 25-43.

McKelvey, Robert. 1985. Decentralized regulation of a common property renewable resource industry with irreversible investment. J. Environ. Econ. Manag., Vol. 12, pp. 287-307.

McKelvey, Robert. 1987. Groundwater: Ramification of Irreversible Investment Decisions Under Common Property Exploitation. (Under preparation).

Nieswiadomy, Michael Louis. 1983. Adjusting to Diminishing Water Supplies in Irrigated Agriculture. Ph.D. Dissertation, Texas A&M, 124 pgs.

Supalla, Raymond J., Robert R. Lansford, and Noel R. Gollehon. 1982. Is the Ogalalla going dry: A review of the High Plains study and its land and water policy implication. Journal of Soil and Water Conservation, Vol. 37, pp. 310-314.

Worthington, Virginia, Oscar R. Burt, and Richard L. Brustkern. 1985. Optimal management of a confined groundwater system. J. of Environ. Econ. and Manag., Vol. 12, pp. 229-245.

PARTICIPANT'S COMMENTS

For those interested in Natural Resource modelling, this paper is of interest in that it uses a set of tools and approaches that are well developed in the fisheries and forestry literature, and examines another renewable resource: groundwater. The groundwater problem would appear to be somewhat unique, in that the renewal rate is very low and is independent of the size of the stock (underground reservoir). However, this is not different from a population with a low natural mortality rate and a low recruitment rate that is independent of stock size.

The second interesting aspect of this paper is the nature of the problem itself: the importance of the irrigation based agriculture of much of the American west is enormous, and the economic impact of the changes discussed by McKelvey are great. Thus, the use of the models to clarify potential alternative trends in groundwater use must be a significant component of the political debate over what to do about groundwater management. This is not an academic exercise but rather one with important real world implications.

Finally, the third part of this paper that is perhaps the most important is the consideration of problems of technological change. While McKelvey's analysis may represent a first attempt at looking at technological change in this one resource problem, it provides a starting point and a stimulus for the same type of analysis in the other, better established areas such as forestry and fisheries. Almost all work in natural resource management assumes a stable industry structure, whether it be product mix in forestry, or gear efficiency in fisheries. Systems analysis could make a very significant contribution to these fields if we shifted our focus a bit from more and more detailed analysis of traditional models with unchanging industry, to the dynamics of transition due to technological change. McKelvey's inclusion of this in the groundwater problem thus sets an example for us all.

Ray Hilborn

This paper is an excellent example of how simple models can provide valuable insights about systems that are spatially complex both in terms of the resource dynamics and the utilization/regulation processes. Especially important points are reviewed about sunk capital affects on the optimum development trajectory, common property effects, and the idea that conservation may be optimum given "knowledge of impending technological change".

I saw three points that deserve further analysis. First, the resource spatial structure and capital age structure may act to either smooth out or make more abrupt temporal transitions from boom to bust (or sustained use); the simple models may be far too optimistic (or pessimistic) in their predictions. Second, it would be wise to see how the optimum transition policy changes when other objectives than rent maximization are used (i.e., who are the beneficiaries of regulation, and how should their preferences be weighted). Finally, there needs to be a more careful look at what the real decision variables are in cases such as the Ogalalla; there are not one but many possible pumping "regulators" at different scales, and it would not be simple to obtain consensus among them.

Carl Walters

GREAT LAKES FISHERIES: ARE EXPLICIT CONTROLS NECESSARY?[1]

George R. Spangler
Department of Fisheries and Wildlife
200 Hodson Hall
University of Minnesota
St. Paul, Minnesota 55455

This is an essay on the current state of fisheries management in the Laurentian Great Lakes. It is presented here to introduce the subject to an audience of control specialists who may have somewhat different theoretical and ecological perspectives on fishery management than do biologists practicing in the Great Lakes region. Following a brief history of Great Lakes fishery problems[2], I will review the corrective measures that have been taken and identify the critical uncertainties surrounding past and present management activities. Where possible, I will identify those processes which may be amenable to significant intervention or control by management agencies.

HISTORICAL GREAT LAKES FISHERIES

For nearly two centuries following the settlement of the St. Lawrence basin by European colonists, the Great Lakes produced an abundance of freshwater fish. The biological production sequence responsible for this fish yield throughout the Great Lakes depended upon an aquatic community of primary producers and microfauna that have subsequently come to be known as "clean-water associates." These included benthic and pelagic species characteristic of clear waters containing near-saturation levels of oxygen. These organisms provided the food resources that supported a complex assembly of Coregonines (whitefishes) which, in turn, supported the top-level piscivorous trout and freshwater cod. The fish most highly prized in the marketplace were members of the Salmonidae, including especially, the shallow-water cisco, *Coregonus artedii*, lake whitefish, *C. clupeaformis*, deep-water ciscoes or chubs, *Coregonus* spp., and lake trout, *Salvelinus namaycush*.

Late in the Nineteenth century, it was recognized that fish production was declining, especially in the lower Great Lakes, and inquiries were initiated to discover the causes. The U.S. Fish Commission, established in 1871, conducted a number of studies of the Great Lakes fisheries (Milner, 1874; Smith and Snell, 1890), and commercial fishing records consisting of landed (or shipped) weights were being

---

[1]Contribution No. 15,046 of the Scientific Journal Series of the Minnesota Agricultural Experiment Station, St. Paul, Minnesota.

[2]Lake-by-lake accounts of changes in the Great Lakes fisheries are presented in case history studies in the proceedings of the SCOL symposium, (J. Fish. Res. Board Can., Vol. 29, No. 6, 1972) and in Tech. Report Nos. 19-23, of the Great Lakes Fishery Commission, Ann Arbor, MI.

kept by some conservation agencies. The U.S. fishery statistics, although incomplete and fraught with reporting errors, clearly indicated that Great Lakes catches had peaked by the turn of the century and were entering a period of somewhat variable, but regular, decline (Baldwin, et al., 1979). Some of the early studies, for example the faunistic survey of Lake St. Clair by Jacob Reighard (1894), resulted in a thorough understanding of the natural history of some of the commercially important species of fish, but little was known of the critical processes needed to sustain valuable fish populations indefinitely.

Throughout the early decades of the present century, Great Lakes fishermen intensified their efforts with improvements in fishing technology and vessels. During this period, some species were swiftly declining in abundance (Koelz, 1926; Van Oosten, 1929; Regier et al., 1969) while others seemed to be maintaining relatively stable yields. The collapse of the once large and valuable stocks of shallow-water cisco in Lake Erie precipitated a persistent controversy over the cause of the decline (Egerton, 1985). On the one hand, Thomas Langlois argued that pollution was the causative agent, whereas John Van Oosten believed overfishing to be the cause. Early in this exchange of views, Langlois (1942) made the important observation that the disppearance of the cisco, dramatic though it was, did not result in a significant overall reduction in the aggregate catch of all commercial species from Lake Erie. Although each of the protagonists in the debate stated their opinions convincingly, the failure to reach unequivocal conclusions regarding the *single most important reason* for collapse of the stocks simply underscored the fact that no one factor could account for changes of the magnitude that were being observed. Elsewhere in the Great Lakes, valuable stocks of fish were beginning to show the cumulative signs of stress that would ultimately result in catastrophic changes in the fish communities.

Collapses of Great Lakes fish stocks did not occur simultaneously across the entire basin in spite of the apparent constancy in total commercial fish yield statistics (Fig. 1). The earliest declines occurred in Lakes Ontario and Erie where the premium salmonid species were essentially eliminated before the 1920's. While salmonids declined steadily in Lakes Huron and Michigan, fishermen in the lower lakes (especially Lake Erie) were turning to other species. The substitution of lower value species (largely non-native marine fish) for the premium salmonids is particularly conspicuous in the case of smelt (*Osmerus mordax*) in Lake Erie during the 1960's, and for alewives (*Alosa pseudoharengus*) in Lake Michigan after 1965. The aggregate effect was, as Langlois had observed earlier, that total fish yield from the entire system appeared to be somewhat stable for the 50-year period since 1920.

During the 1930's, following the expansion of the Welland Canal, parasitic stage sea lampreys (*Petromyzon marinus*) were observed in the three upper Great Lakes (Smith and Tibbles, 1980). Spawning lampreys were first observed in a Lake Michigan tributary in 1936 and populations increased rapidly in abundance during the following decade. Almost immediately, lake trout and other valuable species of Great Lakes fish began precipitous declines in abundance. Sea lamprey wounds (see King, 1980, for illustrations) on dead and dying lake trout implicated the lamprey as the cause of the declines of valuable fish stocks and research was initiated to find a means of controlling this marine invader. Further evidence of the lamprey effect was the sequential repetition of an inverse relationship between lake trout commercial yield and lamprey abundance (Smith, 1968) in each of the upper lakes. Mechanical and electrical barriers on tributaries used by spawning lampreys killed hundreds of thousands of lampreys, but fish stocks continued to decline.

Fortunately for future lamprey control efforts, the initial life history studies (Applegate, 1950) revealed that lampreys undergo an extensive larval life history in tributary streams, generally requiring 3 to 5 years as non-parasitic ammocoetes before transforming into the parasitic (or predatory) form detrimental to fish. Also, the parasitic animals lived only 12 to 18 months in the lake, returning just once to the tributaries where they would spawn and die. In 1958, a halogenated mononitrophenol (3-trifluoromethyl-4-nitrophenol, or TFM) was found to have greater toxicity to larval lampreys than to other fish. TFM was quickly adopted to control sea lampreys (Applegate et al., 1961) by application to tributary streams inhabited by ammocoetes. TFM remains today the single most important method of control.

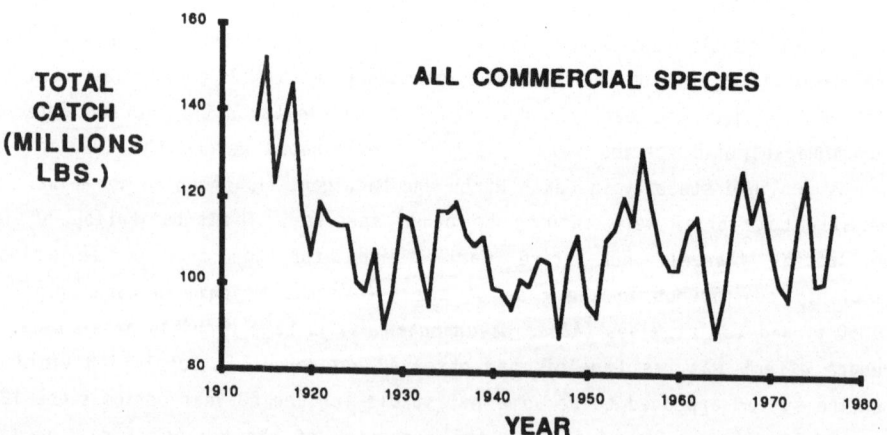

Figure 1.   Reported landings of all commercial species from the Great Lakes since 1914 (Source: Baldwin et al., 1979).

Lamprey control has been a dramatically successful program in the Great Lakes by almost any measure. First applied to Lake Superior tributaries (in order to save remnant lake trout stocks still extant during the early 1960's) and subsequently extended to the other lake basins, there has been a reduction of lamprey abundance of approximately 90% from the pre-TFM control period. In Lake Superior, 6 lamprey assessment barriers on the south shore from 1954 to 1978 showed a marked decline in the number of adult lampreys returning to spawn (Fig. 2). The extended larval life history of sea lampreys allows application of TFM at 2 to 4 year intervals, depending upon the growth of the ammocoetes. While some of the more problematic streams are treated annually, many lamprey-producing tributaries require only infrequent treatment, and some streams have not been recolonized by lamprey for many years following the initial application of TFM (Torblaa and Westman, 1980).

In spite of the dramatic improvements in fish stocks attributed to lamprey control, there has been little progress in identifying the relationship between control activity and a response that can be immediately observed. Ideally, we would like to be able to obtain a direct estimate of reduction in fish mortality as a consequence of intensity of control. Procedures have been developed to estimate the impact of lampreys on prey fish stocks (Spangler, et al., 1980; Youngs, 1980), but these invariably require additional information on lamprey and prey fish abundance in order to relate to lamprey control.

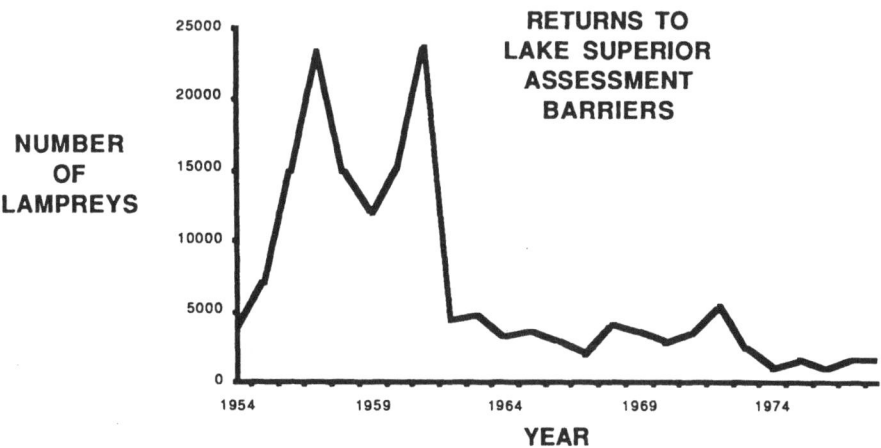

Figure 2.   Numbers of sea lampreys taken in spawning runs to Lake Superior lamprey assessment barriers from 1954-1978 (Source: Smith and Tibbles, 1980, Table 5).

INTERIM MANAGEMENT ACTIVITIES

During the period following collapse of valuable stocks due to lamprey predation, and prior to basin-wide effective lamprey control, i.e., from the mid-1940's until the mid-1970's, major changes occurred in the Great Lakes ecosystem. Water quality improved greatly in the Lake Erie basin due to reductions of plant nutrients in domestic waste effluents and agricultural runoff. Replacement of DDT, dieldrin, and other "hard" pesticides by less persistent toxicants resulted in declining concentrations of some environmental contaminants although there were increases in some others, notably toxaphene and PCBs in the upper lakes, and dioxin in Lake Ontario. In Lake Erie, where anthropogenic eutrophication had resulted in nuisance blooms of filamentous algae, and occasional anoxic conditions in hypolimnetic waters, public resolve to clean up the lake resulted in empirical proof of the reversibility of eutrophication.

In the upper Great Lakes, water quality degradation had generally been conceded to be minimal except near population centers, yet the fisheries for valued species continued their somewhat erratic declines, punctuated occasionally by spectacularly abundant year-classes that would eventually diminish to insignificance. Major changes occurred in the species composition of the offshore fish communities, with smelt and alewives dominating regions of the lakes formerly occupied by deepwater and shallow-water ciscoes. Speculation abounded among biologists as to the reasons for the instability of the fish community, but hypotheses attributing general responses to single factor causes clearly resembled the Langlois-Van Oosten debate in failing to account for much of the empirical evidence and demonstrating little predictive power (see especially Van Oosten et al., 1946; Smith, 1970; and Selgeby et al., 1978).

Finally, a useful synthesis of events emerged from the SCOL Symposium convened in 1971 to examine the singular and joint effects of the three major stresses impinging upon the Great Lakes during the Twentieth Century (Loftus and Regier, 1972). These stresses, fishery exploitation, cultural eutrophication, and the invasion or introduction of exotic species, were judged responsible for the state of the fish community. Among these factors, exploitation was assumed to be controllable, eutrophication had only recently been shown to be reversible, and most biologists were skeptical about the reversibility of effects due to the presence of exotic species. Except for a few locations, sea lampreys were being effectively controlled and survival of the larger species of fish was improving with lamprey reductions (Pycha, 1980; Spangler and Collins, 1980).

Fishery exploitation during this period remained at levels commensurate with harvest of those fish stocks which had survived the period of peak lamprey abundance. In the Lake Huron fishery, whitefish and chubs became the dominant species and the attrition in numbers of commercial fishermen continued to reduce participation in the fishery. In some jurisdictions, commercial fishing licenses were not being renewed. Only a few jurisdictions revoked licenses of *active* fishermen during this period, thus there is little empirical evidence that Great Lakes fishery management agencies would act with resolve to significantly restrict redevelopment of intensive fisheries should the fish populations recover sufficiently to attract attention again. The test of the assumption of controllability of the fisheries would simply have to wait until fishing improved.

Early in this interim period, fish stocking programs were undertaken to mitigate the losses being endured due to lampreys. Lake Superior was increasingly being planted with lake trout fingerlings and yearlings in an effort to bolster stocks and to discover the sizes of fish that would survive best in future planting efforts. During the mid-1960's, a number of exotic salmonids were planted in the Great Lakes in an effort to find edible or recreationally interesting species that could survive the presence of sea lampreys and utilize the enormous biomass of alewives extant in Lakes Michigan and Huron. Coho Salmon (*Oncorhynchus kisutch*), initially intended to establish the species in the Great Lakes (Tody and Tanner, 1966), quickly became the basis for extremely important artificial fisheries (Talhelm and Bishop, 1980) in U.S. waters. These plantings were soon followed with chinook salmon, *O. tshawytscha*, Atlantic salmon, *Salmo salar*, and increases in the numbers of other non-native salmonids that had been stocked irregularly during the past thirty years. In Canadian waters, kokanee salmon (*O. nerka*) and a reproductively viable strain of the splake hybrid (*Salvelinus namaycush X. S. fontinalis*) were planted in selected locations in attempts to establish self-sustaining stocks.

REVITALIZATION OF THE FISHERIES, AND PROSPECTUS

From the late 1970's until the present, the scientific and management questions began to shift from uncertainty about whether or not we could have a productive fishery, to a realization that the Great Lakes could provide major food and recreational benefits through the fisheries in spite of some unanswered questions. Water quality, especially concerns about human health-threatening hydrocarbon contaminants, continued to be a focus for research and control efforts. Substantial declines in DDT, dieldrin, and PCB's were occurring, although new contaminants were being discovered almost daily with improvements in analytical methods and instruments. Lake Erie, once characterized in the popular press as "dead" or "dying," continued to respond to reductions in loadings of agricultural fertilizers and domestic and industrial wastes. Early in the 1980's, Lake Erie produced some of the largest year-classes of perch (*Perca flavescens*) and walleye (*Stizostedion vitreum*) ever observed in the western basin of the lake.

With the realization that the Great Lakes fisheries could be rehabilitated came major new challenges for management agencies that had formerly been concerned almost exclusively with problems of a "biological" nature. As the degree of "artificiality" (i.e., dependence upon external controls) in the fisheries has increased, so too have the resource management agencies increasingly foresaken the traditional "conservation" *raison d'etre*, and assumed active roles in socio-economic reform. Thus, the transformation of the traditional food fish commercial fishery of the Great Lakes into a commercial recreational fishery (selling fishing rather than fish) has been facilitated by management agencies through their selection of piscivorous species for planting. Institutional transformations of this kind, exacerbated by explicit allocation questions such as Native American claims to fish production, further demonstrate that fisheries management problems of the Great Lakes today are predominantly political, cultural and economic. In a curious twist of fate, biologists, who for years have decried political intervention in resource management, now find the shoe on the other foot, firmly implanted in the business of political decision-making (Fraidenburg and Lincoln, 1985)! From these realities have emerged two resource management philosophies that are apparently competing for the distinction of guiding Great Lakes fishery management into the Twenty-First century.

At least one of these philosophies has the issue of control as a fundamental element of its rationalization. The issue is whether or not it is "best" in some sense to re-create (to whatever extent possible) and subsequently manage a fish community consisting of a mix of species somewhat like those of the recent past, i.e., a top-level piscivore such as lake trout subsisting upon a combination of deep-water and pelagic planktivorous species such as the coregonines (whitefishes). Proponents of this approach argue that we could realize catches in perpetuity comparable to those of the 1930's with a minimum of management inputs (e.g., stocked fish). It is assumed that the established exotic species, particularly smelt and alewives, would persist in the system at some level below their historical peak abundances, and that, overall, individual species in the system will fluctuate to a lesser extent than has recently been the case. The article of faith seems to be that we can get a good return from a self-regulating fish community by preventing excessive mortality of the top-level predators (maintaining control over lampreys and the fishery). Additional benefits would include use of the fish community as an indicator of overall health of the ecosystem. Antagonists argue that we cannot "turn back the clock" and must manage with whatever species seem to support interesting fisheries.

The latter alternative is currently seen as a system which is kept "in balance" by carefully controlled inputs of predators such as Pacific salmon. The idea is to prevent massive fluctuations in prey fish abundance while supporting piscivores that are in popular demand by the fisheries. Protagonists cite fishery values in excess of a billion dollars annually through the recreational fishery. Detractors of this approach insist that we do not know enough about hatchery genetic and disease management and about predator-prey relationships to manage for a stable yield, and that external supplies of salmon gametes cannot be guaranteed. Nor do they accept the billion dollar estimate as anything but money that would surface elsewhere in the Great Lakes economy as a result of expenditures in leisure time activities other than fishing.

Both of these objectives, the restoration of a self-perpetuating fish community and the put-grow-take management scheme, assume that we will be forever successful in lamprey control, and they assume that the fishery (in whatever form it takes) can be regulated by the management agencies that have created it. Even ignoring the railings of environmentalists against the chemical approach to pest control, our experience in agriculture clearly denies the former assumption. The redevelopment of excessively intensive Great Lakes fisheries during the past decade clearly denies the latter. The key to realization of either of these scenarios would seem to lie in further understanding the mechanisms that control not only the functioning of the non-human components of the ecosystem, but also those features of social and institutional structure that seek to maintain the *status quo*.

REFERENCES

Applegate, V.C. 1950. Natural history of the sea lamprey (*Petromyzon marinus*) in Michigan. U.S. Fish Wildl. Serv. Spec. Sci. Rep. Fish., Vol. 34, 237 p.

Applegate, V.C., J.H. Howell, J.W. Moffett, B.G.H. Johnson, and M.A. Smith. 1961. Use of 3-trifluoromethyl-4-nitrophenol as a selective sea lamprey larvicide. Great Lakes Fish. Comm. Tech. Rep., Vol. 1, 35 p.

Baldwin, N.W., R.W. Saalfeld, M.A. Ross, and H.J. Buettner. 1979. Commercial fish production in the Great Lakes, 1867-1977. Great Lakes Fish. Comm. Tech. Rep., Vol. 3, 187 p.

Egerton, F.H. 1985. Overfishing or pollution? Case history of a controversy on the Great Lakes. Great Lakes Fish. Comm. Tech. Rep., Vol. 41, 28 p.

Fraidenburg, M.E. and R.H. Lincoln. 1985. Wild chinook salmon management: an international conservation challenge. N. Am. J. Fish. Mgt., Vol. 5, No. 3A, pp. 311-329.

Koeltz, W.N. 1926. Fishing industry of the Great Lakes. U.S. Comm. Fish. Report for 1925, pp. 553-617.

King, E.L, Jr. 1980. Classification of sea lamprey (*Petromyzon marinus*) attack marks on Great Lakes lake trout (*Salvelinus namaycush*). Can. J. Fish. Aquat. Sci., Vol. 37, pp. 1989-2006.

Langlois, T.H. 1942. Two processes operating for the reduction in abundance or elimination of fish species from certain types of water areas. N. Amer. Wildl. Conf. Trans. (for 1941), Vol. 6, pp. 189-201.

Loftus, K.H. and H.A. Regier. 1972. Introduction of the proceedings of the 1971 symposium on salmonid communities in oligotrophic lakes. J. Fish. Res. Board Can., Vol. 29, No. 6, pp. 613-616.

Milner, J.W.   1874.   Report on the fisheries of the Great Lakes: The result of inquiries prosecuted in 1871 and 1872. U.S. Comm. Fish. Report for 1872-73 (Part 2), pp. 1-78.

Pycha, R.L.   1980.   Changes in mortality of lake trout (*Salvelinus namaycush*) in Michigan waters of Lake Superior in relation to sea lamprey (*Petromyzon marinus*) predation, 1968-79. Can. J. Fish. Aquat. Sci., Vol. 37, pp. 2063-2073.

Regier, H.A., V.C. Applegate, and R.A. Ryder.   1969.   The ecology and management of the walleye in western Lake Erie.   Great Lakes Fish. Comm. Tech. Rep., Vol. 15, 101 p.

Reighard, J.E.   1894.   A biological examination of Lake St. Clair. Mich. Fish Comm. Bull., No. 4, 60 p.

Selgeby, J.H., W.R. MacCallum, and D.V. Swedberg.   1978.   Predation of rainbow smelt (*Osmerus mordax*) on lake herring (*Coregonus artedii*) in western Lake Superior. J. Fish. Res. Board Can., Vol. 35, pp. 1457-1463.

Smith, B.R. and J.J. Tibbles.   1980.   Sea lamprey (*Petromyzon marinus*) in Lakes Huron, Michigan, and Superior: History of invasion and control, 1936-78.   Can. J. Fish. Aquat. Sci., Vol. 37, pp. 1780-1801.

Smith, H.M. and M.P. Snell.   1891.   Review of the fisheries of the Great Lakes in 1885. U.S. Fish. Comm., Ann. Rep for 1887, pp. 3-333.

Smith, S.H.   1968.   Species succession and fishery exploitation in the Great Lakes. J. Fish. Res. Board Can., Vol. 25, pp. 667-693.

Smith, S.H.   1970.   Species interactions of the alewife in the Great Lakes.   Trans. Amer. Fish. Soc., Vol. 99, No. 4, pp. 754-765.

Spangler, G.R. and J.J. Collins.   1980.   Response of lake whitefish (*Coregonus clupeaformis*) to the control of sea lamprey (*Petromyzon marinus*) in Lake Huron. Can. J. Fish. Aquat. Sci., Vol. 39, pp. 2039-2046.

Spangler, G.R., D.S. Robson, and H.A. Regier.   1980.   Estimates of lamprey-induced mortality in whitefish, *Coregonus clupeaformis*. Can. J. Fish. Aquat. Sci., Vol. 37, pp. 2146-2150.

Talhelm, D.R. and R.C. Bishop.   1980.   Benefits and costs of sea lamprey (*Petromyzon marinus*) control in the Great Lakes: Some preliminary results. Can. J. Fish. Aquat. Sci., Vol. 37, pp. 2169-2174.

Tody, W.H. and H.A. Tanner.   1966.   Coho salmon for the Great Lakes.   Fish Manage. Rep. Mich. Dep. Conserv., Vol. 1, 38 p.

Torblaa, R.L. and R.W. Westman.   1980.   Ecological impacts of lampricide treatments on sea lamprey (*Petromyzon marinus*) ammocoetes, and metamorphosed individuals. Can. J. Fish. Aquat. Sci., Vol. 39, pp. 1835-1850.

Van Oosten, J.   1929.   Life history of the lake herring (*Leucichthys artedi* Le Sueur) of Lake Huron as revealed by its scales, with a critique of the scale method.   U.S. Bur. Fish. Bull. (for 1928), Vol. 44, pp. 265-428.

Van Oosten, J., R. Hile, and F.W. Jobes.   1946.   The whitefish fishery of Lakes Huron and Michigan with special reference to the deep trapnet fishery. U.S. Fish. Wildl. Serv., Fish. Bull., Vol. 40, pp. 297-394.

Youngs, W.D.   1980.   Estimate of lamprey-induced mortality in a lake trout (*Salvelinus namaycush*) population. Can. J. Fish. Aquat. Sci., Vol. 37, pp. 2151-2158.

PARTICIPANT'S COMMENTS

Fortunately, the organizers of this workshop were successful in securing participants with specialties covering the full range from control theory to resource biology. While sometimes this led to difficulties in communication, the spread proved invaluable. For those towards the control theoretic end of the spectrum, exposure to the complexities and uncertainties of a major resource management problem, such as the Great Lakes fisheries discussed by George Spangler, can be a salutory experience, especially when the exposition may not be neatly placed in a familiar framework.

The jolts begin very early in the paper; in his mildly provocative title George queries whether explicit controls are even necessary at all. He then leads us through the fascinating history of the Great Lakes fisheries management to the present controversy about competing management philosophies. Perhaps somewhat coyly, he does not state his preference, but the wide gulf between the two approaches is made very clear. How these differences may be resolved, and the role that control theory can play is not immediately obvious.

Elsewhere in this volume, the technical concept of controllability is canvassed. In George's paper, we see complementary but rather different concepts of controllability; ones that probably are much more difficult to solve, but are very prevalent in resource management. On the one hand, we have the fact that the managers have the powers to act to control over-exploitation, but in practice they may lack the political will to do so. On the other hand, the mechanics of the resource dynamics are sufficiently ill-understood that it is not known whether a particular species mix can be recreated and maintained. Arguably the system discussed by Anne Blackwell is at least as complex as the Great Lakes system. Perhaps it might be possible to set the latter in a control theoretic framework in a similar way to that achieved by Anne. However, how one can deal with fundamental uncertainties in the form of the resource dynamics is unclear, and similar problems may be experienced in modelling the socio-political influences on management.

Geoff Kirkwood

George Spangler's essay is a fascinating account of the "coevolution" of complex interacting systems (Norgaard, 1981): the Great Lakes aquatic ecosystem has been profoundly affected by, and in turn has exerted its own influence upon, the human economic and cultural structures that have developed around it. Spangler's account reminds us that most bioeconomic analyses, such as optimal harvesting theory, ignore ecosystem effects on the targeted species and assume a static economic environment, including fixed industry structure and technology. However, there has been some attention to trophic communities in modeling multispecies fisheries. See, e.g., Clark, 1976. Plainly, a more long-term and ecologically-comprehensive perspective is necessary in order to understand and manage the Great Lakes fisheries.

It is interesting to realize that the two currently contending strategies for the Great Lakes fisheries have their close analogies in insect pest management (Perkins, 1982), where a "soft path" of "managing with nature" competes with a more aggressive approach aimed at achieving dominant control through large-scale interventions. Working out a balance between these two approaches to biological resource management will be a central theme for decades to come.

References

Clark, C.W. 1976. Mathematical Bioeconomics, John Wiley and Sons.

Norgaard, R.B. 1981. "Sociosystem and ecosystem coevolution in the Amazon". J. Environ. Econ. and Manag., Vol. 8, pp. 238-254.

Perkins, J.H. 1982. Insects, Experts, and the Insecticide Crisis, Plenum Press.

Robert McKelvey

SAMPLING HIGHLY AGGREGATED POPULATIONS WITH APPLICATION TO CALIFORNIA SARDINE MANAGEMENT

Dedicated to the memory of
Philip Morse:
An extraordinary man and scientist
who pioneered bringing the approach
of basic science to operational problems

Marc Mangel
Departments of Agricultural Economics,
Entomology and Mathematics*
University of California
Davis, California 95616

The management of Pacific sardine off the California coast is used to motivate a number of problems associated with egg or larvae sampling. The use of the negative binomial distribution as a model for the spatial distribution of eggs is discussed and inference for the negative binomial by classical and Bayesian methods is introduced. A particular problem, that of presence-absence sampling when not all sampling sites are habitats, is analyzed in detail. The paper closes with a number of open questions, ranging from improvements in the modeling, to prescriptive problems associated with survey design.

INTRODUCTION

The work discussed in this paper was motivated by problems associated with the management of the Pacific sardine (*Sardinops Sagax*) in and near the California current. This particular stock - immortalized by John Steinbeck's Doc, Mac and the boys - is estimated to have peaked at a spawning biomass of more than 11,000,000 metric tons and during the cannery heydays (say 1900-1935) fluctuated between about 2,000,000 metric tons and 9,200,000 metric tons (see Smith, 1978 for details about these estimates). By 1965, the spawning biomass had dropped to less than 10,000 metric tons. Currently, state law requires the California Department of Fish and Game to determine on an annual basis whether or not the spawning biomass exceeds 20,000 short tons (1 short ton = .907 metric tons); if it does, then a modest fishery for sardine may be opened. The problem of estimating such a small biomass by standard methods is a very thorny one (see McCall, 1984a,b; Wolf, 1985 for details) and most of the existing methods simply will not work with any accuracy. Consequently, Wolf and Smith (1985) proposed an "inverse egg production method" (IEPM) for determining the spawning biomass. This method is based on the idea that as the spawning biomass increases, the area in which eggs are found will increase.

---

*Address correspondence to the Department of Mathematics.

Operationally, the method proceeds as follows. One lays down a sampling pattern, as shown in Figure 1. Each dot in Figure 1 represents a station at which eggs are sampled. The region shown in Figure 1 is about five times larger than areas occupied by the Pacific sardine in recent years. Stations are 10 nm (nm = nautical miles; 1 nm = 1853 meters) apart going NW to SE and 4 nm going offshore so that each station represents 40 (nm)^2. The idea behind the IEPM is to estimate a priori the area that a 20,000 short ton spawnings biomass would occupy, then to sample for eggs and determine if this critical area A_c is exceeded. If it is, then it is likely that the spawning biomass exceeds 20,000 short tons. When this method was applied in 1985, 419 stations were sampled. Eleven of these stations had eggs; about 85 eggs were discovered. On the basis of these data and IEPM, a 1000 ton sardine fishery was recommended in 1985 (P. Smith, NMFS, La Jolla, personal communication).

This paper is concerned with various modelling and analytical issues associated with egg surveys similar to the one just described. (These kinds of problems, however, are broader than egg surveys - see Downing (1979) or Resh (1979) for other kinds of applications and motivations.)

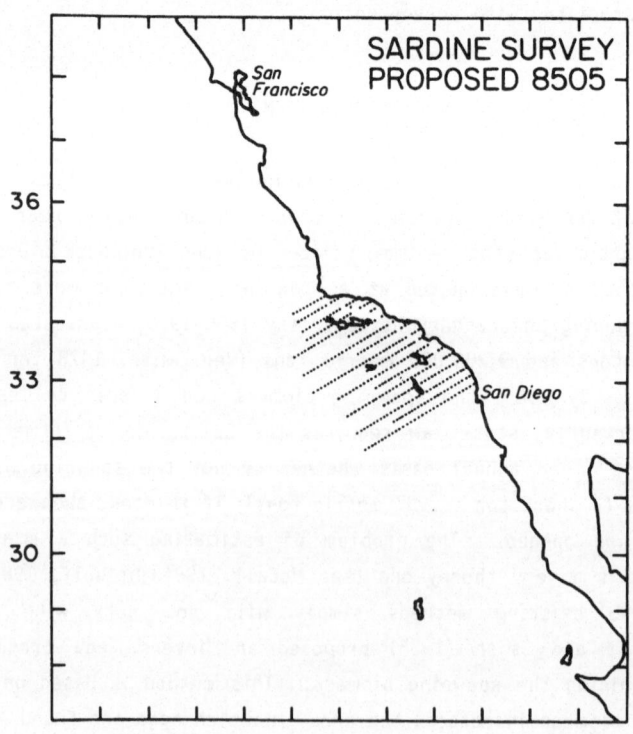

Figure 1.   The sampling sites for the 1985 sardine egg survey proposed by NMFS Scientists (taken from Wolf and Smith (1984)).

Some of these issues are the following: what kinds of models should be used for highly aggregated populations and why; how does one extract the maximum information; how does one deal with data that involve a preponderance of zero's in the samples? In the spirit of a workshop paper, these issues are discussed from the current viewpoint of the author (i.e., subject to possible change) and various untested ideas are presented as a way to probe their usefulness.

MODELING IDEAS AND ISSUES

This section contains a discussion of a number of pertinent questions associated with modeling sampling surveys for highly aggregated populations. To begin, one should think about the spatial scales of interest. These are

Entity	Spatial Scale
individual fish	~ cm
school	~ 100 m
egg patches	~ 1000 m
school groups	~ 10000 m
sampling scale	~ 10000 m

Thus, the sampling scale is large enough to justify the assumption that numbers of eggs taken at different stations are independent random variables. Let X_i be a random variable representing the number of eggs taken in the sample at the ith station, which will henceforth be called a site. Some of the properties of X_i should be the following ones:

$$\Pr\{X_i = 0\} \text{ should be considerable} \tag{1}$$

$$V\{X_i\} \gg E\{X_i\} \tag{2}$$

where $V\{X_i\}$ is the variance of X_i and $E\{X_i\}$ is the expected value (mean of X_i). Properties of (1) and (2) are based on the experimental reality, not any theoretical conceptualization.

For most of this paper, the following model is used: If the ith site is a habitat, the conditional distribution of X_i is a negative binomial with parameters m and k (written NB(m,k)). That is

$$\Pr\{X_i = x \mid \text{site is a habitat}\} = \frac{\Gamma(k+x)}{\Gamma(k)x!} \left[\frac{k}{k+m}\right]^k \left[\frac{m}{k+m}\right]^x \tag{3}$$

where $\Gamma(\cdot)$ is the gamma function. The distribution (3) has the properties (1) and (2) (see Johnson and Kotz (1969) for a discussion of more properties of the NB(m,k) distribution). First

$$Pr\{X_i = 0\} = \left(\frac{k}{k+m}\right)^k \tag{4}$$

which can be considerable even if m is large (see Figure 2). Second

$$\left.\begin{aligned} E\{X_i\} &= m \\[2ex] V\{X_i\} &= m + \frac{m^2}{k} \end{aligned}\right\} \tag{5}$$

so that if k is small, property (2) is satisfied. What value of k should be used? Smith and Richardson (1977) provide the following data

	Spawning Biomass (Millions of Tons)	Estimate of k
	3.9	.14
	2.7	.19
	1.0	.21
	.2	.08
Average	2.95	.155
Coefficient of Variation	1.48	.65

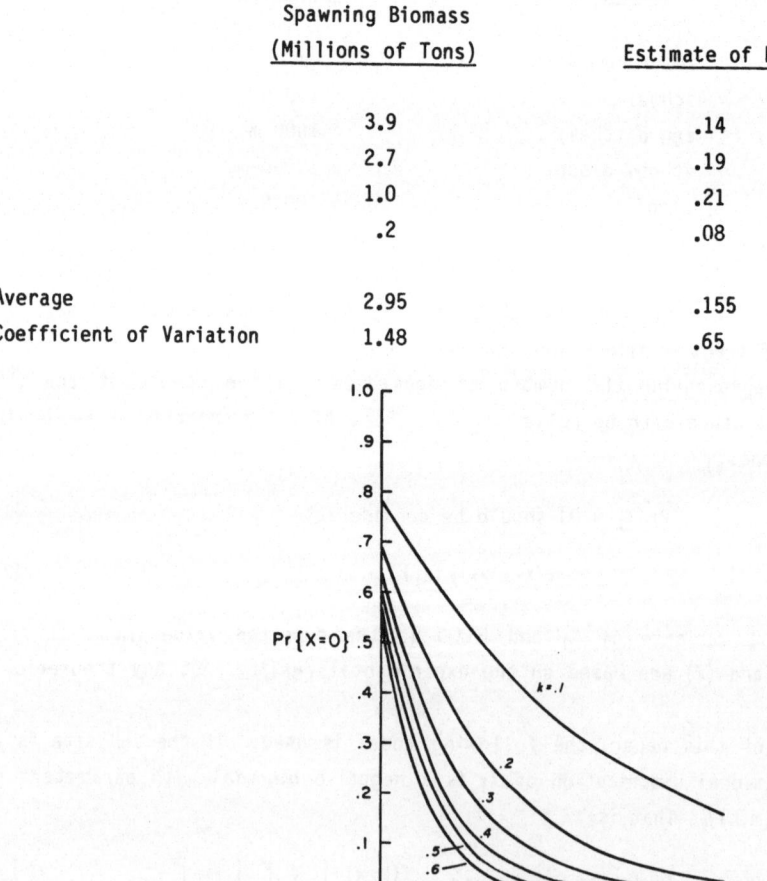

Figure 2.   Likelihood of a zero observation in a NB distribution with parameters m and k.

As the spawning biomass varies over a range of 19.5, note that k varies by a factor of 2.6. For all intents and purposes - and definitely for this paper - k will be treated as a constant, presumed known, in the range of 0.1 to 0.2.

One can legitimately ask if there is a true biological or operational motivation for choosing $X_i \sim NB(m,k)$. Here is one. Let $B(t)$ denote the spawning biomass at time t and assume that it satisfies the following stochastic differential equation:

$$dB = B(t)\{r(1-B(t)/K)dt + \sigma dW\} \tag{6}$$

where $dB = B(t+dt) - B(t)$, r, K and σ are parameters and $dW = W(t+dt) - W(t)$, where $W(t)$ is Brownian motion (see, e.g., Ludwig, 1975; Schuss, 1980, for general discussions of Brownian motion or Mangel, 1985a for a discussion related to natural resource models). If $B(t)$ satisfies (6), then the equilibrium density for B is a gamma density (see Dennis and Patil, 1984, for an elaboration). Assume that, given a value of the equilibrium biomass B_{eq}, the distribution of eggs encountered is a Poisson with parameter $\lambda = \lambda_0 B_{eq}$. Then the unconditional density for the number of eggs is a negative binomial (see, e.g., Mangel, 1985b, for details).

One can also ask if there is a legitimate biological reason for choosing constant k. The answer is, to some extent, yes. In order for eggs to be fertilized, they need to be highly clumped - regardless of the size of the spawning biomass. This will partially justify the use of constant k.

In the analysis which follows in the next section, it is assumed that X_i has the distribution (3) with known k but unknown m and it is assumed that if the spawning biomass exceeds a critical value, then m will exceed a given critical value, m_c.

Before presenting this analysis, however, a number of points need to be cleared up. First, the NB(m,k) model is not the only one with properties (1) and (2). For example, one can use other "contagious" distributions such as the Neyman Type A in which

$$Pr\{X_i = x\} = \sum_{j=1}^{\infty} \frac{e^{-\lambda}\lambda^j}{j!} e^{-j\phi} \frac{(j\phi)^x}{x!} \tag{7}$$

and λ and ϕ are parameters (see Johnson and Kotz, 1969, Chapter 9). In this model

$$
\left.
\begin{aligned}
Pr\{X_i = 0\} &= e^{-\lambda(1-e^{-\phi})} \\
E\{X_i\} &= \lambda\phi \\
V\{X_i\} &= \lambda\phi(1+\phi)
\end{aligned}
\right\} \tag{8}
$$

so that the properties (1) and (2) can be satisfied. The NB(m,k) model is used here, but others are feasible.

Second, note that (3) is conditioned on a site being a habitat. Thus, one needs to append to (3)

$$p_i = \Pr\{i\text{th site is a habitat}\} . \qquad (9)$$

Some choices for p_i are

$$p_i = \begin{cases} p_0, \text{ a constant} \\ p(m), \text{ a function of } m \\ p(i), \text{ a function of site location} \\ p(i,m), \text{ a function of site location and } m \end{cases} \qquad (10)$$

For example, one could use

$$p(m) = 1 - e^{\gamma m} \qquad (11)$$

where γ is a constant.

Using (9) and (3) leads to the following model

$$\Pr\{X_i = 0\} = (1 - p_i) + p_i \left[\frac{k}{k+m}\right]^k \qquad (12)$$

$$\Pr\{X_i > 0\} = p_i \left[1 - (\frac{k}{k+m})^k\right] \qquad (13)$$

This will be the basic model in the next section. Note that, in this model, one still assumes that if eggs are present, they're found. But the cell size is 40 nm^2 and the sampler size is roughly $\sim .05$ m^2 so that even if eggs are present, they could be missed. A way around this problem is discussed in the last section.

Third, one should separate descriptive and prescriptive sampling problems. The prescriptive problems are "survey optimization" ones: how should optimal surveys be connected? These are sexy problems, but often of less use to managers than the descriptive problem of "here's the data, what does it mean?" In the next section, a descriptive problem and its analysis are described. Optimal surveys are discussed in the last section.

PRESENCE-ABSENCE SAMPLING FOR EGGS

In this section, the ideas developed thus far are applied to a problem which is analogous to the sardine egg sampling problem (the details of where the analogy fails are discussed in the next section). The set-up for the problem is this: the ith site is a habitat with probability p_i and one samples for the presence or absence of eggs (ignoring the actual number of eggs, if there are eggs present). This scheme, in operational terms, would allow the survey scientist to bring the samples back in, hold it up to the light and determine the presence or absence of eggs.

As before, X_i is the number of eggs in the sample at the ith site and $X_i = 0$ is called a "negative" sample; $X_i > 0$ is called a positive sample. Equations (12) and (13) give the probabilities that $X_i = 0$ and $X_i > 0$ respectively.

Assume that there are N_n negative samples and N positive samples. Let n denote those sites at which a negative sample was obtained and p denote those sites at which a positive sample was obtained. The likelihood of $\{n,p\}$ is

$$\pounds(n,p|m) = \prod_{i \in n} \{1-p_i+p_i \left(\frac{k}{k+m}\right)^k\} \prod_{i \in p} \{p_i-p_i \left(\frac{k}{k+m}\right)^k\} \tag{14}$$

A number of different kinds of sampling schemes can be derived, based on assumptions about the values of p_i. Some of these will now be discussed.

First consider the case in which all the p_i take the same value, p. Then (14) becomes

$$\pounds(N_n,N|m) = \left[1-p+p \left(\frac{k}{k+m}\right)^k\right]^{N_n} \left[p-p \left(\frac{k}{k+m}\right)^k\right]^N \tag{15}$$

Note the following about (15): the model is now essentially a binomial model, with success probability $p-p(k/(k+m))^k$. Thus, the likelihood in (15) is the unnormalized probability of N successes in $N_n + N$ trials. The normalization constant, a binomial coefficient, is not needed for any of the calculations that follow.

The maximum likelihood value of $m, \hat{m}$ is found by taking the derivative of the logarithm of $\pounds(N_n, N|m)$ with respect to m and setting it equal to 0. This leads to a nonlinear equation for m, which is easily solved on a desktop microcomputer. Once the MLE $\hat{m}$ is known, one can investigate likelihood ratios for other values of m. This approach will be reported elsewhere (Mangel and Smith, 1987).

Instead, consider a Bayesian approach in which one wishes to compute the posterior probability that $m > m_c$, given the data. In order to do this, one needs a prior distribution of m. Two choices are the uniform prior

$$f_0(m) = 1 \qquad 0 < m \leqslant m_m \tag{16}$$

where m_m is a specified value, and the noninformative prior

$$f_0(m) = \frac{1}{\sqrt{m(k+m)}} \tag{17}$$

(The noninformative prior (17) is derived in the appendix.)

These are chosen to represent "ignorance" about the value of m. When the uniform prior (UP) is used, all values of m between 0 and m_m are given equal weighting. When the noninformative prior (NP) is used, data change the position, but not the shape of the posteriori distribution (see Martz and Waller, 1982, for further discussion).

If the uniform prior is used, the posterior probability that m exceeds m_c is given by

$$P_{UP}(m > m_c) = \frac{\int_{m_c}^{m_m} \left[1-p+p(\frac{k}{k+m})^k\right]^{Nn} \left[p-p(\frac{k}{k+m})^k\right]^N dm}{\int_0^{m_m} \left[1-p+p(\frac{k}{k+m})^k\right]^{Nn} \left[p-p(\frac{k}{k+m})^k\right]^N dm} \tag{18}$$

Since m_m may be quite large (say of the order of 1000) it helps to introduce

$$\left.\begin{aligned} w &= \frac{k}{k+m} \\ dw &= -\frac{k}{(k+m)^2} dm = -\frac{w^2}{k} dm \\ w_c &= \frac{k}{k+m_c} \qquad w_m = \frac{k}{k+m_m} \end{aligned}\right\} \tag{19}$$

The integral in (18) becomes

$$P_{UP}(m > m_c) = \frac{\int_{w_m}^{w_c} \frac{[1-p+pw^k]^{Np} [p-pw^k]^N}{w^2} dw}{\int_{w_m}^1 \frac{[1-p+pw^k]^{Np} [p-pw^k]^N}{w^2} dw} . \tag{20}$$

These integrals are easily computed on a desktop microcomputer.

When the noninformative prior is used, one makes the transformation

$$\left.\begin{aligned} m &= k^2 \tan\theta \\ dm &= 2k \tan\theta \, d\theta/\cos^2\theta \\ k + m &= k/\cos^2\theta \\ \\ \theta_c &= \arctan\left[\sqrt{\frac{m_c}{k}}\right] \end{aligned}\right\} \tag{21}$$

and finds that the posterior probability is given by

$$P_{NP}(m > m_c) = \frac{\int_{\theta_c}^{\theta} [1-p+p(\cos\theta)^{2k}]^{Nn} [p-p(\cos\theta)^{2k}]^N \frac{d\theta}{\cos\theta^2}}{\int_0^{\theta_m} [1-p+p(\cos\theta)^{2k}]^{Nn} [p-p(\cos\theta)^{2k}]^N \frac{d\theta}{\cos\theta^2}} \tag{22}$$

Figure 3 shows $P\{m > m_c\}$ as a function of the number of positive samples using both priors. The NP is more "conservative" than the UP.

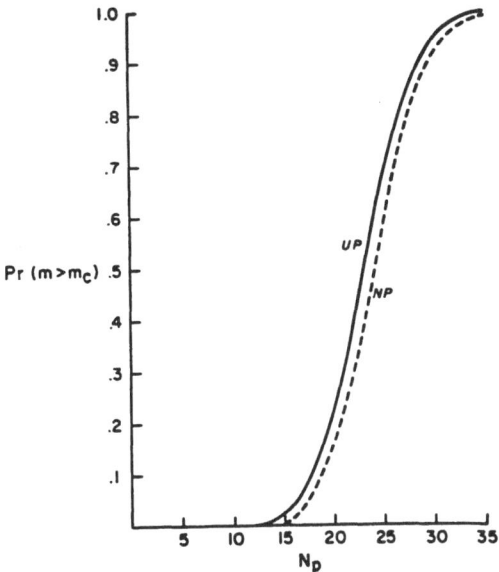

Figure 3.   Probability that m exceeds m_c = 1.14 as a function of the number of positive samples (N) in a total of N = 100 samples.   Other parameters: m_m = 1000, k = .2, p = .8.

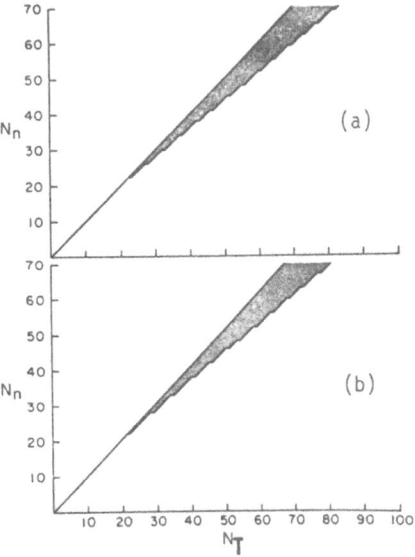

Figure 4.   Sequential sampling charts in which the number of negative samples (N) is plotted against the total number of samples N_T. If the data fall in the shaded region, one can conclude with 99% confidence (Figure 4a) or 95% confidence (Figure 4b) that m < m_c. Other parameters: m_c = 1.14, k = .2, p = .8, m_m = 1000.   The uniform prior was used in the calculations.

Figure 3 is an _ex post facto_ probability statement made after the data are collected. On the other hand, for many situations a sequential sampling plan is often more useful. Figure 4 is a sequential sampling diagram used to compute the probability that $m < m_c$ under the uniform prior. In this diagram, one plots N_n versus $N_T = N + N_n$. If an observation falls in the shaded region, then one can conclude that $m < m_c$ with probability .99 (Fig. 4a) or .95 (Fig. 4b). If the current data point (N_T, N_n) does not fall in the shaded region, then an additional site is sampled.

These same kinds of calculations can be performed when the generalized likelihood (14) is used. For example, when the UP is used, one finds

$$P_{UP}(m < m_c) = \frac{\int_{w_c}^{1} \prod_{i \in n} \{1 - p_i + p_i w^k\} \prod_{i \in p} (p_i - p_i w^k) \frac{dw}{w^2}}{\int_{w_m}^{w_c} \prod_{i \in n} \{1 - p_i + p_i w^k\} \prod_{i \in p} (p_i - p_i w^k) \frac{dw}{w^2}} \tag{23}$$

The only difficulty is that one cannot develop charts similar to Figures 3 and 4. On the other hand, (23) is ideal for use in real-time with a microcomputer. For example, assume that $m_c = 1.14$, $m_m = 1000$, $k = .2$ and let each data point (p_i, X_i) be represented with $X_i = 1$ for a positive sample and $X_i = 0$ for a negative sample. Suppose that the first 10 data points are (1,0), (1,0), (.95,0), (.95,0), (.9,0), (.9,0), (.85,0), (.85,0), (.8,0), and (.8,0). Using (23) shows that $P_{UP}(m < m_c) = .82$. If the next five data points are (1,0), (1,0), (.95,0), (.95,0), and (.9,1), then $P_{UP}(m < m_c) = .86$. If the next five data points are (1,0), (1,0), (.95,0), (.95,0) and (.9,0), then $P_{UP}(m < m_c) = .96$ and sampling can stop if a 95% confidence level is desired.

Two points are worth noting. First, there is a preponderance of zeroes in the data. This kind of result is, in fact, observed in sampling. Second, a large amount of negative information is needed to insure that $m < m_c$ with a high confidence level. One must remember, however, that with the UP, the initial probability that $m < m_c$ is m_c/m_m. So, for example, for the values presented here, the prior probability that $m < m_c$ is 1.1×10^{-3}. In addition, since not every site is a habitat, the effects of negative information on the updated distribution are mitigated (i.e., as $P_i \to 0$, the data have decreasing effects on the Bayesian update).

OPEN QUESTIONS

Since a major purpose of a workshop is to raise questions, a workshop paper can (and perhaps should) end with open questions rather than conclusions. In this spirit, a number of open questions and directions for future work are indicated.

1)   Many Age Classes of Eggs. In the actual sardine survey, four age classes of eggs (<1 day, 1-2 days, 2-3 days old and >3 days old) are sampled, each with a different m and k.   Thus, the data are more complicated, consisting of presence-absence of the four age classes or the actual counts of the four age classes. The question of how to use these data is complex. One could assume, for example, that the four age classes represent completely independent events (probably an unrealistic assumption). The other extreme is one of complete correlation: if any age class is present, then they all are. Reality probably lies somewhere between the two extremes, with a partial correlation. A Bayesian approach to this problem can also be developed (see Mangel et al., 1984, pg. 568) where the correlation level is a user-inputted variable.

For example, let m_j and k_j denote the unknown mean and known aggregation parameter for the jth egg class. A reasonable model is the following one (P. Smith, NMFS, La Jolla, personal communication):

$$
\left.
\begin{aligned}
&k_j = .1j \qquad j = 1,2,3,4 \\
&m_j = s_j m \text{ where m is unknown and} \\
&s_1 = 1 \\
&s_2 = .8 \\
&s_3 = .6 \\
&s_4 = .4
\end{aligned}
\right\} \tag{24}
$$

Finally, let ρ_c denote a correlation parameter, in the following sense: with probability ρ_c, if eggs of one age class are present, then eggs of all age classes are present. With this kind of model, the probability of a negative sample is

$$
\Pr\{X_i=0\} = (1-\rho_c) \prod_{j=1}^{4} \left[\frac{k_j}{k_j+s_j m}\right]^{k_j} + \rho_c \max_j \left[\frac{k_j}{k_j+s_j m}\right]^{k_j} \tag{25}
$$

One can do similar sorts of analyses using (25). Preliminary investigations based on (25) with four age classes of eggs (as in (24)) support the recommendation of opening a small sardine fishery in 1985.

2)   Imperfect Sampling. Another possible extension allows for the chance of imperfect sampling. One way to do this is to use the weighted NB(WNB) model of Bissell (1972). According to that model, if a site is a habitat

$$
\Pr\{X_i=x\} = \frac{\Gamma(k+x)}{x!\,\Gamma(k)} \left[\frac{k}{mW_i+k}\right]^k \left[\frac{mW_i}{mW_i+k}\right]^x \tag{26}
$$

where W_i is a measure of sampling efficiency. (Zweifel and Smith (1981) discuss the interpretation of W_i.) The data now consist of triplets (p_i, W_i, X_i). The methods of the previous sections can be extended to cover this case with essentially no conceptual difficulty and only minor computational difficulty.

Another way to do imperfect sampling is to take into account explicitly the chance that an egg patch might be missed during the samplings. For example, let B_i denote the number of eggs in the cell containing the ith site and let $A(b)$ denote the area of patch with b eggs. One choice is

$$A(b) = A_m(1 - e^{\gamma b}) \tag{27}$$

where A_m and γ are constants. One expects $A_m \ll S$ where S is the 40 nm^2 area of each cell. Finally, assume that

$$Pr\{detecting\ eggs\ |\ B_i = b\} = \frac{A(b)}{S} \tag{28}$$

There are now two ways for $X_i = 0$: no eggs present ($B_i = 0$) or eggs present, but missed. Thus,

$$Pr\{X_i=0\} = Pr\{B_i=0\} + \sum_{B=1}^{\infty} Pr\{B_i=b\}\ Pr(no\ detection|B_i=b)$$

$$= \left(\frac{k}{k+m}\right)^k + \sum_{b=1}^{\infty} Pr\{B_i=b\} \left[1 - \frac{A_m}{S}(1-e^{-\gamma b})\right] \tag{29}$$

After a little algebra, and use of the generating function for a NB distribution, one can develop an explicit formula for $Pr\{X_i=0\}$ in terms of m, k, A_m, S and γ. The development and use of this formula will be given in Mangel and Smith (1987).

3)   Joint Estimation of Habitat Boundaries and m.   An open question, which may require a new formulation of the problem, involves the simultaneous estimation of the habitat boundary and m.   That is, one might consider the joint density that $p_i = 0$ and m takes a certain value.   The approach to this problem is not clear, although a Bayesian formulation seems natural.

4)   Ideal-Free Sardine Eggs.   At the workshop, Mike Rosenzweig pointed out the ecological theory of habitat selection could be used to generate stratified sampling plans.   That is, use habitat theory (see Rosenzweig's article in this volume) to predict the proportion of sites with eggs in habitats of different quality as a function of spawning biomass. If one could identify habitat quality on the basis of oceanographic factors (e.g., satellite photographs of temperature and chlorophyll distributions), then it would be possible to use a sampling scheme based on habitat

quality. Such a scheme might require considerably less information (i.e., fewer samples).

5)   Adaptive Survey Optimization. Of all the possible prescriptive problems, perhaps the most interesting one is the development of an adaptive (i.e., closed loop) algorithm which can be used to guide the survey vessel. That is, based on the sampling history thus far, which site should be visited next.

6)   Economic Modeling. Recall that the purpose of the egg survey is to determine a level of confidence about the biomass and that if the biomass exceeds a critical level, then a complete stock survey will be conducted. One can extend the methods of this paper to include the costs of the egg survey, the cost of the complete stock survey, and the cost of not allowing fishing when the stock exceeds the critical level.

7)   Egg Surveys as Priors. Assuming that one decides to pursue a complete stock survey. The results of the egg survey can be used as a prior density when planning the larger survey. The results presented in the previous section on estimating the extent of the habitat could be especially useful.

ACKNOWLEDGMENTS

This work was partially supported by NSF Grant MCS-81-21659, by the Agricultural Experiment Station of the University of California, and by NOAA through the California Sea Grant Project. I thank Richard Plant, Patti Wolf, Don Ludwig and Carl Walters for useful discussions and Paul Smith for his willingness to share ideas, time and data.

REFERENCES

Anscombe, F.J. 1950. Sampling theory of negative binomial and logarithmic series distributions. Biometrika, Vol. 34, pp. 358-382.

Bissell, A.F. 1972. A negative binomial model with varying element sizes. Biometrika, Vol. 59, pp. 435-441.

Bliss, C.I. 1958. The analysis of insect counts as negative binomial distributions. Proc. Tenth Intl. Cong. Entom., pp. 1015-1032.

Box, G.E.P. and G.C. Tiao. 1973. Bayesian Inference in Statistical Analysis. Addison Wesley, Reading, MA, 588 pp.

DeGroot, M. 1970. Optimal Statistical Decisions. McGraw-Hill, NY, 489 pp.

Dennis, B. and G.P. Patil. 1984. The gamma distribution and weighted multimodal gamma distributions as models of population abundance. Mathematical Biosciences, Vol. 68, pp. 187-212.

Downing, J.A. 1979. Aggregation, transformation and the design of benthos sampling programs. J. Fisheries Res. Board of Can., Vol. 36, pp. 1454-1463.

Feller, W. 1968. An Introduction to Probability Theory and its Applications, Vol. 1, John Wiley, NY, 509 pp.

Gerard, G. and P. Berthet. 1971. Sampling strategy in censusing patchy populations. In G.P. Patil, E.C. Pielou and W.E. Waters (eds.), Statistical Ecology, Vol. 1, pp. 59-68, Pennsylvania State University Press, University Park, PA.

Gunderson, D.R., G.L. Thomas, P. Cullenberg, D.M. Eggers and R.F. Thorne. 1980. Rockfish investigations off the coast of Washington. Report FRI-UW-8021, Fisheries Research Institute, University of Washington, Seattle.

Hewitt, R. 1976. Sonar mapping in the California Current area: A review of recent developments. Cal. COFI Report, Vol. 18, pp. 149-154.

Hewitt, R. 1981. The value of pattern in the distribution of young fish. Rapp. P-v. Reun. Cons. Int. Explor. Mer., Vol. 178, pp. 229-236.

Hewitt, R. 1984. 1984 Spawning biomass of Northern Anchovy. Administrative Report LJ-84-18, Southwest Fisheries Center, National Marine Fisheries Service, La Jolla, CA.

Hewitt, R. and P. Smith. 1979. Seasonal distributions of epipelagic fish schools and fish biomass over portions of the California Current region. Cal. COFI Report, Vol. 20, pp. 102-110.

Hewitt, R., P.E. Smith, and J.C. Brown. 1976. Development and use of sonar mapping for pelagic stock assessment in the California Current area. Fish Bull. US, Vol. 74, pp. 281-300.

Hewitt, R. and P.E. Smith. 1982. Sonar mapping of the California Current area: Some considerations of sampling strategy. Report, Southwest Fisheries Center.

Johnson, N. and S. Kotz. 1969. Discrete Distributions in Statistics. Wiley, NY.

Leaman, B.M. 1981. A brief review of survey methodology with regard to groundfish stock assessment. Can. Spec. Pub. Fish. Aq. Sci., Vol. 58, pp. 113-123.

Lloyd, M. 1967. Mean crowding. J. Anim. Ecol., Vol. 36, pp. 1-30.

Ludwig, D. 1975. Persistence of dynamical systems under random perturbations. SIAM Review, Vol. 17, pp. 605-640.

MacCall, A.D. (ed.) 1984a. Report on a NMFS-CDFG workshop on estimating pelagic fish abundance. Administrative Report LJ-84-40, Southwest Fisheries Center, POB 271, La Jolla, CA 92038.

MacCall, A.D. (ed.) 1984b. Management information document for California coastal pelagic fishes. Southwest Fisheries Center Administrative Report LJ-84-39. Southwest Fisheries Center, POB 271, La Jolla, CA 92038.

Mangel, M. 1985a. Decision and Control in Uncertain Resource Systems. Academic Press, NY.

Mangel, M. 1985b. Search models in fisheries and agriculture. In M. Mangel, (ed.), Proc. of the Ralf Yorque Workshop on Resource Management, Springer Verlag, NY.

Mangel, M. and P.E. Smith. 1987. Presence-absence plankton sampling for fisheries management. Can. J. Fish. Aq. Sci., to appear.

Martz, H. and R. Waller. 1982. Bayesian Reliability Analysis. John Wiley and Sons, NY. 745 pp.

Pennington, M. 1983. Efficient estimators of abundance, for fish and plankton surveys. Biometrics, Vol. 39, pp. 281-286.

Pennington, M. and P. Berrien. 1984. Measuring the precision of estimates of total egg production based on plankton surveys. J. Plankton Res., Vol. 6, No. 5, pp. 869-880.

Pielou, E.C. 1977. Mathematical Ecology. Wiley, NY. 385 pp.

Resh, V.H. 1979. Sampling variability and life history features: Basic considerations in the design of aquatic insect studies. J. Fish. Res. Board Can., Vol. 36, pp. 290-311.

Schuss, Z. 1980. Theory and Application of Stochastic Differential Equations. Wiley, NY.

Smith, P.E. 1978. Biological effects of ocean variability: Time and space scales of biological response. Rapp. P-v. Reun. Cons. Int. Explor. Mer., Vol. 173, pp. 117-127.

Smith, P.E. and S.L. Richardson. 1977. Standard techniques for pelagic fish egg and larva surveys. FAO Fisheries Technical Paper 175, Food and Agriculture Organization of the United Nations, Rome, Italy.

Taylor, C.C. 1953. Nature of variability in trawl catches. Fish. Bull. 83, U.S. Department of the Interior, Vol. 54, pp. 145-166.

Taylor, L.R. 1971. Aggregation as a species characteristic. In G.P. Patil, E.C. Pielou and W.E. Waters (eds.), Statistical Ecology, Vol. 1, pp. 357-377, Pennsylvania State University Press, University Park, PA.

Wald, A. 1947. Sequential Analysis. Dover, NY. 121 pp.

Wolf, P. 1985. Status of the spawning biomass of the Pacific Sardine, 1984-85. Marine Resources Report to the Legislature, California Department of Fish and Game.

Wolf, P. and P.E. Smith. 1985. An inverse egg production method for determining the relative magnitude of Pacific sardine spawning biomass off California. Cal. COFI Report, Vol. 26, pp. 130-138.

Zweifel, J.R. and P.E. Smith. 1981. Estimates of abundance and mortality of larval anchovies (1951-75): Application of a new method. Rapp. P-v. Cons. Int. Explor. Mer., Vol. 178, pp. 248-259.

APPENDIX: DERIVATION OF THE NONINFORMATIVE PRIOR

The approximate noninformative prior for the NB distribution is derived as described by Martz and Waller (1982, pg. 224). Viewing (3) as the likelihood of m given x, the log-likelihood is

$$L(m|x) = -k \log(k+m) + x[\log m - \log(m+k)] + \ell(x,k) \qquad (A-1)$$

where $\ell(x,k)$ contains terms independent of m. The derivatives of the log-likelihood are

$$\frac{\partial L}{\partial m} = -\frac{k}{k+m} + \frac{x}{m} - \frac{x}{m+k}$$

$$\frac{\partial^2 L}{\partial m^2} = \frac{k}{(k+m)^2} - \frac{x}{m^2} + \frac{x}{(m+k)^2} \tag{A-2}$$

Setting $\partial L/\partial m = 0$ shows that the maximum likelihood estimate is $\hat{m} = x$. (For n independent observations, the MLE $\hat{m}$ is easily shown to be the sample mean.) Define

$$J(\hat{m}) = -\frac{\partial^2 L}{\partial m^2}\bigg|_{\hat{m}} = \frac{\hat{m}}{\hat{m}^2} - \bigg/\frac{\hat{m}+k}{(k+\hat{m})^2} = \frac{k}{\hat{m}(k+\hat{m})} \tag{A-3}$$

The approximate non-informative prior is then

$$f_0(m) \propto J(m)^{1/2} \propto m^{-1/2}\,(k+m)^{-1/2} \tag{A-4}$$

PARTICIPANT'S COMMENTS

Mangel's paper addresses a relatively common problem experienced by field biologists, i.e., the non-random distribution of organisms in nature. The approach he is suggesting, use of discrete distribution statistics rather than strict reliance upon the Central Limit Theorem and large numbers of samples to yield good approximations to the Gausian distribution, is likely to become increasingly popular as microcomputers become standard field equipment. The negative binomial distribution has enjoyed some attention among biologists in the past (see Elliott (1977) for examples), but the Pacific sardine example is a particularly interesting case because the presence or absence of eggs in the spawning ground survey determines whether or not the fishery will open in a given year. The 1985 results, 85 eggs at 11 stations, led to the very surprising recommendation to open a 1000 ton sardine fishery! This appears to be very little empirical information upon which to base such a decision, and it immediately raises other technical questions such as the certainty of identification of the eggs, or determination of their ages.

In the Open Questions section, Mangel raises several interesting possibilities for further development of the model. If the negative binomial is really a good model for the underlying distribution of eggs, then the k parameter may be a useful index of dispersion. It is sensitive to size and number of sampling units but within a particular survey it should be useful as a relative index. The four different age classes of eggs (each with different m and k) should show a progression of dispersion with time since spawning. This may allow inference about the number of spawning aggregations present in the region. This hypothesis could be tested with historical data if survey information is available from periods in which there has been some variation in abundance of the spawning stock.

Further consideration of sampling methods appropriate to contagious distributions is long past due in fisheries management. The problem of estimating angler effort in recreational fishing surveys (creel surveys) may be amenable to sampling from satellite imagery where the boats or ice-houses concentrate over known habitats or, more directly, on schools of fish. The degree of aggregation of fishing units may prove to be a useful index of habitat quality or a covariate with the catchability coefficient. Improvements in our ability to estimate the latter statistic would greatly enhance the utility of most of the catch-effort models currently used in fishery management.

Reference

Elliott, J.M. 1977. Some methods for the statistical analysis of samples of benthic invertebrates. Sci. Publ. No. 25, Freshwater Biological Association, Ambleside, Cumbria: 160 pp.

George R. Spangler

The Mangel method promises a revolution in management techniques. As I have indicated in my own chapter of the workshop, it may be possible to refine it by consciously sorting the sampling sites according to habitat quality, and then sampling them in inverse order of their quality (poorest first, etc.).

It is not often that good management practices can actually improve the theory that produces them, but this time that may happen. Practicing the Mangel method will require developing a habitat quality index, $Q(\underset{\sim}{h})$, for the species being managed (where $\underset{\sim}{h}$ is a vector of habitat properties). It also requires knowing at which population density each quality of habitat is added. Let us standardize the densities by dividing them by a constant obtained through sampling (say $\bar{Y}$, the average yield per unit fishing effort). Then the management data will determine the function:

$$Q(\underset{\sim}{h})^* = f_i(N_i/\bar{Y}_i)$$

where $Q(\underset{\sim}{h})^*$ is the habitat quality which is marginally used when the standard density of the i-th species is $N_i/\bar{Y}$.

The subscript i introduces the possibility of a multispecies view, and here is where the need for data collection by managers will enhance the basic science on which they rely. Community ecologists are investigating the fundamental structure of sets of niches (Rosenzweig, in press; P.S. Giller and J.H.R. Gee (eds.), 27th British Ecological Symposium, 1986: Organization of Communities: past and present). Eventually, this knowledge should also benefit the management of exploited populations, but it is still too rudimentary for that. However, obtaining the set of functions, f_i, for a guild of species on one quality index, $Q(\underset{\sim}{h})$, will illuminate what is happening among the species. If, for example, all the f_i are positive functions, then the species rank the habitats similarly. If some are negative, then they have at least two distinct habitat preferences. If they are all unimodal, but peak at various places along the $N/\bar{Y}$ axis, then they all have distinct preferences. Other patterns, of which no one has yet conceived, may emerge. But there is no question that knowing the peaks of the various species and how they relate to each other will advance community ecology and may suggest improvements in management policies.

Michael L. Rosenzweig

THE STRUCTURE OF FISHING SYSTEMS AND THE
IMPLEMENTATION OF MANAGEMENT POLICY

Susan S. Hanna
National Marine Fisheries Service
Northwest and Alaska Fisheries Center

and

Department of Agricultural and Resource Economics
Oregon State University

Management policies that achieve desired results are difficult to design.
This is particularly true in multispecies fisheries where policies are
developed on a single species basis and the economic reasons for existing
operating structures are ignored. This paper discusses the problem of
effort control in the multispecies groundfishery of Oregon, arguing that
an understanding of the economic structure of the fishery as a system is
a necessary precondition for successful management. To reflect the
economic structure of the fishery, functional groupings of landings data
are made by season and operating strategy. Resultant catch mixes
exhibit strong seasonality, distinct differences in species composition,
and a reduced variability in revenues earned. These properties suggest
that catch mixes are an important element of economic structure and
that a data base organized along single-species lines is inappropriate
for management decisions. They further suggest that an explicit
recognition of the economic structure of a fishery will lead to better
anticipation of both management impacts and fishermen's response, and
thereby to more effective policy implementation.

INTRODUCTION

The successful implementation of policies limiting fishery exploitation has
proven to be an elusive goal. Effort control is a particularly vexing problem in
multispecies fisheries. Bioeconomic models used to address resource management
questions offer valuable insight into a system's internal linkages and sensitivity to
perturbation. However, models of necessity present a simplified picture. The
fishery system is comprised of three fundamental elements: biological, represented
by stock dynamics, economic, represented by fisherman/processor dynamics, and
oceanographic. Because basic taxonomic work on the economic structure of a fishery
is sparse, modelers often find themselves in a position of being able to represent
the stock dynamics with a much greater degree of accuracy than the fishermen
dynamics. This point is supported by Hilborn (1985) who argues that it is not a poor
understanding of fish biology that has been responsible for many instances of fishery
collapse, but rather a poor understanding of the dynamics of fishermen.

The author acknowledges helpful comments and suggestions from R. Marasco, J. Terry,
D. Brooks, W. Getz and C. Smith.

One inherent limitation of models built without appreciation for economic structure is an inadequate capacity for anticipating the sensitivity of fishermen's response to management. Managers are often surprised to find themselves faced in the second round with the need to undo unexpected results of policies in the first round.

Walters and Hilborn (1978) distinguish three general types of uncertainty common to fisheries management:

1) Random effects whose future frequency of occurrence can be determined by past experience

2) Parameter uncertainty that can be reduced by research and information acquisition

3) Ignorance about appropriate variables and appropriate model form.

Uncertainties of types (2) and (3) are represented in our ignorance of economic structure and its role in promoting or inhibiting the achievement of management objectives. A more complete taxonomy of fishing systems can mitigate these types of uncertainties. Optimal control policies derived from models built on a realistic economic structure will be better suited to the challenging problem of effective implementation.

The control problem in fisheries management is one of melding a policy from the various goals of fishermen, scientists, and managers that will effectively limit the level of fishing mortality. To accomplish this, the design of management policies must be adapted to the control of fishermen's actions, rather than of stocks of fish. The fundamental questions are: How do fishermen interact with their environment? How do the biological and economic variabilities of this environment shape behavior? Many chronic fisheries management problems can be traced to ignorance of the complex human systems through which policies are filtered.

This paper takes a statistical, rather than a control theory, approach to the problem of managing uncertain renewable resources. A subset of the Oregon groundfishery is used as an example of the importance of system structure to effective management. This subset consists of two components of fishing effort: the choice of fishing area and the mix of landed catch for bottom trawlers fishing out of the port of Astoria. System structure is described by reviewing operating strategies of fishermen and examining the economic properties of the mixes of landed catch. Implications of the economic structure of fishing effort for management are then discussed.

THE STRUCTURE OF FISHING EFFORT: AN EXAMPLE

Logbook data of fishing location, trawl hours, and landed mixes of groundfish catch are analyzed for the period 1977-1983. Over fifty species of fish are categorized as groundfish (Pacific Fishery Management Council, 1985); thirteen species comprise the bulk of the economically valuable (landed) species. Rockfish (Pacific ocean perch, widow rockfish, yellowtail rockfish, a complex of other *Sebastes*), roundfish (sablefish, lingcod, Pacific cod), and flat fish (soles, flounders) are found in various combinations in landed catch.

The biological environment of the groundfishery is characterized by high levels of variability and strong seasonality. A derivative effect of biological variability is economic variability. The economic environment of the fisherman is characterized by an uncertain stream of expected revenues. This uncertain flow of future returns from individual stocks creates an economic incentive for fishermen to diversify catch among more than one stock of fish. Variability in stock abundance means that a fisherman must opportunistically capitalize on a range of species.

Typically, decisions on effort control are made using data aggregations designed for stock assessment, i.e. landed weights of individual species over an entire fleet. For purposes of controlling levels of multispecies catch, a data base constructed along these lines contains arbitrary divisions that do not reflect fishery structure (Silvert and Dickie, 1982). A more promising approach is to base management decisions on data aggregations reflecting natural functional groupings within the fishery.

To isolate functional groupings in this fishery, industry spokesmen (fishermen, processors, and managers of fisherman associations) were asked to identify common fishing patterns. Three types of operation were described as typical: 1) a shallow water "beach" fishery (0-99 fms) for a mix of flatfish; 2) a mid-depth "slope" fishery (100-199 fms) for rockfish and some flatfish; 3) a deep water "deep slope" Dover sole fishery (200+ fms) that includes some rockfish, Pacific ocean perch, and sablefish. These classifications are supported by cluster analyses of recurrent multispecies mixes found in this area (Tyler et al., 1982; Hallowed, Northwest and Alaska Fisheries Center, personal communication, 1985). Three seasonal divisions of the data are made based on the timing of fishing activities: a winter season from December-March reflecting fishing activity on spawning schools; a spring-summer season from April-July; and a fall season from August-November.

For each fishing activity, $k = 1, \ldots, p$ landings are recorded in a matrix of total catch representing weight in pounds of species $i = 1, \ldots, n$, landed together in season $j = 1, \ldots, m$, i.e.:

$$
C_k = \begin{bmatrix} c_{11k} & \cdots & c_{n1k} \\ \cdot & & \\ \cdot & & \\ \cdot & & \\ \cdot & & \\ \cdot & & \\ \cdot & & \\ c_{1mk} & \cdots & c_{nmk} \end{bmatrix} \qquad k = 1, \ldots, p
$$

In our case

n = 13 species

m = 21 seasons (1977-1983)

p = 3 fishing activities

Catch mixes are not an end to themselves, but rather a means to earned income. Multiplying the weight of each species in the landed catch by its ex-vessel price yields a matrix of seasonal gross revenues corresponding to the seasonal catch matrix. The composition of the catch/revenue mixes changes from season to season. The rate of change of each species' contribution to the total catch/revenue mix will contribute to the level of variability. The revenue mix earned by a mix of catch landed together is here called an "economic assemblage." The extent to which the percentage shares (in revenues) of species in an economic assemblage correspond to their shares (in weight) in the catch mix will of course depend on the relative prices of species in the mix. Figure 1 illustrates a catch mix and its corresponding economic assemblage for the shelf fishing activity in the spring-summer season of 1983.

Once the data are partitioned according to separate fishing activities, some interesting patterns emerge. Figure 2 illustrates a strong seasonality to the number of trawl hours applied to each fishing activity, with seasonal distinctions becoming more marked in later years in the series. An increasing seasonal complementarity between fishing activities is also obvious throughout the time period, indicating a diversification of vessels into multiple fishing activities.

Within each fishing activity we see a further diversification over the number of species landed. Table 1 details the increase in the average number of species landed in each fishing activity over the time period. We know that, theoretically, fluctuating resources tend to promote generalist patterns of exploitation (Schoener, 1969; McKelvey, 1983). Observation of the multispecies patterns in this fishery supports this theoretical conclusion.

The benefits of diversification over species can be measured in economic terms by comparing the weighted average variability (measured by the coefficient of variation) of revenues earned by all species in an economic assemblage to the

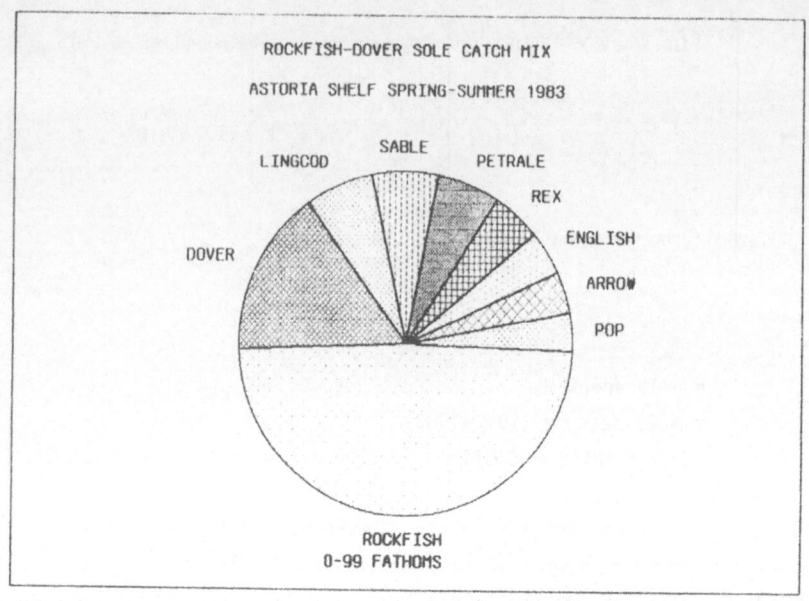

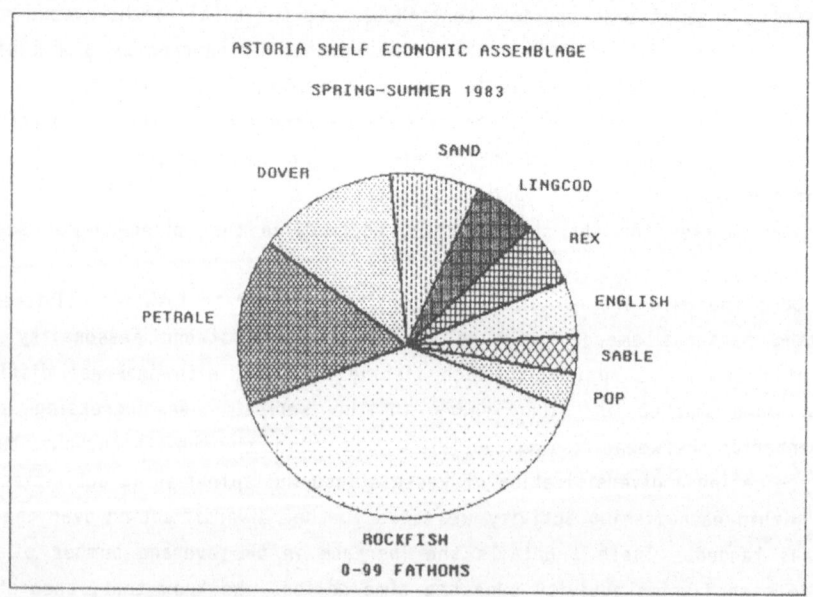

Arrow   = Arrowtooth flounder   POP   = Pacific ocean perch
Dover   = Dover sole   Rex   = Rex sole
English = English sole   Sable = Sablefish
Petrale = Petrale sole

Figure 1. Catch mix and economic assemblage from shelf fishing activity,
           spring-summer 1983

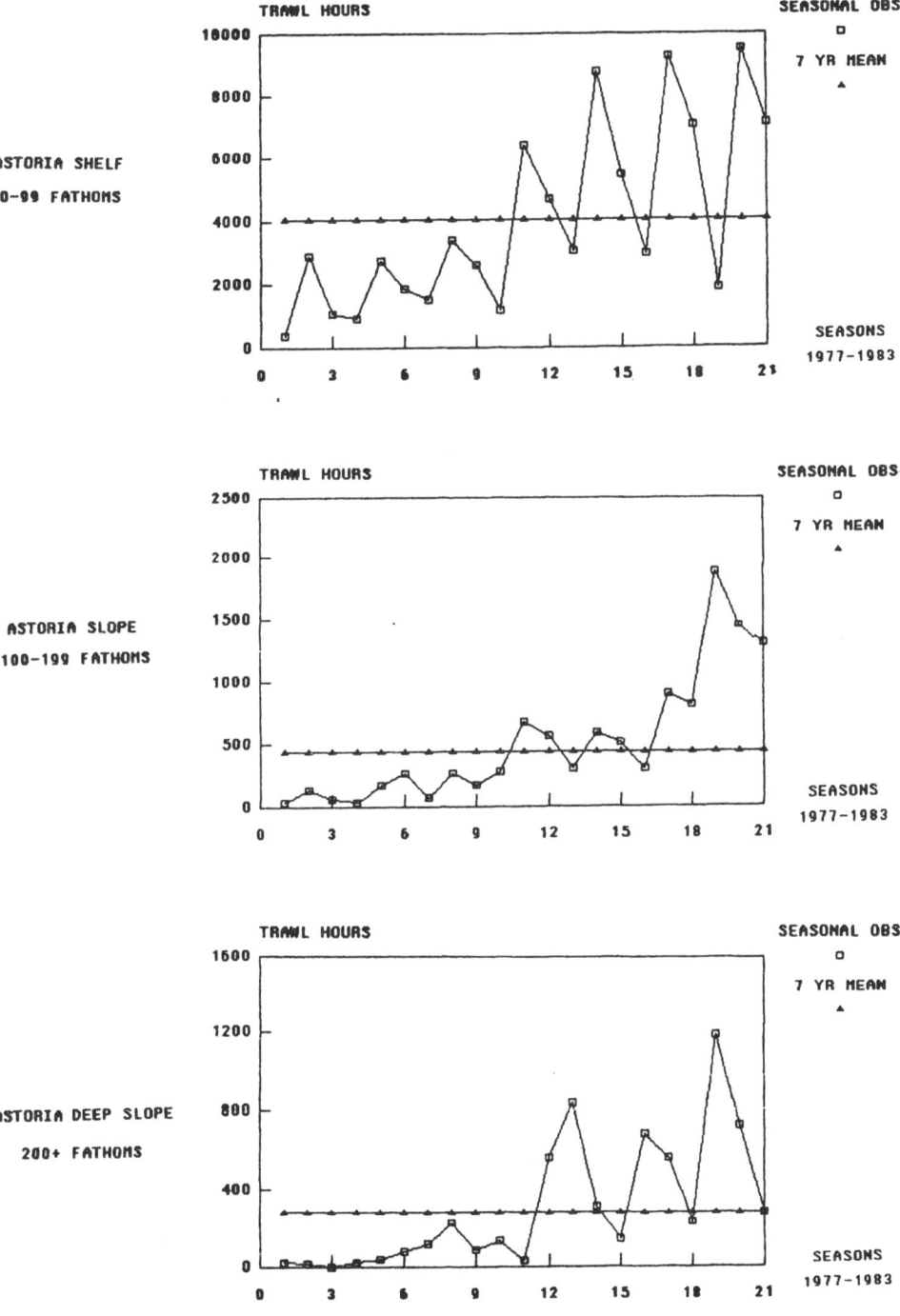

Figure 2. Seasonal trawl hours applied to three fishing activities,
1977-1983

Table 1.   Average number of species in trawl tows, Astoria, Oregon, 1977-1983.

	Shelf	Slope	Deep slope
1977	5.0	3.7	3.4
1978	5.7	5.6	4.1
1979	6.5	5.7	5.3
1980	6.4	5.7	4.4
1981	5.9	3.6	4.2
1982	6.1	5.5	4.8
1983	6.2	6.3	5.2

variability of total revenues earned in an economic assemblage (Table 2).   Sizeable reductions in revenue variability are realized by combining species in landed mixes.

This function of an economic assemblage has important implications for management, but is an aspect of the structure of fishing effort that would be missed by a reliance on data aggregated along single species lines.   The potential for reducing the variability of revenues will provide a strong incentive for fishermen to maintain these or similar patterns of fishing activity.   Although it is admittedly difficult to find many issues about which fishermen agree, it is likely that general agreement could be found on the desirability of a fairly stable floor of earnings (not necessarily yields).

Table 2.   Revenue variability in individual species and economic assemblages, Astoria, 1977-1983.

	Average variability of all species measured as cv (%)	Variability of total economic assemblage measured as cv (%)	% change in variability
Shelf	87.4	72.6	-17
Slope	116.2	81.5	-30
Deep slope	175.2	118.6	-32

*coefficient of variation

IMPLICATIONS OF SYSTEM STRUCTURE FOR MANAGEMENT

In this fishery the economic assemblage is a functional component of system structure. It is clear that fishing takes place on an assemblage basis rather than on a species basis. Fishing activities combine species of fish in ways that reduce the variability of revenues. Although an economic assemblage may have a dominant species, associated species play an economic role as well. It is likely that species associated with the dominant species in a mix serve as "hedges" to counteract the revenue variability of the dominant species. Diversification over several species is a necessary but not sufficient condition for reducing variability in revenues to fishermen. For diversification to provide stability of earnings the variability of individual species must be offsetting. Concurrent highs and lows of landings of major species would only exacerbate the variability of revenues. Economic assemblages provide cases where predation with switching provides a stabilizing effect to the revenues earned by fishermen.

The structure of fishing effort in this fishery is viewed here as an adaptation to a variable environment. This suggests that for management to be effective it must adapt to the way fishing activities are conducted. Management policies that work against fishery structure are also working against the economic incentives that gave rise to that structure and are doomed to meager success if not complete failure.

The management history of the West Coast groundfishery is one of single species techniques. Overall quotas are set for species evaluated to be distressed. The timing of landings of species under quota is controlled by the use of size/-frequency limits on trips. Yellowtail rockfish are one example. As trip limits have decreased in size, the management of yellowtail rockfish has been characterized by an increased frequency of regulation change, an increasing complexity of options, and an increased precision required in estimates of both stock size and weights carried on board.

Economic assemblages illustrate two fundamental ways the current management of this fishery is maladaptive to fishery structure. First, because concern is focused on single species, management policies ignore the association of limited species to other species in the landed mix. Limits placed on a single species within an economic assemblage will create an incentive to discard that species once the limit is reached, thereby having no positive effect on limiting fishing mortality. No economic incentives exist for fishermen to change established fishing patterns.

Second, in focusing on landed weights of species, management policies ignore the revenue earning properties of the mix. This means that managers and fishermen may be focusing on different properties of the same species. Attempts to limit fishing mortality calculated in weight do not address the question of that species' earnings potential in the overall economic assemblage.

Species-specific management regimes are expensive to operate in a multispecies fishery. They are data-intensive, create incentives for the discard of valuable species, impose handling and avoidance costs on fishermen, and create high information costs for fishermen as the allowable numbers become more precise.

The central question for management is whether the marginal benefits exceed or at least equal the marginal costs of management. The structure of the fishing system through which control policies are filtered is a factor of paramount importance to the benefits and costs of various control measures.

Focusing on the behavioral aspects of actors within the system suggests that data bases as they are currently used for single species management are both aggregated over inappropriate units and aggregated at too high a level (annual landings of fleets). Behavioral analysis of system structure also suggests that theories of search, foraging, and predation will offer useful insights into the patterns of human exploitation of resources (cf Caraco, 1981; Clark and Mangel, 1984; Eales and Wilen, 1982; McCay, 1981; Mangel, 1982; Mangel and Clark, 1983).

Economic analysis can benefit from a judicious borrowing of ecological concepts that promote a systematic taxonomy of fisheries. There is common ground in the fields of ecology and economics (Bernstein, 1981; Rapport and Turner, 1977); by exploiting this common ground both our theory and our design of adaptive management tools will be strengthened. This paper has dealt with only a piece of the total system structure of a fishery. The task of adequately accounting for the full scope of biological and economic complexities in multispecies fisheries is a vast one indeed.

REFERENCES

Bernstein, B.B. 1981. Ecology and economics: complex systems in changing environments. Ann. Rev. Ecol. Syst., Vol. 12, pp. 309-330.

Caraco, T. 1981. Risk-sensitivity and foraging groups. Ecology, Vol. 62, No. 3, pp. 526-531.

Clark, C.W. and M. Mangel. 1984. Flocking and foraging strategies: information in an uncertain environment. Amer. Nat., Vol. 123, pp. 626-641.

Eales, J. and J.E. Wilen. 1982. An empirical examination of searching behavior in the Pacific pink shrimp fishery. Department of Agricultural Economics, University of California, Davis, California, 15 pp. (mimeo).

Hilborn, R. 1985. Fleet dynamics and individual variation: why some people catch more fish than others. Can. J. Fish. Aquatic Sci., Vol. 42, pp. 2-13.

Mangel, M. 1982. Search, effort, and catch rate in fisheries. Eur. J. Oper. Res., Vol. 11, pp. 361-66.

Mangel, M. and C.W. Clark. 1983. Uncertainty, search, and information in fisheries. J. Cons. Int. Explor. Mer., Vol. 41, pp. 93-103.

273 —

McCay, B. 1981. Optimal foragers or political actors? Ecological analysis of a N.J. fishery, Amer. Ethnology, Vol 8.

McKelvey, R.W. 1983. The fishery in a fluctuating environment: coexistence of specialist and generalist vessels in a multipurpose fleet. J. Environ. Econ. Manag., Vol. 10, pp. 287-309.

Pacific Fisheries Management Council. 1985. Status of the Pacific coast groundfish fishery through 1985 and recommended acceptable biological catches for 1986. (Document prepared for the Council and its advisory entities.) Pacific Fishery Management Council, 526 SW Mill Street, Portland, Oregon 97201.

Rapport, D.J. and J.E. Turner. 1977. Economic models in ecology. Science, Vol. 195, pp. 367-373.

Schoener, T.W. 1969. Optimal size and specialization in constant and fluctuating environments: an energy-time approach, Brookhaven Symp. Biol., Vol. 22, pp. 103-14.

Silvert, W. and L.M. Dickie. 1982. Multispecies interactions between fish and fishermen. In Multispecies Approaches to Fisheries Management Advice, M. Mercer, ed. Canadian Special Publications of Fisheries and Aquatic Sciences, Vol. 59, pp. 163-169.

Tyler, A., E. Beals, and C. Smith. 1982. Analysis of logbooks for recurrent multi-species effort strategies. In Symposium on Determining Effective Effort and Calculating Yield in Groundfisheries, and on Pacific Cod Biology and Population Dynamics, INPFC Bulletin, Vol. 42, pp. 39-46.

Walters, C.J. and R. Hilborn. 1978. Ecological optimization and adaptive management. Ann. Rev. Ecol. Syst., Vol. 9, pp. 157-188.

PARTICIPANT'S COMMENTS

This paper illustrates a multispecies fishery in which the traditional approach of single species analysis is clearly inadequate. Fishermen will be choosing where to fish based on alternatives that depend upon the mix of fish they catch, and any attempt to look at catch or effort on a single species basis will be misleading. The author makes the very important point that regulations based on single species quotas or efforts are going to have highly unexpected consequences on other species, and almost certainly not achieve their initial goals.

I do question the implication that the fishermen are actually targetting on species assemblages, rather than simply choosing where to fish. A fisherman can only choose where to fish, and the species mix he catches is not a "target" as much as a product of a set of the net in a particular place. Perhaps on a very small spatial scale fine gradings of species mix can be achieved, but I believe it remains to be demonstrated that the choice is of assemblages to fish rather than locations.

Ray Hilborn

Reply

In response to the question of whether fishermen target on species mixes or simply choose a fishing location:

Fishermen in the Oregon groundfishery approach a fishing location with a prior expectation about the mix of species to be found. The strength of the prior and its conformity to the actual likelihood of occurrence of a species mix will *depend on the experience and skill of the individual fisherman as well as on changes in biological and oceanographic factors since the last sampling time at that site.*

Expectations about likely species mixes to be found are formed with knowledge of the ocean depth, ocean bottom characteristics and previous mixes caught at the location. Fishermen do choose a fishing location but this choice is premised on known associations between location, seasons of the year, and biological assemblages. Since these mixes are known with rough approximations of the relative proportion of individual species over fairly large areas, targeting on mixes is indeed possible and does occur.

S. Hanna

Hanna's paper clearly defines the need for an understanding of both the behavior of fishermen, and the reasons underlying the choices they make in prosecution of their fisheries. Multispecies management, long a concern expressed by biologists over conservation of limited fish stocks, is further confounded by the economic and cultural choices confronting the fishermen. Failure to recognize these features of the fisherman's decision-making will condemn to failure any attempts to rationalize and manage such fisheries, even in the presence of "perfect" biological information about the status of the fish stocks. With the single example of the Oregon groundfish fishery, Hanna illustrates a lesson that should be taken seriously by managers responsible for a majority of the world's fisheries. Hanna's concluding paragraph understates the need for a greater understanding of the parallel features of ecology and economics. In addition to the common ground between these disciplines cited in her literature section, biologists would do well to recognize Adam Smith's Wealth of Nations as one of the foundations upon which The Origin of Species rests.

The challenge in modelling fishing behavior is similar to attempting to model the behavior of a predator with respect to prey of different types. The functional relationships that economists choose to describe these dynamics will invariably be subject to the three general types of uncertainly that Hanna describes in her Introduction. A brief extension of Hanna's exploration of analogues in ecology and economics suggests some of the ways in which economic models may be significantly improved with respect to the problem of "appropriate model form." Factors such as "prey image fixation" may be akin to the vested interest fishermen have in the gear and vessels that they traditionally use. "Handling time" is analogous to the time required to set and retrieve the gear and sort the catch. Hanna aptly makes the point that "predation with switching" within the economic assemblages of species serves to provide economic stability to the fishery, just as a suite of alternate prey for carnivores (piscivores) provides a measure of ecological stability to animal communities. Energy return as a function of the particle size of the prey may have economic analogues in processing costs per individual fish in the catch. At the very lowest level of resolution will be the factors the fishermen use in deciding whether or not to continue to fish or to switch (retrain) to other occupations. The latter decision is, perhaps, similar to the choices that we all make in connection with use of leisure time, i.e., going fishing versus playing golf. Less tangible factors, such as culturally inherited mode of operation, may further exacerbate the multi-species management problem.

The management concerns that are raised in the Oregon groundfish fishery are clearly echoed in most other complex fisheries. The question of whether or not we need better parameter estimates for the single species management models is clearly moot when it is recognized that regulations formulated on the basis of these estimates, no matter how precisely known, may bear no relationship to the fishermen's behavior, or may simple encourage wastage or "illegal" activity. Hanna points out that attempting to limit fishing mortality by weight does not address the overall earnings potential of the economic assemblage. This is comparable to the biological conundrum that a unit of weight means different things to the stock depending upon the status of the stock. Thus, quota management approaches, popular with today's high seas fisheries and some Great Lakes fisheries, and the daily bag limits promulgated in most North American recreational fisheries will fail to adequately control fishery exploitation unless accessory regulations can be developed that will directly control participation in the fisheries.

George Spangler

SPATIAL MODELS OF TUNA DYNAMICS IN THE WESTERN PACIFIC: IS INTERNATIONAL MANAGEMENT NECESSARY?

Ray Hilborn
Tuna and Billfish Assessment Programme
South Pacific Commission
Post Box D-5
Noumea CEDEX
New Caledonia

Two explicit spatial models are presented to analyze the interaction between tuna fisheries in adjacent jurisdictions. Given the movements of skipjack tuna, and the large EEZ size of Western Pacific countries, there would appear to be little effect of skipjack fisheries outside of the EEZ on the catch inside the EEZ of most countries. This is less true for yellowfin tuna which live longer. The models presented, and the data and analysis required to estimate parameters necessary for these models, provides a quantitative framework for the analysis of interaction and the need for international catch regulation.

INTRODUCTION

At a conference on natural resource management that is jointly sponsored by national science organizations in Australia and the U.S., it is appropriate to talk about tuna, since tuna in the Pacific Ocean are perhaps the only renewable natural resource that is jointly exploited by Australia and the U.S. Indeed the conventional view of tuna, that is of highly migratory species that slosh back and forth across the oceans, suggests that there may be some interaction between U.S. and Australian fishing on Pacific tuna stocks. This is especially true since the U.S. high seas tuna fleet now fishes throughout the Western Pacific, and at times enters Australian waters.

The two dominant species in the Pacific Ocean (and world-wide) tuna harvests are skipjack tuna (*Katsuwonus pelamis*) and yellowfin tuna (*Thunnus albacares*), and for the rest of this paper I will consider only these two species. Both species are officially classified as "highly migratory" under the United Nations Law of the Sea convention, a distinction shared with other tuna and billfish, as well as some marine mammals and sharks. However, there is no exact definition of "highly migratory", and in addition, there is considerable international disagreement about fishing rights for these species. In particular, the U.S. does not recognize the coastal state jurisdiction over tuna within EEZs, while every other major fishing nation in the world does recognize the right of countries to regulate tuna harvests in their EEZ. The matter is further complicated because the U.S. does recognize jurisdiction over billfish (members of the tuna family), particularly the marlin prized by U.S. sportsfishermen in Hawaii and Florida.

It appears to be widely accepted that, because tuna have been known to make long individual movements, international agreement to prevent overexploitation is required. Harvests are often monitored by international agencies such as the Inter-American Tropical Tuna Commission (IATTC) in the Eastern Pacific, the International Commission for the Conservation of Atlantic Tunas (ICCAT), the South Pacific Commission (SPC) in the Western Pacific, and the Indo-Pacific Tuna Program (IPTP) in the Indian Ocean and part of the Pacific. Two of these agencies (IATTC and ICCAT) have some regulatory mandate.

No one appears to have analyzed the movement rates of tunas, and then examined the consequences of different management regimes given the size and shape of various countries' EEZs. In this paper, I will look at data on tuna movement rates derived from several tagging studies, and then present two spatial models of tuna harvesting. The consequences of alternative international management practices will be examined for these two models.

TUNA MOVEMENT RATES

The international organizations that monitor tuna stocks often conduct tagging studies to determine movement rates and directions. One of the most ambitious of these programs was the Skipjack Survey and Assessment Programme of the SPC, which tagged over 150,000 tuna, mostly skipjack, between 1977 and 1981 (Kearney, 1983). Over 6,000 of these tagged fish were recovered. This provides an excellent data base on movements of skipjack.

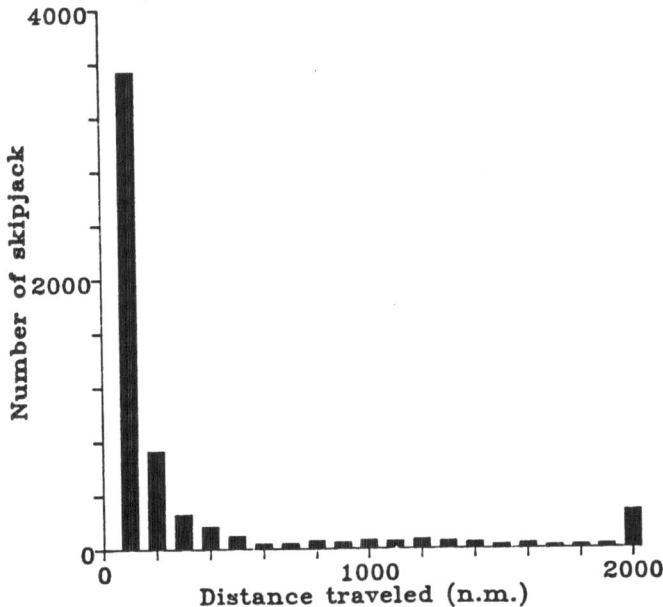

Figure 1.   Frequency distribution of distance moved by skipjack recovered in SPC Skipjack Programme.

Figure 1 shows a frequency distribution of distance moved for skipjack tuna. Almost all fish were recovered within a few hundred kilometers of their tagging site, and indeed approximately 85% of skipjack were recovered in the EEZ of the country in which they were tagged. Thus, we can readily see that far from sloshing back and forth across the ocean, skipjack tuna are generally not moving very far or very fast. Even after a year at large, most tags were recovered close to the site of tagging. This high residence rate is, to some extent, an artifact of the fact that fish were tagged where abundance was high, which also tended to be sites of high fishing pressure. However, analysis of movement rates indicates most individuals move quite slowly.

Figure 2 shows data on average distance traveled vs. time at large for yellowfin tuna tagged in West Africa (Cayre et al., 1974). Again, the average distance traveled is not particularly large, and these data show an apparent annual migration, thought to be associated with seasonal changes in ocean temperatures.

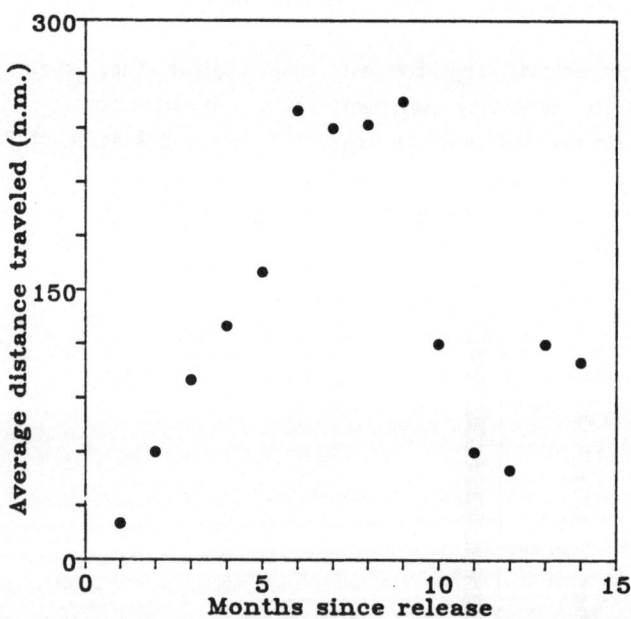

Figure 2.   Distance traveled vs. time at large for yellowfin tagged in the Eastern Atlantic (from Cayre et al., 1974).

The relative rarity of long distance movement in the skipjack data presented above could be due to at least three causes, 1) the fish do not move great distances, 2) the fish move long distances and then return to near their tagging site where they are recaptured, and 3) the distribution of fishing effort causes us to underestimate the movement of marked fish. The second mechanism can be discounted because no evidence for annual migratory movements has been found in these skipjack data (as opposed to the yellowfin data of Figure 2); most fish were recovered close to the site of tagging no matter what time since tagging had elapsed. The third mechanism is harder to refute, certainly tagging took place in known concentrations, generally where there was a commercial fishery. However, in several places where there were commercial fisheries reasonably close, very few tags were recovered in the adjacent fishery.

If an EEZ size is small, as occurs in West Africa and parts of Central America, then movements of a few hundred kilometers may mean the fish are moving quite rapidly from one EEZ to another. However, in the Western Pacific, EEZ sizes are very large, often several million square kilometers, and movements of a few hundred kilometers will leave most fish in the EEZ of the same country.

SPATIAL MODELS

The question we are most interested in is how does the fishing mortality in one location affect the catch rates in another location. The first model will consider two countries, taken in isolation, that exchange fish across their border; we will look at how the fishing pressure in one country affects the other country as a function of movement rates and survival rates. In the second model, we consider an isolated island country with an economic zone of 200-mile radius, with intense fishing taking place outside of its fishing zone. We will examine how the movement rate and the survival rate affect the interaction between catches inside and outside of the EEZ.

Some Tuna Biology

Skipjack and yellowfish tuna are found throughout the tropical oceans of the world, and their larvae are also distributed world-wide, suggesting spawning takes place in essentially all tropical waters. A major exception appears to be the Eastern Pacific, where few skipjack larvae have been found (Argue et al., 1983; Far Seas Fisheries Research Laboratory, 1985). The fish are extremely fast growing, the average skipjack caught is thought to be about 1 year old (2-5 kg), and the purse seine caught yellowfin (5-20 kg) are thought to be 1-3 years old. They also have very high natural mortality rates; I will use the value of 20% per month for skipjack (Kleiber et al., 1983), and 6% per month for yellowfin (Lenarz and Zweifel, 1979). Clearly, the extent of interaction between adjacent countries will depend on both the movement rate and the natural mortality rate, so we will expect to find more interaction between yellowfin stocks than skipjack stocks.

Little is known about the reproductive biology of skipjack and yellowfin, but they are highly fecund; each female releases millions of eggs with each spawning, which takes place year round. Therefore, for simplicity, and in the absence of any evidence to the contrary, I will assume that there is a constant recruitment rate. A rough estimate of the standing stock of skipjack is 0.1 tonne per square kilometer (Kleiber et al., 1983), which with no fishing, and 20% natural mortality per month, would imply a recruitment of 0.02 tonnes per square kilometer per month.

Two Adjacent Countries

Consider two countries with tuna stocks that move between the countries but to no other country. We will use the following model to describe the dynamics in the two countries:

$$N_{1,t+1} = N_{1,t} (s - qE_1 - d_{12}) + N_{2,t}d_{21} + R$$

$$\text{(1)}$$

$$N_{2,t+1} = N_{2,t} (s - qE_2 - d_{21}) + N_{1,t}d_{12} + R$$

where $N_{i,t}$ is the biomass of tuna in country i at time t, s is the survival rate (1 - natural mortality rate), E is the fishing effort (assumed constant at equilibrium), q is a catchability coefficient that is the proportion of the biomass harvested per unit of fishing effort, d_{ij} is the movement rate from country i to country j, and R is the recruitment (also assumed constant). This model is quite similar to one proposed by Beverton and Holt (1957), and used by Sibert (1984) to analyze exchange rates between countries.

The economics can be modeled for each country by assuming a value per tonne of fish caught (v), and a cost of operation for each unit of effort (c). Catch, from equation (1), is the product of E, N, and q.

Clearly, if all parameters are the same in both countries, then the exchange of fish doesn't matter; the biomass will be the same on both sides of the border and the net flow will be zero. The equations become interesting when there is asymmetry. Perhaps the most interesting is asymmetry in management objectives. In international waters one would expect the fishery to go to the point of bioeconomic equilibrium. Some countries may wish to maximize employment or catch for foreign export earnings, which would also put the fishing effort at bionomic equilibrium. This, of course, assumes the countries would not choose to lose money to increase catch. Other countries may wish to maximize the access fees paid by other nations for access to their EEZ, which should mean trying to maximize total profits from the fishery.

One could explore many of the above variations, particularly with various assumptions about asymmetric prices and costs. However, for the purposes of this paper, I will only look at the impacts of one country fishing at bionomic equilibrium (maximize catch so long as profits are not negative), upon the second country attempting to maximize profits.

It is easily demonstrated for the above model that catch will always increase with increased effort, therefore the bionomic equilibrium will occur when profits are zero. The biomass at bionomic equilibrium N^* will be:

$$N^* = \frac{c}{qv} \tag{2}$$

If we assume that area 2 will be held at some specific stock size (N_2), then we can calculate the equilibrium biomass for area 1 as a function of the fishing effort in area 1 as follows:

$$N_1 = \frac{R + (d_{21} N_2)}{(1 - s + d_{12} + qE_1)} \tag{3}$$

The profits in area 1 are:

$$P = E_1 \left[\frac{qv(R+d_{21}N_2)}{1-s+qE_1+d_{12}} - c \right] \tag{4}$$

which is maximized at

$$E_1 = \frac{(-b+sqr(ab/c))}{q} \tag{5}$$

where:

$$a = (r + d_{21} N_2) qv$$

and

$$b = 1 - s + d_{12}$$

When $d_{12} = d_{21} = d$, then profits can be more simply calculated as

$$P = \frac{(sqr(Rqv + cd) - sqr(c(1-s)+cd))^2}{q} \tag{6}$$

Catch can also be explicitly calculated as:

$$Catch = \frac{(Rqv + cd - sqr(c(Rqv+cd)(1 - s+d)))}{qv} \tag{7}$$

The following parameters were used assuming each country has 1,000,000 km^2 of EEZ, and a time step of one month: R = 5000, q = 0.00625, s = -0.9, c = \$200,000, and v = \$1,000. The recruitment and survival rate values (R and s) were chosen as an aggregate for yellowfin and skipjack. The price of \$1,000 per tonne is considerably higher than the recent price, but below historical high prices. The cost was derived from some unpublished purse seine costs available to the SPC.

Using the above parameters, the bionomic equilibrium for each country will be 32,000 tonnes, with an unfished stock size of 50,000 tonnes. Assuming area 2 is managed down to bionomic equilibrium, and area 1 is managed to maximize profits, Figure 3 shows the relationship between the catch and profits in area 1, and the exchange rate (the d parameters) between the two countries. Note that catch does not change too much, but profits are highly sensitive.

While it is difficult to generalize from this analysis, we can see that differences in the objectives of different countries will have a major impact on the interaction of their fisheries, an influence that will, in turn, depend upon the exchange rate of fish. Estimates of exchange rates made by Sibert (1984) for movements between Papua New Guinea and the Solomon Islands, are on the order of 0.01 to 0.05 per month. Thus, there should be little impact of fishing in one country upon the other, although one must always be careful to recognize that profits will be much more sensitive than catch.

The only case in which two countries (or one country and international waters) will interact strongly is where there are different objectives and significant exchange rates. To look at this in more detail, we use the following model.

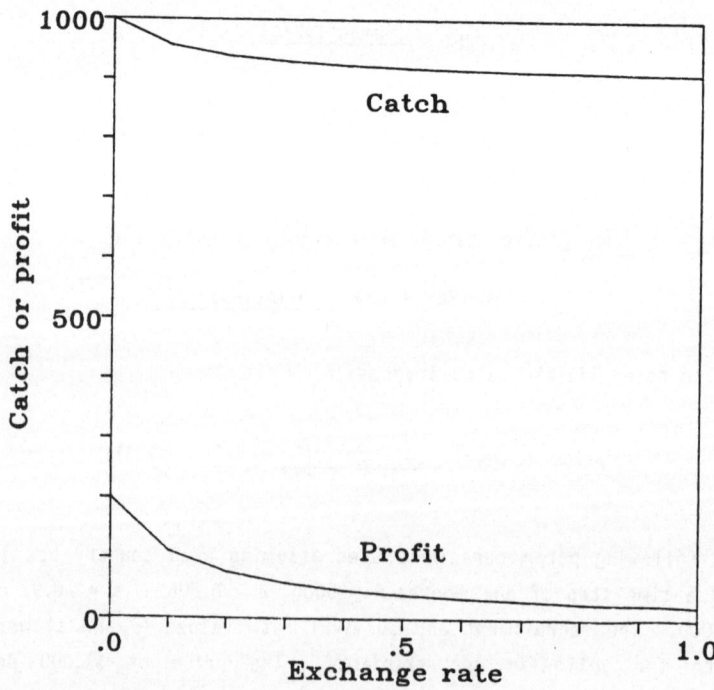

Figure 3.   Catch and profits in a country being managed to maximize profits as a function of exchange rate with a country managed to bionomic equilibrium.

<u>Single Country Model</u>

Imagine a country with an economic zone of radius x, set amidst an endless international high seas zone. Inside of the country's EEZ there is a small artisanal fishery that harvests 1% of the fish each month; outside of the EEZ there is a marauding, unregulated international fishery that is harvesting very hard at 10% per month. If we assume the recruitment and harvest dynamics of the previous model, and imbed this in a diffusion model for the movement, we can then solve quite simply for the effects of diffusion rate and EEZ size using the following model for dynamics at any point.

$$\frac{dN}{dt} = R - (1 - s) N - hN + D \left[\frac{d^2N}{dx^2} + \frac{d^2N}{dy^2}\right] \tag{8}$$

where h is the harvest rate at the specific point and D is the diffusion coefficient. This model is essentially the same as that of Clark's (1976) inshore-offshore fishery model except it takes place in two dimensions.

If we assume there is a harvest rate due to artisanal fishermen (1% per month) inside the EEZ, and a harvest rate of 10% per month outside the economic zone, we can see how much the artisanal catch will be reduced by comparing the equilibrium artisanal catch to that which would occur when the harvest rate was 1% per month outside the economic zone. This will depend upon the mortality rate of the fish and the size of the economic zone.

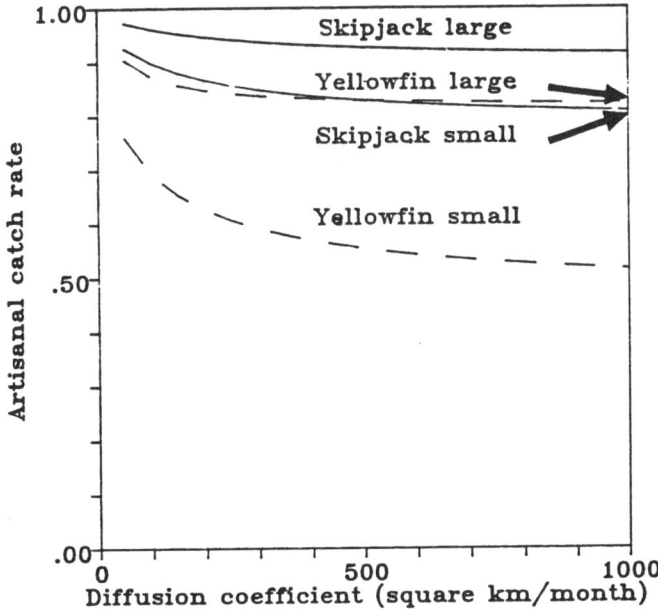

Figure 4.   Artisanal catch rate within a 200-mile EEZ as a function of the diffusion rate. Area outside of EEZ is harvested at 10 times the rate of the area inside the EEZ.

Figure 4 shows the relative artisanal catch rate as a function of diffusion rate using natural mortality rates of 0.2 per month for skipjack and 0.06 per month for yellowfin. This graph is drawn for a small EEZ with a radius of 200 nautical miles (360 kilometers or an area of 407,000 km²), and for a large country with a radius of 1000 km (an area of roughly 3.1 million km²). For a large country, there is very little impact on either species, and for a 'small' country only yellowfin would appear to be strongly affected.

Most of the Pacific Island countries that have major tuna fisheries, Federated States of Micronesia, Papua New Guinea, Solomon Islands for instance, have large EEZs because they consist of archipelagos rather than an individual island. In addition, the concern is not so much about artisanal catch rates, but catch rates of commercial fisheries in their waters. Both of these factors mean the impact of fishing rates outside the EEZ will likely be small. A cautionary note should perhaps be sounded in that while catch rates may not drop very much, the profitability of the fishing operation might be more seriously affected.

All of this depends, to some extent, upon the movement dynamics. Preliminary analysis of skipjack movements indicates that most individuals move very little and may be modeled by a diffusion model with a diffusion coefficient of between 100 and 500 km² per day. A small proportion of the fish (perhaps 10%) seem to move in a more directed manner, and probably cannot be well described by diffusion. More detailed analysis of movement dynamics is a very high priority.

We can therefore say with some certainty that there will be very little effect on skipjack yields by fisheries outside of the EEZ, but in the absence of reliable estimates of yellowfin movement rates we need to be a little more cautious. It is possible that 20-30% reductions in yellowfin catch rates could be experienced because of fisheries beyond the EEZ.

CONCLUSIONS

The traditional assumption that tuna are highly migratory, and what goes on in one country totally depends upon the fishing policies of its neighbors, needs to be discarded. If the EEZs of the countries are large, the tuna stocks can be considered resident and countries can quite nicely manage their tuna stocks without international agreements to limit harvests. International cooperation on data collection will continue to be very important since it may be difficult for any individual country to conduct a research or data collection program on its own, and long-term and large-scale trends will be more easily detected from data aggregated over very large areas, than from data collected within one individual country.

The data base on skipjack movement rates gives us reasonable confidence about our predictions on skipjack. We can be less confident about yellowfin. The IATTC has conducted considerable yellowfin tagging, but has never analyzed the data in a form usable for estimating movement rates. Widespread yellowfin tagging in the Western Pacific is required to accurately assess the interaction between fisheries in different countries and international waters.

I have presented two models, each of which can be used for different purposes. There are a number of game theory extensions that can be performed with the first model, looking, for instance, at reaction curves and Nash equilibria. The second model could be extended by considering multiple classes of fish, each moving at different rates.

ACKNOWLEDGMENTS

This work builds upon a study of tuna movement initiated by Dave Fournier and John Sibert, which, due to circumstances beyond their control, was never completed. Much of the analysis depends upon the valuable body of tagging information collected by the SPC Skipjack Programme, and I am indebted to all the participants in that program for developing and maintaining an excellent data base. John Sibert provided valuable help and ideas at many stages. Simon Levin provided very helpful comments and the explicit derivation of optimal effort and profits in the two-area model. Geoff Kirkwood and an anonymous referee also provided helpful comments.

REFERENCES

Argue, A.W., F. Conand and D. Whyman. 1983. Spatial and temporal distributions of juvenile tunas from stomachs of tunas caught by pole-and-line gear in the central and western Pacific Ocean. Tuna and Billfish Assessment Programme Technical Report No. 9. South Pacific Commission, Noumea, New Caledonia, 47 pp.

Beverton, R.J.H. and S.J. Holt. 1957. On the dynamics of exploited fish populations. U.K. Min. Agric. Fish., Fish. Invest. (Ser 2), Vol. 19, 533 p.

Cayre, P., Y. LeHir, and R. Pianet. 1974. Marquage et migrations des albacores dans la région de Pointe-Noire. Office de la recherche scientifique et technique outre-mer, Documents Scientifique du Centre Pointe Noire Nouvelle Série No. 37.

Clark, C.W. 1976. Mathematical Bioeconomics. John Wiley and Sons, New York, 352 pp.

Far Seas Fisheries Research Laboratory. 1985. Average distribution of larvae of oceanic species of Scombroid Fishes, 1958-1981. Shimizu, 424 Japan, S Series No. 12.

Kearney, R.E. 1983. Assessment of the skipjack and baitfish resources in the central and western tropical Pacific Ocean: A summary of the Skipjack Survey and Assessment Programme. South Pacific Commission, Noumea, New Caledonia, 37 pp.

Kleiber, P., A.W. Argue, R.E. Kearney. 1983. Assessment of skipjack (Katsuwonus pelamis) resources in the central and western Pacific by estimating standing stock and components of population turnover from tagging data. Tuna and Billfish Assessment Programme Technical Report No. 8. South Pacific Commission, Noumea, New Caledonia, 38 pp.

Lenarz, W.H. and J.R. Zweifel. 1979. A theoretical examination of some aspects of the interaction between longline and surface fisheries for yellowfin tuna, (*Thunnus albacares*). Fish. Bull., Vol. 76, pp. 807-825.

Sibert, J.R. 1984. A two-fishery tag attrition model for the analysis of mortality, recruitment and fishery interaction. Tuna and Billfish Assessment Programme Technical Report No. 13. South Pacific Commission, Noumea, New Caledonia, 27 pp.

PARTICIPANT'S COMMENTS

The considerable value of this paper -- and the usefulness in the general analysis of resource systems -- is that a common assumption, which everybody "knows", is successfully challenged. That is, previous perceptions regarding the management of tuna have been based on the idea that tuna are so migratory that international management is a must. Entire organizations exist on the basis of this concept. Hilborn's first contribution is to challenge the basic assumption. Although the evidence that he presents is not perfect, it certainly is compelling and forces one to question some of the underlying precepts of the usual management policies. Hilborn is thus forcing us to think about tuna in ways that we have not previously done and his analyses should be motivation to look at the entire problem again. From the more technical side, the problem of estimating the diffusion coefficient and drift for models of tuna motion is not a completely trivial one and I cannot find a solution in the literature. Thus, Hilborn's second contribution is to suggest meaningful areas of research.

If one reads the literature on problem solving, a recurring theme is that of "conceptual blockbusting". To provide novel solutions to problems, one needs to be able to look at them in novel ways. Often this involves questioning the basic premise on which so much of the usual answer hinges. This paper provides an excellent example of conceptual blockbusting in a fisheries setting. It is, of course, crucial to ultimately know if tuna are migratory or not. Regardless of the answer, a paper such as this one provides a valuable contribution in getting us to question hidden assumptions.

Marc Mangel

EXAMINATION OF INSTITUTIONAL STRUCTURES IN MULTIPLE RESOURCE, MULTIPLE MANAGEMENT SYSTEMS: A CONTROL THEORETIC APPROACH

Ann Lowes Blackwell
Mechanical Engineering Department
University of Texas
Arlington

The constraints which management institutions impose upon resource productivity is a common concern to resource analysts and has been addressed through various means. In this paper, an approach is presented which utilizes control theoretic concepts to determine the effect of multiple institutional constraints upon the achievement of management targets. The method is illustrated by application to a scenario of anadromous fishery and water resource management.

INTRODUCTION

A number of papers presented at this conference dealt with developing optimal strategies for managing a single resource regulated by a single manager (Lesse, Walters) or by multiple managers organized in a centralized management configuration (Sluczanowski) or by multiple managers in a strictly competitive configuration (Hilborn, Nicol). Still others were concerned with strategies for managing multispecies resources regulated by a single manager or by a centralized management configuration (Cohen, Hanna, Spangler).

Another scenario occurring often in resource management is one in which the responsibilities for the management of exploited resources of an ecosystem are allocated among agencies which derive their authority from different governmental levels. When both state and federal agencies are involved, the linkage pattern among the management authorities cannot readily be classified as strictly centralized, cooperative or competitive. The explicit nature of that linkage must be determined by examination of both the authorizing and enabling legislation and the policies and practices of managers.

Although not unique to it, the paradigm of multiple resource, multiple manager interaction is evident in the allocation of water resources and the stewardship of fishery resources. For fisheries involving anadromous species (e.g., salmon) or estuarine dependent species (e.g., penaeid shrimp) the quality of freshwater habitat has an impact upon species production. The effect of habitat variability has been included in fisheries management models as variable environmental conditions (Walters, 1975; May et al., 1978; Swartzman et al., 1983; Tang, 1985; Moussalli and Hilborn, 1986). In these models, habitat quality may be modeled explicitly or be implicit in the fish population dynamics parameters. Fishery management strategies are developed which involve manipulation of the control variables for harvesting and/or stocking of the fish resource rather than control variables for habitat manipulation, i.e. management reflects a strategy within an immutable habitat condition. While this may be appropriate when habitat variability is dominated by natural (i.e. uncontrollable) forces, it is less suitable when habitat condition is largely dominated by human management actions. In the latter case, optimal management strategies are more appropriately developed using models which account for the population dynamics of the biological resource, the explicit manner in which biological and habitat controls affect the population dynamics and the relationships between the biological and habitat control variables.

The common occurrence of this resource management paradigm is reflected in the development of various types of models (e.g., descriptive, conceptual, simulation) by researchers in disciplines concerned with the problem, such as law, economics, and water resources. In the legal literature, an historical, legislative model of multiple resource exploitation and multiple authority management of resources is presented by Wilkinson and Conner (1983). They provide an in-depth review of the development of the management jurisdictions which affect the life history of the chinook salmon in the Columbia River Basin. Noting the involvement of as many as 17 habitat and fishery jurisdictional authorities, they discuss the legislative, policy and operational efforts which have been made to achieve effective coordinated management of the water resources of the Columbia river system and the related fish and wildlife programs. From the perspective of resource economics, Wandschneider (1984a,b) has discussed the nature of the management structure for the Columbia-Snake River system and has developed conceptual models of streamflow management and the interactions of various agencies on the law-making, policy-making, and operational decision-making levels. To include institutional analyses in water resources planning, the U.S. Fish and Wildlife Service (USFWS) has developed a Legal/Institutional Analysis Model to assist in defining the roles and powers of agencies related to water development projects. Other planning and analysis tools such as habitat suitability index models and instream flow suitability curves (e.g., Raleigh et al., 1985; USFWS, 1985) can be used in conjunction with this model to evaluate project impacts upon instream uses such as fish habitat.

The purpose of this paper is to illustrate how a multiple resource exploitation, multiple authority management scenario can be formulated in a control theoretic framework with the expectation that, by examining certain properties of the mathematical model, insights into the nature of the overall system behavior may appear that may not be as readily observed through other approaches. This understanding should expedite determination of how the achievement of goals proposed or adopted by the policy-makers are constrained by the dynamics of the resources and by the roles and powers of the management agencies.

MULTIPLE RESOURCE MANAGEMENT IN CALIFORNIA:
WATER DEVELOPMENT AND THE SALMON FISHERY

The river systems of the North Coast (Klamath - Trinity) and the Central Valley (Sacramento and the San Joaquin) of California are highly developed for water supply purposes. The U.S. Federal and California State governments operate, respectively, the Central Valley Project (CVP) (initiated in the 1930's) and the State Water Project (SWP) (initiated in the 1960's). The major features of the CVP and the SWP are shown in Fig. 1. The water projects and urban developments in the drainage basin of these rivers have altered not only the flow patterns in the rivers and through the Delta and San Francisco Bay to the Pacific Ocean but the water quality regime as well.

LEGEND

━━━ State Water Project

— — Central Valley Project

•••• Joint Use Facilities

Figure 1.   California chinook salmon fishery management system boundaries (-··-): (1) San-Joaquin River Basin, (2) American River Basin, (3) Feather River Basin, (4) Upper Sacramento River Basin, (5) Trinity-Klamath River Basin, (6) Ocean fishery zone. Existing and authorized major features of the State Water Project and Central Valley Project of California are indicated. Map adapted from Layperson's Guide to the Delta, courtesy of Western Water Education Foundation.

These river systems are also the habitat of the second largest population of chinook salmon (*Oncorhynchus tshawytscha*) in the Pacific. Chinook salmon are an anadromous fish species which spawn in fresh water but spend the greater portion of their life cycle in the marine environment. The fall run (season of adult immigration from the ocean) is the most numerous and the major focus of management efforts at this time. The watersheds in which these stocks occur are shown in Fig. 1. Due to the construction of impassable barriers (dams), not all of the streams in the extent of watershed shown are available to the spawning stock.

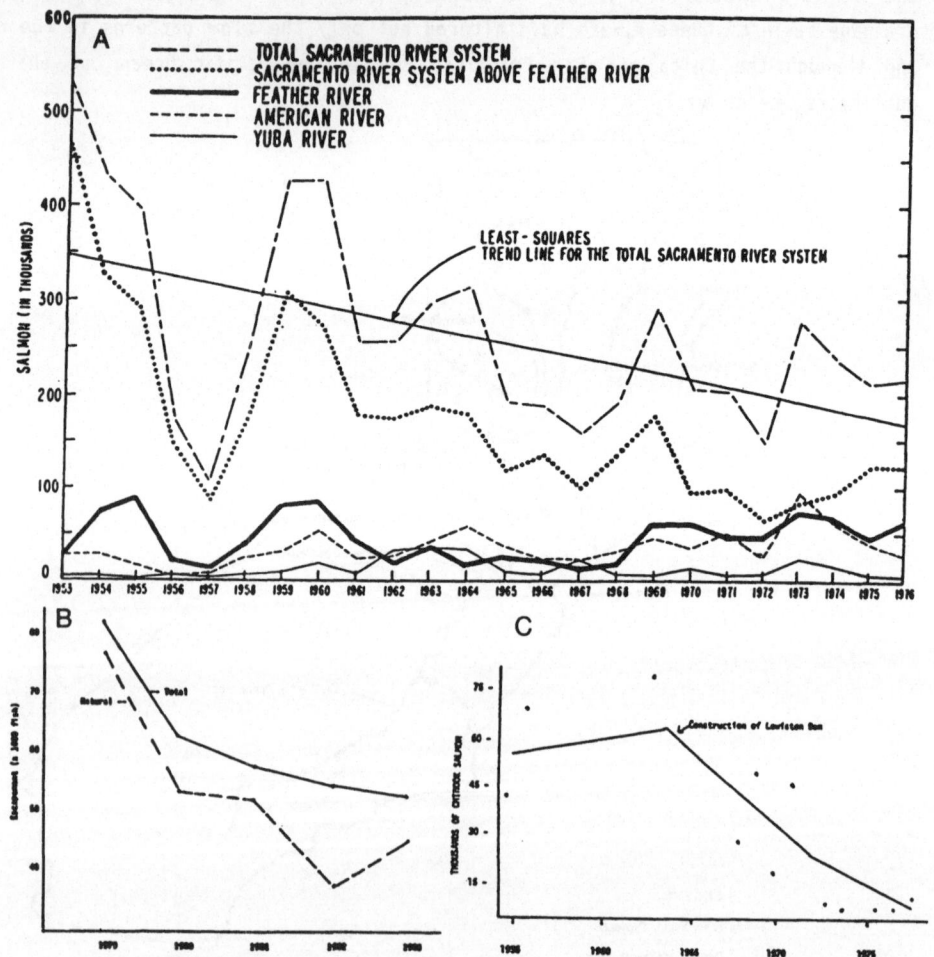

Figure 2. (A) Sacramento River system (Sacramento River system above the Feather River, Feather River, Yuba River and American River) fall spawning chinook salmon populations, 1953-1976 (From Hallock, 1978); (B) Upper Sacramento River system (above the Feather River) fall spawning chinook salmon populations (natural and total (natural plus hatchery)), 1979-1983 (From Fletcher, 1984); (C) Chinook salmon spawner escapement estimates in the Trinity River, 1955-1978 (From U.S. Fish and Wildlife Service, 1980).

In the San Joaquin River the entire spring chinook salmon run has been lost and the fall run is only 10 percent of its historical size. The chinook salmon run in the Sacramento River is about 40 to 50 percent of its size in 1953. (See Fig. 2A and 2B.) An estimated 80 to 90 percent decline in fall run Trinity River spawners has occurred since the Trinity River Division of the CVP was placed in operation in 1963. (See Fig. 2C.) It is believed that the water projects, in addition to overfishing, are responsible for the declines in abundance of salmon runs in these river systems. Previous studies have been conducted to develop tools to assess fish and wildlife resource management in the Sacramento River Basin (H. Chadwick, C.J. Walters, personal communication). However, the analytical tools required to assess the relative impact of each resource use on the salmon spawning population decline have not previously existed.

The evolution of the management system which regulates the salmon resource and its habitat in California reflects the social, economic and political development of the state and region as well as the development of scientific understanding of the environmental needs of the salmon in its various life history stages. Federal and State anadromous fishery managers regulate production and harvesting of fish in different habitats. In general, State regulation of harvesting prevails in State waters and Federal regulation prevails in marine waters beyond the territorial sea. Both the State and the U.S. are engaged in the artificial production of chinook salmon to mitigate the impacts of the water projects. Both governmental entities are also involved in the management of the harvesting of the salmon in the marine ecosystem for commercial and recreational purposes.

Habitat managers manipulate various aspects of habitats not only for purposes of fish production but for utilization by other sectors of the economy, e.g., utilization of land for urban development or timber and mineral production, diversion of water for agricultural and municipal uses, utilization of fresh or marine waters for disposal of wastes, energy production and wildlife sanctuaries. Both habitat managers and fishery managers derive their authority from different jurisdictional levels, creating a complex management structure without a distinct hierarchy. A summary of major exploited ecosystem resources and the major Federal and State management agencies involved is shown in Fig. 3.

Further development of both the water resources and the fishery resources is planned by both the State and the U.S. Proposed and adopted spawner management goals for the Sacramento and Klamath Rivers are illustrated in Table I. Historical and projected water development for the CVP and SWP are illustrated in Figs. 4 and 5. It has become evident to both State and Federal agencies that the complex relationships among the agencies in both governmental levels must be identified, and perhaps modified, in order to ensure the achievement of the joint water and fishery resource development goals.

o **Multiple ecosystem resources - biological** - salmon
                                          other anadromous fish species

                          non-biological renewable - agricultural & municipal water
                                                      diversions hydropower

                          non-renewable - oil & gas exploration, urban land development
                                          mining, agricultural & silvicultural land

o **Multiple institutions -**   **Federal** - DOI - OCS leasing
                                                  BuRec - CVP water system
                                                  USFWS - hatchery, spawning channels
                                                  BIA - Indian lands
                                          USDA - USFS - national forest lands
                                          DOC - NMFS (PFMC) - Pacific fisheries
                                          OOCRM - Marine sanctuary
        No distinct hierarchy
                                      State - cities - urban land uses
                                              counties - subdivision
                                              CFG - fishing, hatcheries
                                              CDWR - SWP water system
                                              SWRCB - water appropriation, quality
                                              BCDC - San Francisco Bay
                                              CCC - Cal coast

Figure 3.   The major exploited ecosystem resources related to chinook salmon
             fisheries and salmon habitats and the Federal and California State
             institutions involved in the management of those resources in the
             Central Valley and Pacific Coastal ecosystems of California. BCDC -
             San Francisco Bay Conservation and Development Commission; BIA - Bureau
             of Indian Affairs; BuRec - Bureau of Reclamation; CCC - California
             Coastal Commission; CDWR - Development of Water Recources; CFG -
             Department of Fish and Game; CVP - Central Valley Project; DOC -
             Department of Commerce; DOI - Department of the Interior; NMFS -
             National Marine Fisheries Service; OCS - Outer Continental Shelf; OOCRM
             - Office of Ocean and Coastal Resource Management; PFMC - Pacific
             Fishery Management Council; SWP - State Water Project; SWRCB - State
             Water Resources Control Board; USDA - Department of Agriculture; USFS -
             Forest Service; USFWS - Fish and Wildlife Service.

Table I. Examples of Management Goals for Fall Chinook Salmon Considered or Adopted by the Pacific Fishery Management Council

Sacramento River Fall Chinook

The Council considered three alternative management goals for Sacramento River fall chinook before it adopted Option 3.

Option 1: Achieve a spawning escapement goal of 99,000 natural and 9,000 hatchery chinook of upper Sacramento River origin by 1988 given average environmental conditions and contingent upon solving the problems associated with the Red Bluff Diversion Dam. A specific schedule to achieve the goal is not included in this option.

Option 2: Achieve an average 20 percent increase in spawning escapement every four years until the long-term goal of 99,000 natural spawning chinook is attained, contingent upon solving the problems at the Red Bluff Diversion Dam. The rebuilding schedule listed below is expressed as spawning escapement except for a small in-river harvest.

1983-86	65,800	1991-94	94,800
1987-90	79,000	1995-98	99,000

Option 3 (adopted by the Council): Achieve a single river spawning escapement goal range of 122,000 to 180,000 Sacramento River chinook. Within this range annual escapements can be expected to vary. Separate goals for the upper and lower Sacramento stocks are not established. The California Department of Fish and Game has provided the following information on state distribution goals and the rationale for this option:

California Department of Fish and Game Distribution Goals for Sacramento River Fall Chinook Salmon[1]

Upper-River:	Natural	99,000
	Hatchery	9,000
Total Upper		108,000
Lower River:		
Feather -	Natural	27,000
	Hatchery	5,000
Yuba -	Natural	10,000
American -	Natural	24,000
	Hatchery	6,000
Total Lower-River		72,000
Total Sacramento		180,000

[1]Distribution goals will not be used as a basis for ocean management. These will be used as management goals by agencies having in-river management responsibilities. Until passage problems at the Red Bluff Diversion Dam are corrected, the up-river distribution goals are not expected to be achieved.

Klamath River Fall Chinook

The Council adopted a rebuilding schedule for Klamath River fall chinook which extends the time beyond 1988 that the long-term escapement goal will be met. Under this rebuilding schedule, Klamath escapements will be increased by an average of 20 percent every four years until the long-term goal is met.

Goals for the Klamath River are expressed as in-river escapement until in-river Indian and recreational harvest allocations are established. Once these harvest allocations are agreed upon, spawning escapement goals will be set.

The rebuilding schedule is to achieve the following in-river run sizes (natural and hatchery combined) for the Klamath River:

1983-86	68,900	1991-94	99,200
1987-90	82,700	1995-98	115,000+[2]

[2]The long-term escapement goal of 115,000 chinook is spawning escapement to which in-river harvest must be added to calculate the ocean escapement goal.

The Klamath River escapement goal may be adjusted in the future upon evaluation of habitat quality, spawner success, and contribution of natural spawning stocks. Also, if in the future an allocation for Indian harvest is set at a level that, when combined with recreational needs and the spawning escapement goal, would require an in-river escapement goal that would result in under-utilization of other stocks in the ocean, the escapement goal may be reevaluated. Such changes would be made by an amendment to the FMP.

(From Pacific Fishery Management Council 1984.)

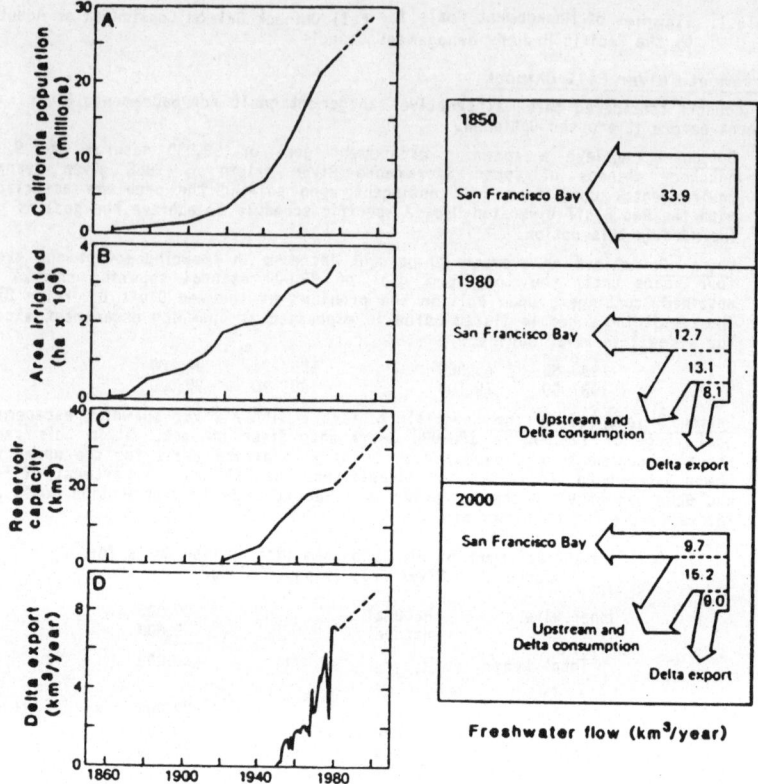

Figure 4.   (left) (A and B) California population growth and land under irrigation; (C) water storage capacity of reservoirs in the Sacramento and San Joaquin River Basin exclusive of those smaller than 8000 ha. Dashed line represents estimated increase in storage; (D) historic and projected export of water from the Sacramento-San Joaquin Delta by the CVP and SWP. (From Nichols et al., 1986. Copyright 1986 by the AAAS. Reprinted by permission.)

Figure 5.   (right) Deposition of Central Valley Runoff Exclusive of transpiration and other natural losses) before about 1850, in 1980, and projected for the year 2000. From Nichols et al., 1986. Copyright 1986 by the AAAS. Reprinted by permission.)

FORMULATION OF THE DYNAMICAL SYSTEM MODEL

There are two distinct components of a fishery/habitat, multiple manager model: the population dynamics model

$$\dot{\underline{x}} = \underline{f}(\underline{x},\underline{p},\underline{u},\underline{v}) \tag{1}$$

containing process models which link the vectors of habitat control variables ($\underline{v}$), fishing and stocking control variables ($\underline{u}$) and life stages ($\underline{x}$) to determine the parameter ($\underline{p}$) dependent transfer rates between life stages, andthe institutional linkage model

$$\underline{g}(\underline{x},\underline{p},\underline{u},\underline{v}) = \underline{0}. \tag{2}$$

describing the relationships among the management agency control variables. Fig. 6 illustrates the relationship between these two components of the system model.

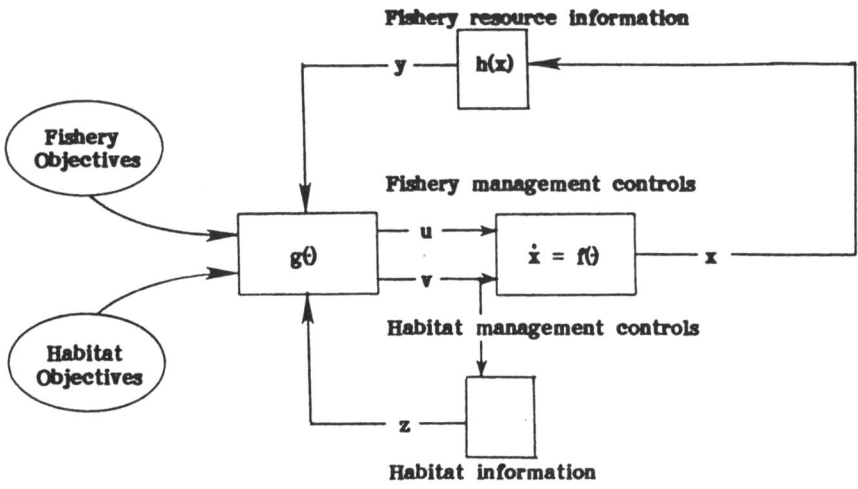

Figure 6. The general structure of fishery-habitat management model. The control variables, $\underline{u}$ and $\underline{v}$, which are determined by fishery and habitat objectives, fishery resource and habitat information and the constraints imposed by the institutional linkage model, $\underline{g}(\cdot)$, are inputs to the population dynamics model, $\dot{\underline{x}} = \underline{f}(\cdot)$.

Population Dynamics

Parsimony in model structure requires a judicious balance between necessary detail and existing knowledge of specific functions and parameters (cf. Ludwig and Walters, 1985). Nevertheless, in a modeling and aggregation scheme, consideration must be given to uniqueness of life stage functions, life stage habitats, and life stage/habitat institutional controls.

Consider a model representing the dynamics of a single cohort of the population.

$$\dot{x}_{ki} = \sum_j f_{kij}(\underline{x}, \underline{p}, \underline{s}(.),\underline{v}) + \sum_l f_{kil}(\underline{x}, \underline{p}, \underline{s}(.),u_{ki}) \tag{3}$$

$f_{kij}(\underline{x}, \underline{p}, s_{ki}(.), \underline{v})$ is the jth process function (dependent upon habitat controls) which regulates the rate of change of the ith life stage of the kth watershed stock and $f_{kil}(\underline{x}, \underline{p}, s_{ki}(.), u_{ki})$ is the lth process function (dependent upon fishing/-stocking controls) which regulates the rate of change of the ith life stage of the kth watershed stock. $s_{ki}(x_{ki}, \underline{p})$ represents the size of individuals of the ith life stage of the kth watershed stock and is used to indicate that for some processes (e.g., harvesting), size of the fish is a controlling factor rather than (or in addition to) age.

For the California chinook salmon fishery eight life stages (i) involving three habitat controlled rate processes (j) and one fishing/stocking controlled rate process function (l) for five watershed stocks (k) are required. The eight life stages are (1) eggs and alevins in the stream spawning segments; (2) fry migrating downstream in the river system; (3) juveniles rearing in and/or migrating through the estuarine system; (4) juveniles in the nearshore habitat; (5) immatures in the pelagic habitat; (6) adults migrating through the estuarine system; (7) adults migrating upstream in the river system; (8) effective spawning females in the spawning grounds. The habitat-controlled rate functions are gains via (1) recruitment, and losses from (2) mortality, and (3) development into the succeeding life stage. The fishing/-stocking-controlled rate function is harvesting/planting. The five watershed stocks are San Joaquin River, American River, Feather River, Upper Sacramento River, and Klamath River (Fig. 1).

Institutional Linkages

The vector function $\underline{g}(\underline{x},\underline{p},\underline{u},\underline{v})$ represents the relationships among the control (i.e. management) variables available to each of the agencies whose actions impact the abiotic and biotic habitat and the fisheries associated with the California chinook salmon.

The functions $g(.)$ are determined by (1) the control authority which the enabling legislation allows each agency in managing the salmon habitat or fishery and (2) the degree of coordination versus independence of the management agencies. These are derived from Federal statutes and Code of Federal Regulations, California statutes and codes, and other documents such as coordinated operating agreements, memoranda of understanding, etc. The functions $g(.)$ constrain the freedom of the control variables, $\underline{u}$ and $\underline{v}$, and hence may limit the ability to reach the target sets (i.e. the goals set by the policy makers). Although the functions $g(.)$ are written as static relationships, the recent formation of joint task forces and reformulation of agency policies may imply that a dynamic function, $g(., t)$, will be required.

EXAMINATION OF THE DYNAMICAL SYSTEM MODEL PROPERTIES

In order to determine whether the targets are achievable, it is necessary to examine the properties of the system model (namely, controllability) along and about the time history trajectories of the state variables as they have been prescribed by the goals of the policy makers. An outline of the procedure to do so follows.

Reference Trajectory Expansion

Expanding (3) in a Taylor's Series about a reference trajectory $\underline{x}|_{ref}$, $\underline{u}|_{ref}$, $\underline{v}|_{ref}$ and retaining only the first order terms in the increments (Elgerd, 1967; and Aoki, 1976), results in a set of non-homogeneous equations of the form

$$\dot{\tilde{\underline{x}}} = A(t)\, \tilde{\underline{x}} + B_1(t)\, \tilde{\underline{u}} + B_2(t)\, \tilde{\underline{v}} + \underline{f}(\underline{x}, \underline{v})|_{ref} + \underline{f}(\underline{x}, \underline{u})|_{ref} - \dot{\underline{x}}|_{ref} \qquad (4)$$

where the elements of matrices $[A]$, $[B_1]$, and $[B_2]$ are the partial derivatives of f_{kij} and f_{kil} with respect to x_{ki}, u_{ki}, and v_{km}; $|_{ref}$ means evaluation along the reference trajectory; and $\tilde{\underline{x}}$, $\tilde{\underline{u}}$, and $\tilde{\underline{v}}$ are incremental values about the reference trajectory.

The Reference Trajectory: Goals and the Management Structure

By setting the reference functions in equation (4) equal to zero

$$\underline{f}(\underline{x}, \underline{v})|_{ref} + \underline{f}(\underline{x}, \underline{u})|_{ref} - \dot{\underline{x}}|_{ref} = \underline{0} \qquad (5)$$

and substituting the goals which have been set for $\underline{x}|_{ref}$, $\underline{u}|_{ref}$, and $\underline{v}|_{ref}$, into (5), the remaining unspecified reference trajectories can be determined (see Elgerd, 1967).

For example, in setting spawning escapement goals for each watershed system, the values of $x_{k8}|_{ref}$ are prescribed. (See Table I.) In setting ocean escapement goals, the values of $x_{k6}|_{ref}$ are prescribed. If hatchery production rates are set, the values of $u_{k2}|_{ref}$ and/or $u_{k3}|_{ref}$ are determined depending upon at what age (fry or fingerlings) and at what geographic location the young are planted in the system.

The nature of the interagency relationships and constraints, $\underline{g}(.)$, determine the relationships among $\underline{u}|_{ref}$ and $\underline{v}|_{ref}$. The solution of (5) must be examined not only to see if biologically realistic values of $\underline{x}|_{ref}$ and $\underline{u}|_{ref}$ are required for solution but also to determine if politically realistic relationships among $\underline{u}|_{ref}$ and $\underline{v}|_{ref}$ exist for the solution.

Determination of Controllability

In order to determine if the states are controllable about the projected reference trajectories, the controllability of the perturbation part of (4)

$$\dot{\tilde{\underline{x}}} = A(t) \; \tilde{\underline{x}} + B_1(t) \; \tilde{\underline{u}} + B_2(t) \; \tilde{\underline{v}} \tag{6}$$

must be determined, i.e. determine if it is possible to find a vector $[\tilde{\underline{u}}(t)|\tilde{\underline{v}}(t)]$ that will transfer any initial state $\underline{x}(t_0)$ to any final state $\underline{x}(t_f)$ in a finite time interval $t_0 < t < t_f$. The incremental control variables, $\tilde{\underline{u}}$ and $\tilde{\underline{v}}$, may also be constrained by biological and physical feasibility and by legislative authority or interinstitutional agreements if the constraints, $\underline{g}(.)$, are active in the perturbation region about the reference trajectory.

Alternatively, the multiresource, multimanager dynamical model structure allows consideration of questions such as: what reference trajectories are feasible given the existing institutional constraints, or what reallocation of institutional authority may be required to achieve controllability along a selected reference trajectory (cf. Aoki and Canzoneri, 1979).

DISCUSSION

The ability to determine certain properties of the mathematical model (e.g., controllability about the reference trajectory, biological feasibility of potential reference trajectory solutions, etc.) is dependent upon knowledge of the structural forms of the process models in the population dynamics model (3), their associated parameters, and the institutional linkage model constraint equations (2). Previous work has involved aggregation of the equivalents of the functional relationships of the state equations for x_1 through x_4 into the well-known stock/recruitment models of Ricker (1954) and of Beverton and Holt (1957). Both of these models involve only two parameters into which are incorporated all effects in the life history between a spawning stock and recruitment from this stock. These parameters are calculated from historic data. As Moussalli and Hilborn (1986) point out, "stock and recruitment models are useful tools ... when mechanisms acting on intermediate life history stages cannot be predicted or controlled... [The] parameters are assumed constant from year to year; thus, stock and recruitment models do not allow incorporation of expected changes in life history events." The form of the population dynamics model (3) follows their philosophy that the factors which control these parameters should be separated by life history stage and, where possible, be explicitly related to anthropogenic effects rather than, as has been assumed, solely to "serially uncorrelated environmentally induced variations."

Getz (1984) states that "recruitment represents the primary source of variability since egg and larval survival rates are much more dependent on environmental factors than are the survival rates of the older fish." The functions $f_{kij}(.)$ and $f_{kil}(.)$ of (3) for i = 1, 2, 3, 6, 7, 8 represent those processes which are most dominated by human manipulation of the abiotic environment. The greatest anthropogenic habitat effects upon the stock/recruitment relationship are reflected in $f_{kij}(.)$ for i = 1, 2, 3, and 8.

The chinook salmon fishery constitutes an important economic resource in California and elsewhere in the Pacific Northwest. Recognition of the value of the resource and the impact of habitat modification upon its productivity has led to recent extensive studies intended to organize existing knowledge and to quantify, with maximum possible precision, the habitat variables which affect the resource (e.g., Kjelson et al., 1982; Aceituno et al., 1985; Raleigh et al., 1985). The data which are being gathered and analyzed concerning habitat-related survival and growth will establish the age-dependent process functions which are important to the dynamics of the separate life history stages in the stock/recruitment functions. This author is currently synthesizing the results of these studies, together with California fall run chinook salmon population dynamics models being developed by others (R. Kope and L. W. Botsford, University of California, Davis; L. B. Boydstun, California Department of Fish and Game; W. L. Kimmerer, Biosystems Analysis, for National Marine Fisheries Service (1985)), into the specific format required by the population dynamics model discussed in this paper.

The literal bases for the functions which comprise the institutional linkages model (2) are being developed by comprehensive investigations conducted by the Pacific Fishery Management Council (PFMC) and the U.S. Bureau of Reclamation (BuRec). Documents issued by PFMC and BuRec will describe (1) the agencies involved in managing the salmon resource and its habitat, (2) the details of the pertinent State and Federal legislation, and (3) the management goals of these agencies (PFMC, 1985; L. Kaufman, BuRec, personal communication). The realizations of these linkage functions from descriptive form into mathematical form are currently under development by the author.

Lastly, it should be noted that many participants at this workshop have expressed the need to accommodate, in modeling and management, the extent of uncertainty which exists, and may continue to persist, in our knowledge of the processes and parameters which govern the dynamics of bioresource systems. In robust control methods, such as those presented at this workshop by Lee and

Leitmann, controllability of the nominal system is assumed. In order to apply these promising control methods to bioresource management (e.g., Corless and Leitmann, 1983), it is evident that issues of constraints upon system controllability, such as those illustrated in this paper, must be addressed.

CONCLUSION

Much research into the development of resource management strategy has been conducted assuming either a centralized or a competitive management structure. In many instances, neither of these assumptions may be compatible with prevailing organizational structures. Resource management institutions, linked directly by political structures and indirectly by their mutual ecosystem impacts, affect the productivity of one ecosystem resource due to exploitation of another. Through examination of the mathematical properties of dynamical system models of multiple manager, multiple resource systems analogous to the Federal/State fisheries and water resource scenario developed in this paper, the precise nature of these interactive effects can be elucidated and the means by which these conflicts might be resolved can be determined.

REFERENCES

Aceituno, M.E., A. Hamilton, M. Hampton, and R. Brown. 1985. Annual report: Trinity River flow evaluation study. U.S. Fish and Wildlife Service, Division of Ecological Services, Sacramento, CA. 27 p.

Aoki, M. 1976. Optimal control and system theory in dynamic economic analysis. North-Holland, NY. 400 p.

Aoki, M, and M. Canzoneri. 1979. Sufficient conditions for control of target variables and assignment of instruments in dynamic macroeconomic models. Int. Econ. Rev., Vol. 20, pp. 605-616.

Beverton, R.J., and S.J. Holt. 1957. On the dynamics of exploited fish populations. Fisheries Investigation Series II, Vol. 19, Ministry of Agriculture, Fisheries, and Food, London.

Corless, M., and G. Leitmann. 1983. Adaptive control of systems containing uncertain functions and unknown functions with uncertain bounds. J. Optimization Theory and Applications, Vol. 41, pp. 155-168.

Elgerd, O. 1967. Control Systems Theory. McGraw-Hill, NY. 562 p.

Fletcher, R.C. 1984. Memorandum dated 3/14/84 to Pacific Fishery Management Council re: Sacramento River salmon escapement goal in 1984. In Final framework amendment for managing the ocean salmon fisheries off the coasts of Washington, Oregon and California commencing in 1985. Pacific Fishery Management Council, Portland, OR.

Getz, W.M. 1984. Production models for nonlinear stochastic age-structured fisheries. Math. Biosci., Vol. 69, pp. 11-30.

Hallock, R.J. 1978. Status of the Sacramento River system salmon resource and escapement goals. p. S-1-Cs-S-25-Cs In Reference documents. Prepared for the comprehensive salmon management plan of the Pacific Fishery Management Council. Pacific Marine Fisheries Commission, Portland, OR.

Kjelson, M.A., P.F. Raquel, and F.W. Fisher. 1982. Life history of fall-run juvenile chinook salmon, _Oncorhynchus tshawytscha_, in the Sacramento-San Joaquin estuary, California. p. 393-411 in V.S. Kennedy (ed.) Estuarine Comparisons. Academic Press, NY.

Ludwig, D. and C.J. Walters. 1985. Are age-structured models appropriate for catch-effort data? Can. J. Fish. Aquat. Sci., Vol. 42, pp. 1066-1072.

May, R.M., J.R. Beddington, J.W. Horwood, and J.G. Sheperd. 1978. Exploiting natural populations in an uncertain world. Math. Biosci., Vol. 41, pp. 159-174.

Moussalli, E., and R. Hilborn. 1986. Optimal stock size and harvest rate in multistage life history models. Can. J. Fish. Aquat. Sci., Vol. 43, pp. 135-141.

National Marine Fisheries Service. 1985. Value of chinook salmon habitat in the Sacramento River Basin. Request for proposal. NOAA, Seattle, WA. 34 p.

Nichols, F.H., J.E. Cloern, S.N. Luoma, and D.H. Peterson. 1986. The modification of an estuary. Science, Vol. 231, pp. 567-573.

Pacific Fishery Management Council. 1985. PFMC strategy for development of comprehensive salmon management. Portland, OR. 9 p.

Raleigh, R.F., W.J. Miller, and P.C. Nelson. 1986. Habitat suitability index models and instream flow suitability curves: Chinook salmon. U.S. Fish Wildl. Serv. Biol. Rep., Vol. 82(10-122), 64 pp.

Ricker, W.E. 1954. Stock and recruitment. J. Fish. Res. Board Can., Vol. 11, pp. 559-623.

Swartzman, G.L., W.M. Getz, R.C. Francis, R. Haar, and K. Rose. 1983. A management analysis of the Pacific whiting fishery using an age structured stochastic recruitment model. Can. J. Fish. Aquat. Sci., Vol. 40, pp. 524-529.

Tang, Q. 1985. Modification of the Ricker stock recruitment model to account for environmentally induced variation in recruitment with particular reference to the blue crab fishery in Chesapeake Bay. Fish. Res., Vol. 3, pp. 13-21.

U.S. Fish and Wildlife Service. 1980. Final environmental impact statement on the management of river flows to mitigate the loss of the anadromous fishery of the Trinity River, California. Vol. I., Sacramento, CA.

U.S. Fish and Wildlife Service. 1985. Flows needs of chinook salmon in the lower American River. Final report on the 1981 lower American River flow study. Division of Ecological Services, Sacramento. 21 p.

Walters, C.J. 1975. Optimal harvesting strategies for salmon in relation to environmental variability and uncertain production parameters. J. Fish. Res. Board Can., Vol. 32, pp. 1777-1784.

Wandschneider, P.R. 1984a. Control and management of the Columbia-Snake River system. XB 0937 1084. Agricultural Research Center, Washington State University.

Wandschneider, P.R. 1984b. Managing river systems: centralization vs. decentralization. Natural Resources Journal, Vol. 24, pp. 1043-1066.

Wilkinson, C.F., and D.K. Conner.   1983.   The law of the Pacific salmon fishery: conservation and allocation of a transboundary common property resource.   Univ. Kans. Law Rev., Vol. 32, pp. 17-109.

PARTICIPANT'S COMMENTS

Since complex models in renewable resource management are usually impossible to analyze, most published studies, as in these Proceedings, deal with subsystems delineated by clear boundaries. The solution of real-world problems is more difficult than these studies usually suggest.

Numerous individuals and institutions are involved in the decision making process. The question of whether their energy and resources are well directed is important, particularly since many objectives remain unfulfilled.   The possible causes include the following:

* each institution's model of the overall real-world system may be incorrect or incomplete, particularly with regard to their understanding of the relationship and dynamics of other participants.

* individual goals, combined with others', may not be achievable simultaneously.

* achievement of each institution's goals may require use of controls outside its influence.

This problem can only be resolved with the development and understanding of a metasystem taking account of all the interacting subsystems at the same time.

Ann Blackwell has developed the structure of a model describing an important Californian fishery and the institutions involved in managing it.   She also describes a technique which could determine its controllability. The model's necessary scope, complexity and resolution highlight the magnitude of the task facing administrators and researchers and adds an important perspective on studies concerned with details of the system.

Since correct identification and parameter estimation of such a complex model is probably impossible, the question remains as to which of the diverse objectives are achievable and how the instutional structure will adapt as greater understanding and outside pressures evolve.

Philip R. Sluczanowski

This paper deals with the formulation of a multiple resource exploitation (fish and watershed resources) subject to multiple authority management (agencies of the Federal and State governments).

While related to other papers in the Conference, it differs in the multiple-multiple aspects of the situation considered. Some conference papers deal with management of single resources by a single manager or by competitive multiple managers; others deal with management of multiple resources by a single manager.

The author proposes a control-theoretic framework with the institutional linkages accounted for by constraints. Both the dynamical model and the constraints are discussed in general terms, as is the goal of the multiple management.

The first question posed, but not yet answered, concerns the system's controllability; i.e., can the system be managed to achieve the desired goal in the light of the institutional linkage constraints?

Since a problem must be posed properly before an attempt can be made to solve it, this paper constitutes an important first step toward the rational management by often conflicting jurisdictions of a variety of interdependent resources.

George Leitmann

OPTIMAL HARVESTING OF A YEAR-CLASS OF PRAWNS

Philip R. Sluczanowski
Department of Fisheries
Box 1625, G.P.O. Adelaide
South Australia, 5001

The optimal closure policy for a year-class of prawns is derived using dynamic programming. The decision to fish or not to at a certain time depends on whether the number of prawns exceeds a certain critical level, represented by a "switching curve", whose shape is determined by the model's input parameters: individual growth, mortality, seasonal relative vulnerability to fishing gear, size specific prawn prices and costs and limits on effort. The switching curve offers a convenient tool for analyzing the sensitivity of the optimal harvesting periods to the input parameters, thereby identifying those parameters which research should aim to estimate most accurately. The paper presents an algorithm for computing such an optimal switching curve.

INTRODUCTION

A number of prawn fisheries around the world are characterized by the annual harvesting of single year-classes of prawns. The seasonal closure of such fisheries intended to maximize the yield or value per recruit is a major concern of fisheries managers (FAO, 1985) and procedures for determining appropriate times are under development (Somers, 1985). The dynamics of each year-class are typically described by an extension of the Beverton-Holt model taking account of natural and fishing mortality, growth and the effects of fishing effort on fishing mortality. Clark (1976) and Goh (1980) described the optimal solution of such models, while Clark et al. (1973) and Sluczanowski (1976) determined the pulse fishing strategy which maximizes the biovalue yield from such a model which also included costs. Dudley and Waugh (1980) used a dynamic programming technique to derive the optimal control solution to the same problem.

In practice, however, the problem of determining when to harvest a year-class for maximum economic return is further complicated by the effects of additional parameters which make a realistic model even more complex and difficult to solve than those referred to above. The most important of these are changes in the relative vulnerability of prawns to fishing gear during different seasons and the effects on profitability of altered economic factors such as size-specific prices and limits and costs of fishing effort. Even where these parameters are well established, the decision when to harvest is not obvious, because it may happen, as in the case below, that the period during which the year-class reaches its maximum natural biovalue corresponds to the prawns' low vulnerability to fishing gear, giving rise to a clash between high revenue and high costs.

Sluczanowski (1984) estimated the parameters of a population dynamics model of the South Australian gulf prawn fishery, added economic components, and then determined optimal control policies which took account of parameter uncertainty, system noise and varying economic conditions. This paper now presents previously unpublished details of the dynamic programming procedure used to determine the optimal closure strategy derived from that model. The optimizing procedure is efficient (storage requirements of control parameters increase linearly with increases in resolution) and can be applied to any fishery where a single year-class can be modelled as a bilinear system in which the control variable appears linearly in the objective function.

An extensive research program is underway to estimate the biological parameters necessary to model the dynamics of the Spencer Gulf prawn stocks more accurately than has been previously possible (Carrick, 1982), and so to provide a basis for management decision making in the fishery. In particular, a method is required for determining closure periods which maximize profits, and the dynamic programming solution given here provides the necessary tool, once the key parameters have been estimated with sufficient confidence.

The solution described also offers a useful method of measuring the sensitivity of the optimal harvesting periods to the various associated parameters, thereby permitting an assessment of the potential value and relative importance of research aimed at refining their accuracy. It also can be used to determine the relative economic consequences of a range of possible closure regimes.

THE MODEL AND DATA

The model describes a single year-class of western king prawns (*Peneaus latisulcatus*) in Spencer Gulf over 74 weeks. It is based on continuous time functions for numbers, weight and relative vulnerability, which are transformed into discrete time equivalents for the optimization stages. For convenience, recruiting prawns are assigned arbitrary age 0 corresponding to the beginning of January and no further recruitment to the stock is assumed after this time.

In the unfished population, the number $N(a)$ of prawns of age a depends on the starting population N_1 and instantaneous rate of natural mortality M.

$$N(a) = N_1 \exp(-Ma), \qquad\qquad 0 < a < 74. \qquad (1)$$

The function describing mean weight (gm) of individuals $w(a)$ is a piecewise linear function summarizing the progression of size frequency modes measured by King (1978):

$$
\begin{aligned}
w(a) &= 9.940 + 0.900a, & 0 < a < 26 & \\
&= 26.280 + 0.272a, & 26 < a < 52 & \qquad (2) \\
&= -7.360 + 0.918a, & 52 < a < 74. &
\end{aligned}
$$

The relative vulnerability of prawns to capture by fishing $v(a)$ varies through the seasons according to a function estimated by Sluczanowski (1981):

$$
\begin{aligned}
v(a) \quad &= 1.00 \quad , &0 < a < 13 \\
&= 0.001747a^2 - 0.1066a + 2.090 \quad , &13 < a < 48 \\
&= 1.00 \quad , &48 < a < 52 \qquad (3) \\
&= v(a-52) \quad , &52 < a
\end{aligned}
$$

The discrete time bioeconomic model was constructed by appending components representing fishing mortality and economic factors to discrete time versions of (1)-(3). A weekly time interval was chosen, with subscript i denoting discrete variables.

The population of age i prawns at the start of week i, N_i is given by

$$
N_{i+1} = N_i \exp(-M-qv_i E_i) , \qquad\qquad i = 1,2,\ldots,74 \qquad (4)
$$

where the product $qv_i E_i$ represents fishing mortality. A factor of proportionality q relates fishing effort to fishing mortality, E_i is the fishing effort in week i, and v_i is the discrete time equivalent of $v(a)$. The number of vessels, their fishing power and the maximum number of fishing hours they can expend during a week were assumed constant and therefore E_i has an upper bound $\bar{E}$. We impose the further constraint that during any week either the fleet fishes at maximum capacity or no fishing occurs:

$$
E_i = 0 \qquad \text{or} \qquad \bar{E} . \qquad\qquad (5)
$$

Singular solutions are not allowed because they would be too difficult to implement practically. The model does, however, allow the possibility of multiple seasons. Integration of the continuous functions $v(a)$ and $w(a)$ given by equations (2) and (3) gave v_i and w_i.

Larger prawns fetch higher prices. Table 1 shows the prices paid for prawns in commercial size gradings corresponding to specified weight categories. The function $p(w_i)$ is shown as a function of time i in Figure 1, which also presents the input time series N_i, w_i and v_i described above.

The variables illustrated in Figure 1 can be combined as various products to describe the unfished population available for exploitation. The product $N_i w_i$ is the unfished biomass of a year-class, reaching its peak at about 22 wk, as shown in Figure 2. Multiplying the biomass by the size specific price function yields the unfished biovalue curve $N_i w_i p(w_i)$, which has a peak later in the year at about 24 wk because of the higher price of larger prawns.

Whereas the biomass and biovalue curves represent the state of the population in the sea, the catch rates experienced by fishermen or researchers are affected by the seasonal relative vulnerability of prawns to fishing gear. The "apparent" biomass $v_i N_i w_i$ and "apparent" biovalue $v_i N_i w_i p(w_i)$ curves represent the kg/hr and \$/hr experienced by fishing vessels, reaching their peak values at about 13 wk and 15 wk respectively. It is clear in this situation that catch rates alone are not a good indicator of the state of the stock available for exploitation. In fact the highest biovalue occurs at a time of low and declining "apparent" biovalue.

Table 1. Input parameter values for the optimal harvesting model of a year-class of prawns.

NOTATION	MEANING	VALUE	UNITS
M	Instant. rate of natural mortality	0.0302	wk^{-1}
q	Factor proportionality btw. E and F	0.0095	
	Maximum age of prawns in time intervals	74	wk
N_1	Mean population at age 0	700	million prawns
	Standard deviation of N_1	150	" "
$\bar{E}$	Maximum fishing effort per week	2.282	1000 hr
c	Unit cost of fishing effort	0.391	A$mil/1000 hr

Prices $p(w_i)$ in commercial size gradings

Grades (no/lb)	<10	11/15	16/20	21/25	26/30	31/35	36/40	41/50	51/60	>60
Low bound (gm)	45.36	30.24	22.68	18.14	15.12	12.96	11.34	9.07	7.56	0.00
Price ($/kg)	8.65	8.60	7.90	6.11	4.81	4.48	3.37	2.85	2.00	2.00

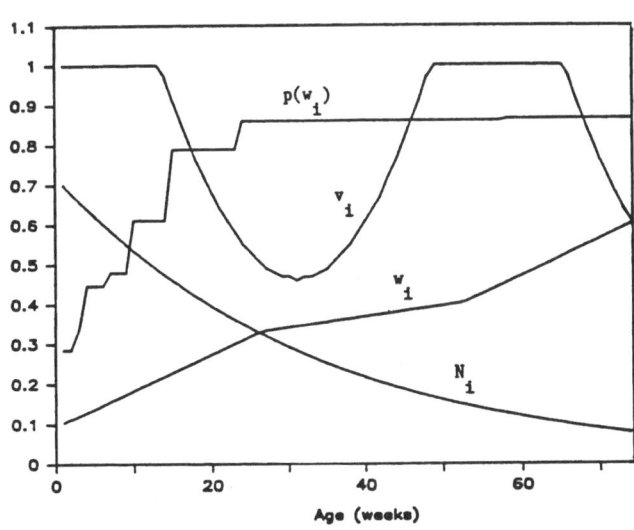

Figure 1. Time series data for a year-class of prawns, used to determine the optimal harvesting policy. A unit on the dependent axis represents 1000 million prawns for numbers N_i, 100 gm for weight w_i, 1.0 for vulnerability v_i and A$10.00/kg for price $p(w_i)$.

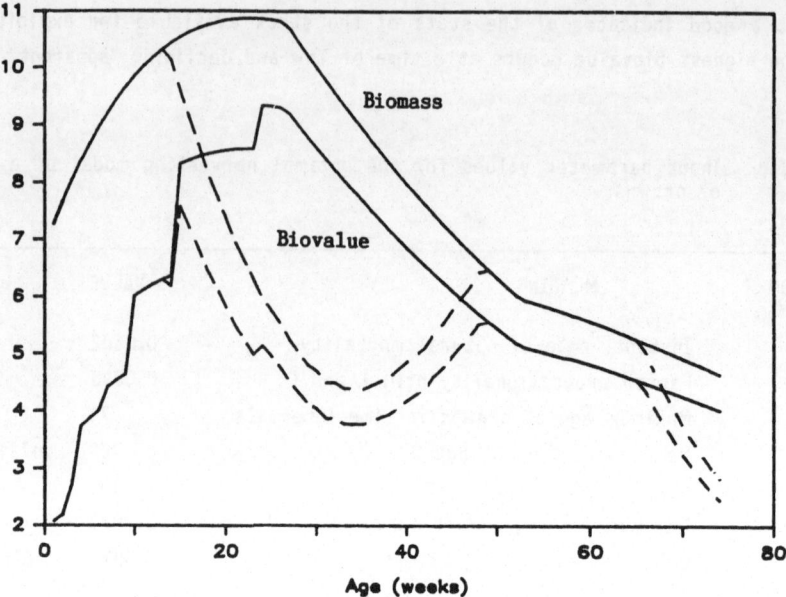

Figure 2.   The biomass $N_i w_i$ (1000tns) and biovalue $N_i w_i p(w_i)$ (A$mils) of a year-class of prawns in an unfished stock. The dashed lines show the "apparent" levels, $v_i N_i w_i$ and $v_i N_i w_i p(w_i)$, sampled by gear influenced by seasonal relative vulnerability of prawns.

The optimal exploitation strategy of the stock can be easily derived for extreme cases. If the costs of effort are negligible and the upper bound on fishing effort $\bar{E}$ is extremely high, then one should remove most of the stock when the biovalue curve reaches its peak at about 24 wk. For lower values of $\bar{E}$, one should still fish during the same period, but start sooner and finish later. On the other hand, a level of unit fishing cost c exists such that profits can only be made if the "apparent" biovalue exceeds (say) 7 in Figure 2. In that case, optimal fishing would occur only at about 14 wk. As shown later (Figure 4b), ten weeks difference in fishing period can have a significant effect on overall profit. Clearly, non-biological parameters such as $\bar{E}$, c and $p(w_i)$ have a determining effect on the optimal closure policy. The algorithm presented in this paper presents a general framework for deriving this solution.

The profit P_i obtained from fishing during week i was expressed as the difference between revenue and cost as follows:

$$P_i = E_i[q v_i w_i p(w_i) g_i N_i - c] \tag{6}$$

where c is the unit cost of fishing effort and g_i is a correction factor necessary to make the discrete time equation (6) exactly equivalent to that which would result from integrating one derived from the continuous time model (1) and is given by the following expression:

$$g_i = [1 - \exp(-M-qv_iE_i)]/(M+qv_iE_i) \tag{7}$$

For simplicity, the unit value of prawns at age i was defined as

$$u_i = w_ip(w_i) \tag{8}$$

Equations (2) and (3) contain numerical estimates necessary for evaluating $w(a)$, w_i, $v(a)$ and v_i. Trial values for the other model parameters M, q, N_1, $\bar{E}$, c and $p(w_i)$ based on previous estimates are given in Table 1.

DYNAMIC PROGRAMMING SOLUTION OF THE DISCRETE MODEL

Because we are interested in the effects of varying recruitment, the problem to be solved includes finding the optimal solution as a function of the initial state N_1. It can be stated as follows:

State equation: $\qquad N_{i+1} = N_i\exp(-M-qv_iE_i), \quad i=1,2,\dots,n \tag{9}$

Control constraint: $\qquad E_i = 0 \text{ or } \bar{E} \tag{10}$

Objective function term: $\qquad P_i = E_i[N_iqv_ig_iu_i - c] \tag{11}$

$$\text{where } g_i \text{ is given by (7)}$$

Objective function from time i: $\qquad J_i(N_i) = J_i(N_i,E_i,E_{i+1},\dots,E_n) \tag{12}$

$$= \sum_{j=i}^{n} P_j \tag{13}$$

Find optimal solution: $\qquad J_1^*(N_1) \text{ and } \{E_i^*(N_1), i=1,2,\dots,n\} \tag{14}$

such that: $\qquad J_1^*(N_1,E_1^*,E_2^*,\dots,E_n^*) = \max_{E_1,E_2,\dots,E_n} J_1(N_1) \tag{15}$

We use the fact that P_i is linear in the state variable N_i together with the knowledge that (9) is exactly equivalent to the continuous model derived from (1) to solve the problem using dynamic programming techniques. Guided by an understanding of the system dynamics, we seek an optimal strategy in the form of a switching curve $\hat{N}_i$. This not only solves the problem for initial condition N_1, but increases our understanding of the general form of the optimal solution. Whereas we have assumed the parameters M, $\bar{E}$ and c to be constant for all i, it is trivial to extend the model to allow them to vary.

The expression for the 'objective function until the end' $J_i(N_i)$ given by (13) can be combined with (11) to give the following recursive relationship:

$$J_i(N_i,E_i) = E_i(N_iqv_ig_iu_i - c) + J_{i+1}(N_{i+1}), \quad J_{n+1}(N_{n+1}) = 0 \qquad (16)$$

where N_{i+1} is given by (9). If the '*' denotes optimal solution [see (14-15)], then because the optimal solution over each subinterval i is part of the overall solution, the following is certainly true:

$$J_i^*(N_i) = E_i^*(N_iqv_ig_i^*u_i - c) + J_{i+1}(N_{i+1})$$

where $J_{n+1}^*(N_{n+1}) = 0$, $g_i^* = g_i(E_i^*)$, $N_{i+1} = N_iexp(-M-qv_iE_i^*)$.

Following the general technique of dynamic programming (Bellman, 1967), we find the optimal effort E_i^* in each interval i by solving for each N_i:

$$\max_{E_i=0 \text{ or } \bar{E}} J_i(N_i,E_i) = \max_{E_i=0 \text{ or } \bar{E}} \left[E_i(N_iqv_ig_iu_i - c) + J_{i+1}^* \{N_iexp(-M-qv_iE_i)\}\right] \qquad (17)$$

Equation (17) can be used to work backwards through time periods to find the optimal strategy. For each interval i, two items are computed as functions of the state at the beginning of the interval N_i: the optimal control $E_i^*(N_i)$ and the optimum of the objective function from then on $J_i^*(N_i)$. The assumption of 'bang-bang' controls together with our understanding of the physical system lead us to expect an optimal control strategy $E_i^*(N_i)$ stored in the form of a switching curve $\hat{N}_i$ such that

$$E_i^*(N_i) = 0 \qquad \text{for } N_i < \hat{N}_i \qquad (18a)$$

$$E_i^*(N_i) = \bar{E} \qquad \text{for } N_i > \hat{N}_i \qquad (18b)$$

Indeed we construct a solution of this form.

Before describing the procedure for finding the optimal control strategy, we show that the optimal objective function from time i, $J_i^*(N_i)$, can be stored for this model as a piecewise linear function of N_i as follows:

$$J_i^*(N_i) = a_i^jN_i + b_i^j \text{ for } N_i \in (d_i^j, d_i^{j+1}), \qquad j = 1,2,...,n_i \qquad (19)$$

$$\text{where} \qquad n_{n+1} = 1, \quad a_{n+1}^1 = b_{n+1}^1 = d_{n+1}^1 = 0, \quad d_{n+1}^2 = +\infty$$

We show by induction that if controls over each interval are 'bang-bang', then a piecewise linear representation for $J_i^*(N_i)$ is exact.

Suppose that $J_{i+1}^*(N_{i+1})$ is exact and piecewise linear in N_{i+1}. That is:

$$J_{i+1}^*D(N_{i+1}) = a_{n+1}^j N_{i+1} + b_{n+1}^j \quad \text{for} \quad N_{i+1} \in (d_{n+1}^j, d_{i+1}^{j+1}), \quad j = 1,2,\dots,n_{i+1} \quad (20)$$

Then (17) implies that

$$J_i^*(N_i) = E_i^*(N_i q v_i g_i^* u_i - c) + a_{i+1}^j N_i \exp(-M - q v_i E_i^*) + b_{i+1}^j \quad (21a)$$

where
$$N_i \exp(-M - q v_i E_i^*) \in (d_{i+1}^j, d_{i+1}^{j+1}), \quad j = 1,2,\dots,n_{i+1}. \quad (21b)$$

This form is equivalent to the following exact and piecewise linear expression for $J_i^*(N_i)$:

$$J_i^*(N_i D) = a_i^j N_i + b_i^j \quad \text{for} \quad N_i \in (d_i^j, d_i^{j+1}) \quad (22)$$

where
$$a_i^j = E_i^* q v_i g_i^* u_i + a_{i+1}^j \exp(-M - q v_i E_i^*) \quad (23a)$$

$$b_i^j = b_{i+1}^j - E_i^* c \quad (23b)$$

$$d_i^j = a_{i+1}^j \exp(M + q v_i E_i^*) \quad (23c)$$

Since $J_{n+1}^*(N_{n+1}) = 0$ is exact and piecewise linear, it follows by induction that $J_i^*(N_i)$ is exact and piecewise linear.

Having established the form of the optimal solution, we are now in a position to derive an algorithm to compute it from given data.

Given $\quad J_{i+1}^*(N_{i+1})$ in the form of $\{a_{i+1}^j, b_{i+1}^j, d_{i+1}^j, \quad j = 1,2,\dots,n_{i+1}\}$,

one can iterate backwards to evaluate:

i) $\hat{N}_i$, the switching curve which yields $E_i^*(N_i)$ as given by (18)

ii) $J_i^*(N_i)$

Figure 3 illustrates details of the relationship between variables at each backstep, in particular the relationship between the indices k and m.

i) We define

$$DJ_i(N_i) = J_i(N_i, \bar{E}, E_{i+1}^*, E_{i+2}^*, \dots, E_n^*) - J_i(N_i, 0, E_{i+1}^*, E_{i+2}^*, \dots, E_n^*) \quad (24)$$

$$= \bar{E}(N_i q v_i \bar{g}_i u_i - c) + a_{i+1}^k N_i \exp(-M - q v_i \bar{E}) + b_{i+1}^k - a_{i+1}^m N_i \exp(-M) - b_{i+1}^m \quad (25a)$$

where
$$N_i \exp(-M - q v_i \bar{E}) \in (d_{i+1}^k, d_{i+1}^{k+1}) \quad (25b)$$

and
$$N_i \exp(-M) \in (d_{i+1}^m, d_{i+1}^{m+1}) \quad (25c)$$

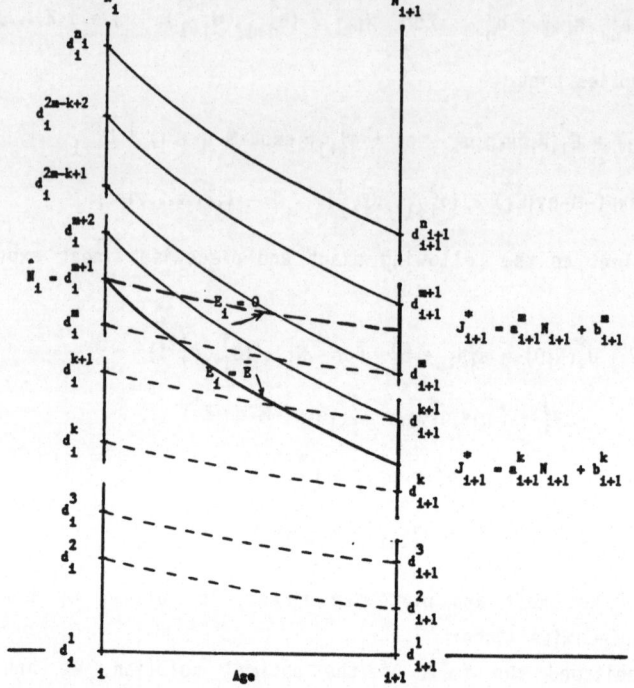

Figure 3. Details of the dynamic programming backstep between intervals i+1 and i. The alternative trajectories from the switching point N_i are used to derive $J_i^*(N_i)$ from $J_{i+1}^*(N_{i+1})$.

For each value of N_i, there are two possible trajectories to N_{i+1} (using $E_i = 0$ or $E_i = \bar{E}$), ending respectively in the intervals m and k as in equation (25). If $\hat{N}_i$ is the switching curve then:

$$DJ_i(N_i) > 0 \text{ for } N_i > \hat{N}_i \text{ and } E_i^*(N_i) = \bar{E} \qquad (26a)$$

$$DJ_i(N_i) < 0 \text{ for } N_i < \hat{N}_i \text{ and } E_i^*(N_i) = 0 \qquad (26b)$$

$$DJ_i(\hat{N}_i) = 0 \qquad (26c)$$

Given $J_{i+1}^*(N_{i+1})$ it is simple to step through pairs of intervals (k,m) for all possible values of N_i. The algorithm for finding $\hat{N}_i$ does this, at each pair of intervals solving the linear equation (25a) for $\tilde{N}_i$ which would make $DJ_i(\tilde{N}_i) = 0$. If $\tilde{N}_i$ satisfies (25b) and (25c) then $\hat{N}_i = \tilde{N}_i$, otherwise we move on to the next pair (k,m). Note that $\hat{N}_i$ may be zero or infinite.

$DJ_i(N_i)$ is a well defined piecewise linear function of N_i that represents the gain or loss to the objective function due to using either of the two possible controls. It measures the relative loss of profit (penalty) due to making an incorrect choice of control at (i, N_i). This function can be used to determine the periods during which it is important to impose correct optimal controls and those during which one can afford to be indifferent.

ii) Having found a unique $\hat{N}_i$ and given $J^*_{i+1}(N_{i+1})$, $J^*_i(N_i)$ is easily evaluated. We need to consider two cases, depending on whether $N_i < \hat{N}_i$ or $N_i > \hat{N}_i$, respectively. In the first case, $N_i < \hat{N}_i$. The interval boundaries d and linear coefficients (a,b) are updated according to equation (23) with $E^*_i(N_i) = 0$. $\hat{N}_i$ becomes an interval boundary, i.e., $d^{m+1}_i = \hat{N}_i$. We must still compute $J^*_i(N_i)$ for $N_i > \hat{N}_i$. The factors d and (a,b) are updated as in the first case, but using $E^*_i = \bar{E}$ instead of 0 in equation (23). The number of intervals n_i increases by (m+k-1) from n_{i+1} and the system's behaviour ensures that problems of an 'expanding grid' do not arise.

The procedure described was translated into a formal algorithm which provided a basis for a computer program to carry it out automatically.

RESULTS

The dynamic programming procedure was used to find the optimal switching curve $\hat{N}_i$ for the model described above. The resulting curve is shown in Figure 4a. The optimal sequence of fishing seasons and closures was then derived for the mean population at age 0, N_1, resulting also in an optimal trajectory $\bar{N}_i$ based on the mean recruitment N_1. The figure also contains the threshold number of prawns necessary to ensure that revenue from the catch exceeds the costs of catching it. It was derived by setting $P_i = 0$ in equation (11) and solving for N_i. Once the switching curve $\hat{N}_i$ was established, the procedure was repeated, this time evaluating $DJ(\bar{N}_i)$, the possible penalty at each time period due to an incorrect choice of control (Figure 4b).

Table 2 lists further results for the model. The total profit of 3.84 (A$mil) evaluated as the sum of the profit terms in each week matched that derived from the piecewise function for $J^*(N_1)$. The operation of the program was verified further by running the model for extreme values of the various parameters, when the results comformed to those expected. Furthermore, the switching curve behaves qualitatively as expected for changes in parameter values about their means.

Examination of Figure 4 reveals some of the characteristics of the optimal solution. The virtually asymptotic entry of the switching curve at about nine weeks is clearly the dominant effect in the determination of optimal closures. This means that the opening date of the first fishing season is almost the same for any level of recruitment.

Figure 4b highlights those times of year when it is important whether or not to fish. Whereas shifting the dates of closures a week or two either side has little effect in most cases, it is important not to start the first opening too soon.

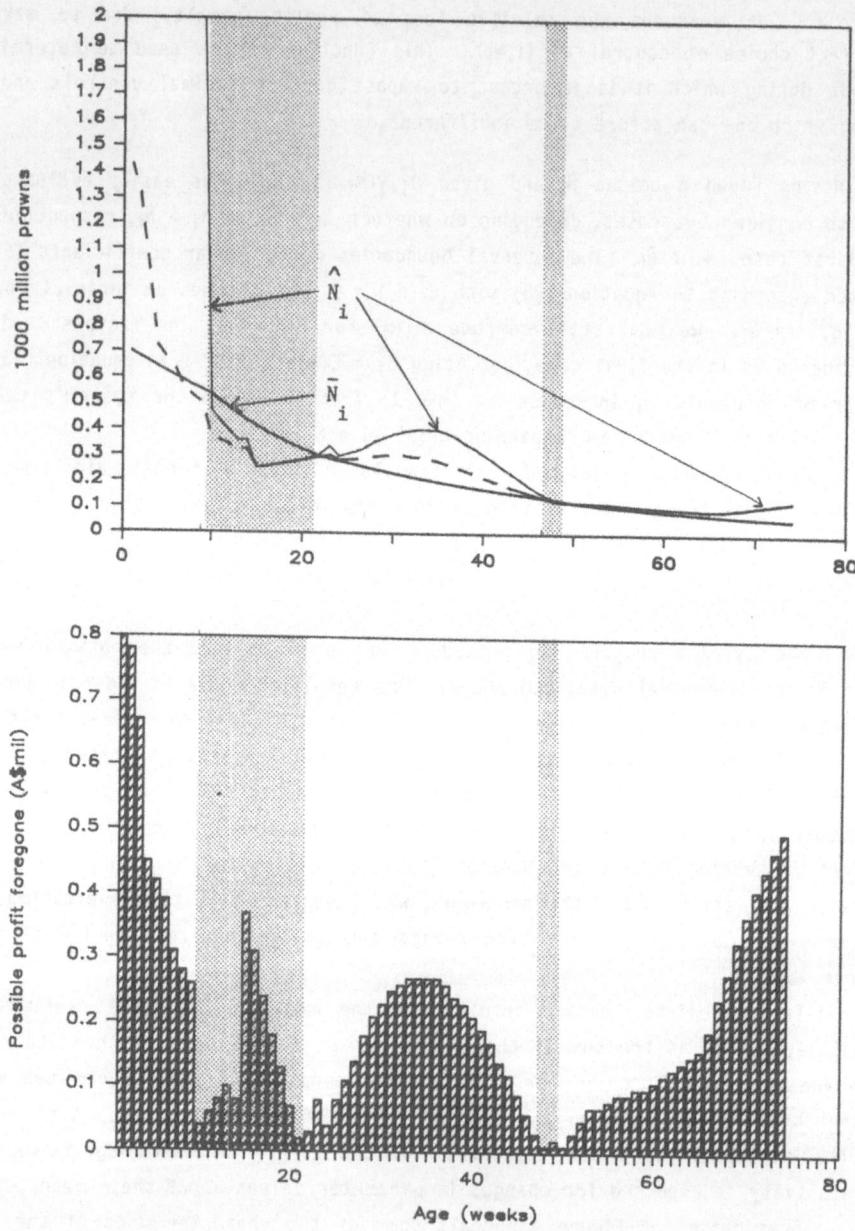

Figure 4.   a) The optimal switching curve $\hat{N}_i$, mean trajectory of number of prawns $\bar{N}_i$, the profit threshold level (dashed) and the resulting optimal fishing seasons (shaded).   b) $DJ_i(N_i)$, the possible return foregone at each interval by applying the incorrect control E_i while on the mean trajectory $\bar{N}_i$.

Table 2.   Optimal output values for the optimal harvesting model of a year-class
            of prawns with mean recruitment value.

	OPTIMAL VALUE	UNITS
Total catch	2.279	1000t
Total fishing effort	31.95	1000hr
Average catch rate	71.35	kg/hr or t/1000hr
Total revenue	16.33	A$mil
Total cost	12.49	"
Total profit	3.84	"

This is because the penalties in terms of profit foregone are likely to be severe.
In the face of parameter uncertainty, it is probably better to extend this date to be
more assured of not making a bad error. The figure also shows that a substantial
portion of profits is realized towards the latter stages of the life cycle and that
fishing in the final season should not stop too early. This suggests that the
results are sensitive to the maximum expected age of the prawns and that future
practical application of the method should examine the effects of this parameter on
optimal closure times.

DISCUSSION

The paper shows that if one restricts the pattern of fishing a year class of
prawns to a series of weeks when either full fishing effort or no fishing occur,
then the range of optimal solutions can vary considerably, depending on different
model parameters. The solution is in the form of a sequence of closed and open
seasons, whose pattern may not be obvious at the outset. The optimal solution for a
given set of parameter values can be succinctly summarized for all levels of
recruitment by means of an optimal switching curve, whose shape also offers further
insight into the dynamics and sensitivity of the particular system under analysis.

There is a clear relationship between the optimal switching curve and the
profit threshold level. Further studies have shown that the two curves are almost
concurrent for this model over the range of ages greater than that when fishing
first commences. However the profit threshhold does not offer guidance as to the
start of the optimal first fishing season, the most critical decision.

An examination of the behavior of the switching curve in response to changes in parameter values reveals the sensitivity of optimal closures and the consequences of incorrect decisions. The profits lost through basing closure decisions on a model having incorrect values of each parameter can be calculated and then used by managers to weigh the importance of more accurate estimates, thereby allowing a better allocation of research resources.

The model and optimization procedure presented above is fairly general and can easily be extended to account for more complex situations. For example, social discount rates or proportional crew payments can simply be included in time varying functions u_i and c_i. It can be used to determine the economic effectiveness of short term closures imposed during full moon to minimize the supposed effects of reduced vulnerability and poor prawn condition during such periods.

An obvious shortcoming of the model is its deterministic nature, whereby the behavior of a heterogeneous population is represented by that of the "mean individual". The fact that a real population has a constantly changing size distribution has a smoothing effect on the curves, which appear so ragged in the model used as an example above. The practical user of this model would need to remedy this shortcoming.

REFERENCES

Bellman, R. 1967. Dynamic Programming. Princeton University Press, Princeton, New Jersey.

Carrick, N.A. 1982. Spencer Gulf prawn fishery - surveys increase our knowledge. SAFIC, Vol. 6, No. 4, pp. 3-32. Department of Fisheries, Adelaide.

Clark, C.W. 1976. Mathematical Bioeconomics: The Optimal Management of Renewable Resources. Wiley-Interscience, New York.

Clark, C.W., Edwards, G. and Friedlander, M. 1973. Beverton-Holt model of a commercial fishery: Optimal dynamics. J. Fish. Res. Board Can., Vol. 30, pp. 1629-1640.

Dudley, N. and Waygh, G. 1980. Exploitation of a single-cohort fishery under risk: A simulation-optimization approach. J. Environ. Econ. Manage., Vol. 7, pp. 234-255.

FAO. 1985. Report of the FAO/Australia Workshop on the management of Penaeid shrimp/prawns in the Asia-Pacific region. Kooralbyn Valley, Queensland, Australia, 29 October-2 November 1984. FAO Fisheries Report 323.

Goh, B.S. 1980. Management & Analysis of Biological Populations. Elsevier, Amsterdam.

King, M.G. 1977. The biology of the western king prawn Peneaus latisulcatus Kishinouye and aspects of the fishery in South Australia. M.Sc. Univ. of Adelaide.

Sluczanowski, P.R. 1976. Implementable policies for improving the biomass yield of a fishery. In Optimization Techniques: Modelling and Optimization in the Service of Man, Pt. 2, J. Cea (ed.), pp. 182-206. Lecture Notes in Computer Science 40, Springer-Verlag, Berlin.

Sluczanowski, P.R. 1984. Modelling and optimal control: A case study based on the Spencer Gulf prawn fishery for *Peneaus latisulcatus* Kishinouye, J. Cons. int. Explor. Mer, Vol. 41, pp. 211-225.

Somers, I.F. 1985. Maximizing value per recruit in the Gulf of Carpentaria banana prawn fishery. Proc. of the Second Australian National Prawn Seminar Kooralbyn 1984. CSIRO, Cleveland, Australia.

PARTICIPANT'S COMMENTS

One of the main objectives of the workshop was to bring together researchers in resource management, biological systems theory, and control theory, so that a variety of techniques could be focused on the topic of renewable resource management. The paper under discussion is a prime example of this theme.

The paper presents a very practical application of an optimization and optimal control technique (dynamic programming applied to a model of a real resource management system. The paper begins by presenting a continuous-time model for a particular prawn fishery, along with experimentally determined relationships for various quantities involved in the model. The model is then simplified by transforming to a discrete-time model, the control variable is simplified to discrete levels (minimum or maximum fishing effort), and then the dynamic programming technique is applied to determine the optimal economic harvesting policy. Complete details of the procedure are presented, including both a proof that the "optimal return" function is piece-wise linear and a development of the optimal switching function for the problem. The paper closes with a discussion of the practical aspects of the results, points out where the results are sensitive to various parameters in the model, and indicates some areas for further investigation.

Hopefully, in the future, the results of the paper will be implemented and we will learn not only one effective optimal control policies may or may not be, but also something about possible implementation procedures and difficulties.

Walter Grantham

The paper by Philip Sluczanowski embodies many of the ideas appropriate to a Workshop on Renewable Resource Management. The author has chosen a specific biological system, that of western king prawns in Spencer Gulf, and used mathematical modelling and optimal control techniques to provide a basis upon which a sensible management policy can be developed. Although the model itself is extremely simple and one might argue (as the author does) that it suffers from being deterministic rather than stochastic in nature, it nevertheless provides valuable insight into the behavior one might expect from a more realistic model.

An interesting result from the model is that the opening date of the fishing season is almost independent of recruitment although, of course, the length of the season may be highly correlated with recruitment. I believe that the actual recruitment for this fishery takes place over about a three-month period. The model assumes that recruitment has ceased before fishing commences. Is this a valid assumption, and if not, what effect is this likely to have on the model predictions?

The key to the algorithm for solving the optimization problem is that the objective function from a given time can be expressed as a linear function of the population at that time. This property will, of course, be dependent on the complexity of the state equation, which is a discrete analogue of the simple Beverton-Holt model. In the discussion, the author mentions that the model and procedure can easily be extended to deal with more complex situations. If, for example, the model includes density dependent mortality, then the objective function will no longer be linear in the state variable suggesting that an alternative procedure for solving the optimization problem will have to be developed.

Mike Fisher

Your source for advances in theoretical biology and biomathematics

Journal of

Mathematical Biology

ISSN 0303-6812

Title No. 285

Editorial Board: K. P. Hadeler, Tübingen; S. A. Levin, Ithaca (Managing Editors); H. T. Banks, Providence; J. D. Cowan, Chicago; J. Gani, Santa Barbara; F. C. Hoppensteadt, East Lansing; D. Ludwig, Vancouver; J. D. Murray, Oxford; T. Nagylaki, Chicago; L. A. Segel, Rehovot

For mathematicians and biologists working in a wide variety of fields – genetics, demography, ecology, neurobiology, epidemiology, morphogenesis, cell biology – the **Journal of Mathematical Biology** publishes:

● papers in which mathematics is used for a better understanding of biological phenomena
● mathematical papers inspired by biological research, and
● papers which yield new experimental data bearing on mathematical models.

The following selection of articles from recent issues reflects the **Journal of Mathematical Biology's** range and scope:

Subscription information:
To enter your subscription, or to request sample copies, contact Springer-Verlag, Dept. ZSW, Heidelberger Platz 3, D-1000 Berlin 33, W. Germany

Springer-Verlag
Berlin Heidelberg New York
London Paris Tokyo

Bio-mathematics

Managing Editor: S. A. Levin

Editorial Board: M. Arbib,
H. J. Bremermann, J. Cowan,
W. M. Hirsch, J. Karlin,
J. Keller, K. Krickeberg,
R. C. Lewontin, R. M. May,
J. D. Murray, A. Perelson,
T. Poggio, L. A. Segel

Springer-Verlag
Berlin Heidelberg New York
London Paris Tokyo

Springer

Lecture Notes in Biomathematics